Lecture Notes in Computer Science 6164

Commenced Publication in 1973
Founding and Former Series Editors:
Gerhard Goos, Juris Hartmanis, and Jan van Leeuwen

Finn Arve Aagesen
Svein Johan Knapskog (Eds.)

Networked Services and Applications – Engineering, Control and Management

16th EUNICE/IFIP WG 6.6 Workshop, EUNICE 2010
Trondheim, Norway, June 28-30, 2010
Proceedings

 Springer

Volume Editors

Finn Arve Aagesen
Norwegian University of Science and Technology (NTNU)
Department of Telematics
O.S. Bragstads plass 2B, 7491 Trondheim, Norway
E-mail: finnarve@item.ntnu.no

Svein Johan Knapskog
Norwegian University of Science and Technology
Centre for Quantifiable Quality of Service in Communication Systems (Q2S)
O.S. Bragstads plass 2E, 7491 Trondheim, Norway
E-mail: knapskog@q2s.ntnu.no

Library of Congress Control Number: 2010929191

CR Subject Classification (1998): C.2, E.3, D.4.6, K.6.5, E.4, K.6

LNCS Sublibrary: SL 3 – Information Systems and Application, incl. Internet/Web
and HCI

ISSN   0302-9743
ISBN-10   3-642-13970-1 Springer Berlin Heidelberg New York
ISBN-13   978-3-642-13970-3 Springer Berlin Heidelberg New York

springer.com

© IFIP International Federation for Information Processing 2010
Printed in Germany

Typesetting: Camera-ready by author, data conversion by Scientific Publishing Services, Chennai, India
Printed on acid-free paper   06/3180

Preface

The EUNICE (European Network of Universities and Companies in Information and Communication technology) (http://www.eunice-forum.org) mission is to jointly develop and promote the best and most compatible standard of European higher education and professionals in ICT by increasing scientific and technical knowledge in the field of ICT and developing their applications in the economy. The EUNICE Workshop is an annual event. This year the workshop was sponsored by IFIP TC 6 WG 6.6: Management of Networks and Distributed Systems.

Eight years ago, the seventh edition of the EUNICE workshop took place in Trondheim with the topic "Adaptable Networks and Teleservices." Since then "adaptability" has become a topic which is found in most ICT conferences. The concept teleservices, which is a telecommunication domain concept from the 1980s, has been lifted out of the telecom community and is now found with new and sometimes mysterious names such as service–oriented architecture and cloud computing.

This year's workshop title, "Networked Services and Applications – Engineering, Control and Management," was more generic than the 2002 topic. Networked services and applications have developed from being important research topics within the telecom and computer network communities, respectively, to become one of the core drivers for the whole ICT domain. From being services either extending basic telephony functionality applied by telephony customers or applications used by computer professionals, common networked services and applications are now used by almost everyone related to almost every business. The impact of networked services and applications as an important part of society infrastructure is increasing. EUNICE 2010 addressed research issues of services and applications as considered through disciplines such as architecture, engineering, security, performance and dependability as well as through service and application frameworks and platforms.

After the review process, 24 papers were accepted for presentation in technical sessions. In addition, 15 posters were allocated for a poster session during the conference. Extended abstracts for six of these posters were accepted to be included in these proceedings. Every submission received at least three reviews from the members of the Technical Program Committee and/or external reviewers. Our gratitude goes to all the reviewers for their efforts.

We would like to take this opportunity to express our thanks to the technical and financial sponsors of the 16th EUNICE Workshop: Department of Telematics NTNU; Q2S - Centre for Quantifiable Quality of Service in Communication Systems at NTNU; Euro-NF, European Network of Excellence; UNINETT; IFIP TC6 WG 6.6; and Norwegian University of Science and Technology (NTNU).

June 2010

Finn Arve Aagesen
Svein Johan Knapskog

Organization

EUNICE 2010 was co-organized by ITEM (Department of Telematics (http:// www.item.ntnu.no)) and Q2S (Centre for Quantifiable Quality of Service in Communication Systems (http:// www.q2s.ntnu.no)) at the Norwegian University of Science and Technology.

Technical Program Committee Co-chairs

Finn Arve Aagesen	ITEM, NTNU
Svein Johan Knapskog	Q2S, NTNU

Technical Program Committee

Finn Arve Aagesen	NTNU, Trondheim, Norway
Sebastian Abeck	University of Karlsruhe, Germany
Rolv Bræk	NTNU, Trondheim, Norway
Jörg Eberspächer	Technical University of München, Germany
Olivier Festor	INRIA, Nancy, France
Markus Fiedler	Bleking Institute of Technology, Sweden
Edit Halász	Budapest University of Technology and Economics, Hungary
Jarmo Harju	Tampere University of Technology, Finland
Poul Heegaard	NTNU, Trondheim, Norway
Bjarne E. Helvik	NTNU, Trondheim, Norway
Yuming Jiang	NTNU, Trondheim, Norway
Yvon Kermarrec	TELECOM Bretagne, France
Svein Johan Knapskog	NTNU, Trondheim, Norway
Paul Kühn	University of Stuttgart, Germany
Øivind Kure	NTNU, Trondheim, Norway
Xavier Lagrange	TELECOM Bretagne, France
Maryline Laurent-Maknavicius	TELECOM SudParis, France
Ralf Lehnert	TU Dresden, Germany
Stefan Lindskog	University of Karlstad, Sweden
Chris Mitchell	Royal Holloway, University of London, UK
Maurizio Munafó	Politecnico di Torino, Italy
Elie Najm	ENST Paris, France
Miquel Oliver	Universitat Pompeu Fabra, Barcelona, Spain
George Polyzos	Athens University of Economics and Business, Greece
Aiko Pras	University of Twente, The Netherlands
David Ros	TELECOM Bretagne, France
Sebastian Sallent	UPC-BARCELONA TECH, Spain

Gwendal Simon	TELECOM Bretagne, France
Burkhard Stiller	University of Zürich, Switzerland
Robert Szabo	Budapest University of Technology and Economics, Hungary
Samir Tohmé	University of Versailles Saint Quentin en Yvelines, France
Arne Øslebø	UNINETT, Trondheim, Norway

Referees

Finn Arve Aagesen	Ralph Lehnert
Gergely Biczók	Stefan Lindskog
Máté J. Csorba	Patrick Maillé
Jörg Eberspächer	Chris Mitchell
Martin Eian	Maurizio Munafò
Gabor Feher	Georgios Pitsilis
Olivier Festor	George Polyzos
Markus Fiedler	Aiko Pras
Edit Halász	David Ros
Jarmo Harju	Gwendal Simon
Poul Heegaard	Bilhanan Silverajan
Bjarne E. Helvik	Vidar Slåtten
Tamas Holczer	Burkhard Stiller
Shanshan Jiang	Gábor Szücs
Yuming Jiang	Robert Szabo
Yvon Kermarrec	Geraldine Texier
Svein Johan Knapskog	Attila Vidacs
Lill Kristiansen	Benedikt Westermann
Øivind Kure	Otto Wittner
Paul Kühn	Arne Øslebø
Xavier Lagrange	Harald Øverby
Maryline Laurent	

Invited Talks

Danilo Gligoroski: *Swiss Army Knife in Cryptography and Information Security—Cryptographic Hash Functions*
Paul Kühn: *Modeling Power Saving Strategies in ICT Systems*
Aiko Pras: *Research Challenges in Network and Service Management*
CEO Wireless Trondheim Thomas Jelle: *Wireless Trondheim Living Lab*

Technical Sponsors

Euro-NF, European Network of Excellence (http://euronf.enst.fr/en_accueil.html)
International Federation for Information Processing (IFIP) TC6 WG 6.6: Management of Networks and Distributed Systems (http://www.simpleweb.org/ifip/)

Sponsoring Institutions

Department of Telematics at NTNU (http://www.item.ntnu.no/)
Centre for Quantifiable Quality of Service in Communication Systems
  (http://www.q2s.ntnu.no/)
NTNU (http://www.ntnu.no/)
UNINETT (http://www.uninett.no/)

Table of Contents

Congestion Control

Monitoring and Filtering

Dependability

Adaptation and Reconfiguration

Poster Session

On the Performance of Grooming Strategies for Offloading IP Flows onto Lightpaths in Hybrid Networks

Rudolf Biesbroek[1], Tiago Fioreze[1], Lisandro Zambenedetti Granville[2], and Aiko Pras[1]

[1] University of Twente, Design and Analysis of Communication Systems (DACS)
Enschede, The Netherlands
[2] Federal University of Rio Grande do Sul, Institute of Informatics
Porto Alegre, Brazil

Abstract. Hybrid networks take data forwarding decisions at multiple network levels. In order to make an efficient use of hybrid networks, traffic engineering solutions (*e.g.*, routing and data grooming techniques) are commonly employed. Within the specific context of a self-managed hybrid optical and packet switching network, one important aspect to be considered is how to efficiently and autonomically move IP flows from the IP level over lightpaths at the optical level. The more IP traffic is moved (offloaded), leaving the least amount of traffic on the IP level, the better. Based on that, we investigate in this paper different strategies to move IP flows onto lightpaths while observing the percentage of offloaded IP traffic per strategy.

Keywords: Grooming strategies, IP flows, lightpaths, ns-2, hybrid networks.

1 Introduction

The need for a separation between heavy applications and the normal Internet traffic over a shared network infrastructure has increased the importance of hybrid networks. Through the use of hybrid network infrastructures, backbone networks are able to provide better performance by means of faster delivery and more reliable data transmission. In such a hybrid environment, IP flows can traverse a hybrid network through either a lightpath or a chain of routing decisions. Moving large amounts of data from the IP level to the optical level enables flows to experience faster and more reliable transmissions with optical switching than with traditional IP routing. Meanwhile, the regular IP routing level is offloaded and can serve smaller flows better. Moreover, transmitting data flows at the optical level is cheaper than transmitting them at the IP level [11].

In order to configure a hybrid network and create lightpaths for IP flows, a management mechanism is required. Currently, GMPLS signaling and conventional management are important solutions for that [3]. GMPLS coordinates the creation of lightpaths by employing signaling messages that are exchanged between adjacent nodes along the path from source to destination node of a flow [12]. In the conventional management, on the other side, a central manager individually configures each node in the transmission path. Both GMPLS and conventional management rely on human decisions in

F.A. Aagesen and S.J. Knapskog (Eds.): EUNICE 2010, LNCS 6164, pp. 1–10, 2010.

order to select which flows would remain at the IP level and which other flows should be offloaded to the optical level. As expected, the human intervention turns the whole process slow and error-prone.

Based on the aforementioned state-of-the-art for the management of hybrid networks, it would be interesting to have a decision making process that could be automated in order to minimize human intervention. Having that in mind, a new management approach for hybrid networks is under investigation at the University of Twente, namely self-management of hybrid optical and packet switching networks [7,9,8]. One of the main challenges in such an investigation is to find out appropriate lightpath setups in which the available capacity of optical wavelengths is consumed in an optimal manner. For example, through the multiplexing of many flows into a single wavelength. Techniques for that, while considering certain design conditions (*e.g.*, minimum cost), are generally referred as *traffic grooming* [6,14].

In this context, we pose the following research question to be answered in this paper: *what traffic grooming strategy offloads the highest percentage of IP traffic to the optical level?* Depending on the grooming strategy employed, the percentage of offloaded traffic could differ significantly. At the optical level, each wavelength has a fixed amount of available bandwidth. In most cases, the sum of the offloaded flow rates will not fill the fully available wavelength capacity, leaving some of the capacity unused. Therefore, grooming techniques should strive to minimize the amount of unused capacity, which increases the possible offload percentage.

In this paper we evaluate the performance of some grooming strategies. These strategies have the purpose of grooming many IP flows, regardless the granularity of the IP flows, over the available lightpaths. The list of strategies that we investigate here is inspired by an earlier research on strategies and related algorithms for achieving dynamic routing of data flows for global path-provisioning [13]. Whereas the authors of the previous research have investigated the blocking probability while observing different offloading strategies to accommodate LSPs (Label Switched Path) on established lightpaths, we observe the percentage of IP traffic that can be offloaded to the optical level.

Through the use of simulation we evaluate the performance of grooming strategies while observing the percentage of traffic that is offloaded by each one of them. For that, we employ three different strategies: *dedicated*, *spreading*, and *packing*. As a side research, we also observe the energy consumption of each strategy. In order to do that, we look at the number of in-use wavelengths while accommodating the offered flows that need to be offloaded to the optical level.

The remainder of this paper is structured as follows. In Section 2 we review the current status in the field of traffic grooming in hybrid networks. In Section 3 we describe our simulation model and present a network topology used in our evaluation scenarios. In Section 4 we discuss the simulation results and finally, in Section 5, we close this paper with final remarks and perspectives for future work.

2   Related Work

Our research is inspired by the research performed by Sabella *et al.* [13] who focus on a solution for an online-routing function, which allows the network to promptly react to traffic changes. The authors have proposed a strategy and related algorithms to achieve dynamic routing of data flows. To accommodate new traffic requests, they have proposed the use of two algorithms: (i) a routing algorithm, to find a route for the requested traffic, and (ii) a grooming algorithm, to assign for any link of the route the traffic to an optical channel. Looking at the latter, the authors concluded that, by choosing the right grooming strategy, a reduction from two up to about four time of refused bandwidth for a network load of 70% and 55%, respectively, can be achieved. Moreover, the authors have argued that the gain of the proposed strategy (*packing* strategy) is greater when the average granularity of LSP's are coarser, and have remarked that this gain tends to diminish when the network becomes uniformly congested.

An important difference between our work and of Sabella *et al.* [13] is the goal of the grooming function; where Sabella *et al.* have aimed to maximize non-blocking probability when multiplexing LSPs into the given wavelengths, we aim to achieve maximum percentage of offloading IP traffic when sending high amount of traffic reaching up to 100% of the total bandwidth.

Drummond *et al.* [5] carried out a similar investigation through the use of simulation. They showed that the NS-2 simulator, combined with the OWns package, is able to simulate grooming capabilities of IP flows into wavelengths. Despite the use of simulation to observe different grooming strategies, Drummond *et al.*, did not consider the performance of such strategies as we consider in this paper.

Operational research aims at providing analytic methods to structure and understand complex situations as well as to use this understanding to improve and predict a system's behavior [10]. Based on that, our work is aligned with the operational research field, since we aim at formulating a model that enables us to analyze and understand the behavior of our system by means of simulation. The result of our simulation enables us to analyze the system behavior regarding the formulated method, which leads to the best performing system (*e.g.*, highest offload percentage).

3 Simulation Model

In this section we describe the model we use to simulate the offloading strategies considered in this paper. We then present: a network topology (subsection 3.1), the flow handling (*i.e.*, starting, offloading, and termination) (subsection 3.2), the evaluated criteria (subsection 3.3), and the scenarios (subsection 3.4). This simulation model enables the evaluation of the performance of our system in terms of percentage of offloaded IP traffic.

3.1 Topology

Our topology (Figure 1) consists of two routers being logically connected via an OC-192 link, which actually comprises of eight OC-24 links. The unidirectionally transmitted data (IP flows) is sent from Router 1 to Router 2. These IP flows vary in rate between 1 Mbps to 500 Mbps. In order not to exceed the overall optical link bitrate, the total bandwidth of the transmitted flows is limited to 9.952 Gbps (the equivalent of an OC-192 link). All the flows generated by Router 1 are initiated at the IP level and they stay at such a level until the offload procedure moves them to the optical level.

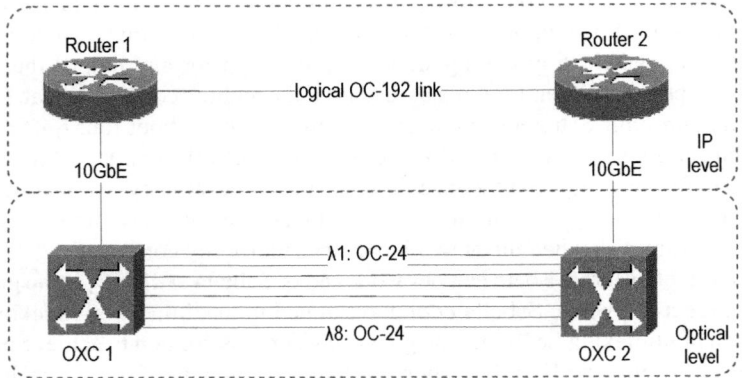

Fig. 1. Our simulation topology

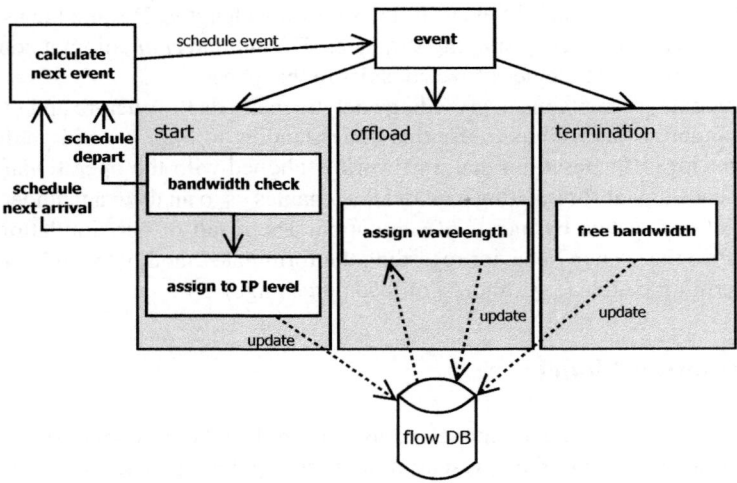

Fig. 2. Sketch of the flow handling (start, offload, and termination) during the simulation

3.2   Start and Termination of Flows

The start and termination of flows is regulated as described in Figure 2. When an arrival event occurs, a new flow is started and the next flow arrival event is scheduled.

For the inter arrival of flows we assumed a negative exponential distribution ($\lambda = 0.1234175251$). For the termination of a flow, we assume a Weibull distribution ($\lambda = 0.0190299$, $k = 0.17494315$). These assumptions are based on the analysis performed on the IP data collected from the University of Twente network.

Upon starting the flow, a bandwidth check is performed to ensure the available bandwidth in the network and to prevent packet loss. The total available capacity C^A is found by taking the used bandwidth C_i^U per wavelength i, and the used bandwidth C_{IP}^U on the IP level, together with the total link capacity L_c (=9.952 Gbps).

$$C^A = L_c - \left(\sum C_i^U + C_{IP}^U \right) \tag{1}$$

If enough bandwidth has been determined, a depart event is scheduled and the flow is assigned to the IP level. The offload event is triggered in a fixed interval of one second. This event will cause the offload process of moving IP flows from the IP level to the optical level. It is important to mention that the order of offloading the flows is determined according to the rate of the flow: *flows with the highest rates are offloaded first*. When a flow has been offloaded and assigned to a wavelength, it will stay on this wavelength until it terminates. When a departure event is triggered, the associated flow will be terminated. At the moment of its termination, a flow can reside at the IP level or at the optical level. In both cases, the bandwidth associated with the flow will be released. In case this event will cause a wavelength to become empty, this wavelength will be torn down in order to save energy.

3.3 Evaluation Criteria

During the simulation, flows are generated and, whenever possible, offloaded to the optical level. The evaluation of the simulation is done with the goal of finding the best performing offload strategy (*e.g.*, the strategy that has the highest percentage of offloaded IP traffic). Thus, we take the amount of traffic that resides at the IP level and compare it with the amount of traffic at the optical level. We use the percentage of offloaded traffic as a measurement to determine the performance of the offloading strategy.

$$\text{offloaded}(\%) = \frac{\sum C_i^U}{C_{total}^U} \times 100 \tag{2}$$

Where C_{total}^U is formulated as:

$$C_{total}^U = \sum C_i^U + C_{IP}^U \tag{3}$$

We also evaluate the energy consumption at the optical level by monitoring the number of wavelengths used during our simulation. The power consumption values of each optical element is depicted in Figure 3. The 8 x 1 Gbps transponders represent the 8 x OC-24 lightpaths connecting the OXC (Optical Cross-Connect) with the WDM terminal, and a 10 Gbps transponder connecting the WDM terminal with a demux. This demux connects to its counter-part, with amplifiers in between. Then, all the optical elements aforementioned repeat themselves in inverted order until OXC2. It is important to highlight that the transponders are switched on and off on-demand. They are automatically switched on when there is data to be transmitted and switched off when there is no data transmission (to save energy). The minimum energy consumption (*e.g.*, no data is sent) comprises of the energy consumption of the OXCs + the WDM terminals + 10GTx/Rx packs + WDMs + amplifiers on both sides. The amount of energy consumption when data is transmitted is the minimum amount of energy consumption *plus* the corresponding link's transponders of the in-use wavelengths.

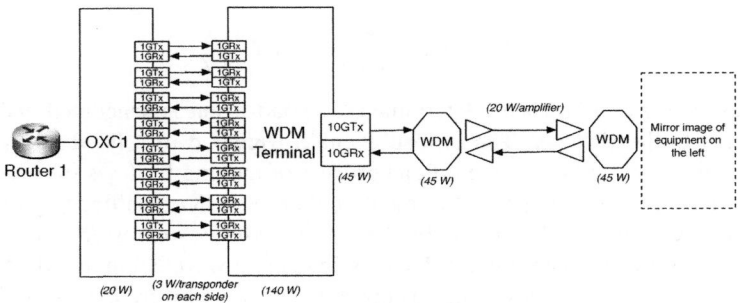

Fig. 3. Energy consumption per element (watt) at the optical level

3.4 Scenarios

When IP flows are offloaded to the optical level, and more than one wavelength is able to serve the offloaded flow, a criteria is needed to select the proper wavelength. In our attempt to find the best performing strategy for offloading IP flows from the IP level to the optical level, three different strategies are considered: *dedicated*, *spreading*, and *packing*. For the sake of performance evaluation we also investigate the effects when offloading only the biggest flow per offloading event, instead of offloading all possible IP flows. Furthermore, we look into the effects of allowing more coarser flows by limiting the maximum flow rate to the equivalent of an OC-24 link (*i.e.*, 1.244 Gbps).

Dedicated: This strategy offloads one flow per unused wavelength. In practice, this means that a maximum of eight flows (one per wavelength) are offloaded to the optical level. The other remaining flows stay at the IP level until one of the offloaded flows ends, releasing thus the wavelength. When a new offload event occurs, the biggest flow at the IP level is offloaded, and subsequently, dedicatedly assigned to one wavelength.

Spreading: This strategy aims for an equally distributed utilization (in terms of bandwidth) of the wavelengths. By choosing the least loaded wavelength for assigning the offloaded flow, an attempt is made to equally divide the flows over the available wavelengths. This offloading algorithm chooses the wavelength with the most bandwidth capacity available to serve the flow to be offloaded.

Packing: Opposite to the *spreading* strategy, this strategy chooses the wavelength which is most loaded, *i.e.*, by checking all the available wavelengths and selecting the most loaded one able to serve the requested flow. Since the other wavelengths are kept less loaded, the chance to serve a flow with a large bandwidth request increases. By using this strategy an attempt is made for maximizing the probability to find a wavelength that is able to serve a flow with a large bandwidth demand.

3.5 Biggest Flow, Highest Priority

While non-offloaded flows are kept at the IP level due to not enough bandwidth available, all other flows at the IP level are offloaded to the optical one. For the sake of our performance evaluation, we also observed the performance when we offload only the flow with the highest rate at the IP level. If this flow cannot be offloaded, all flows are kept at the IP level until the next offload event is triggered, serving the highest possible priority to the biggest flow.

4   Simulation Results

In this section we discuss the simulation results. For the simulation of flows over an optical connection, we made use of the NS-2 simulation tool [1]. This tool does not simulate WDM networks [5] [4] without an additional module such as Optical WDM Network Simulator (OWns) [2]. Some features needed for our simulation, such as grooming multiple flows into the same wavelength, are not available in the OWns module. Just like NS-2, the OWns module is based on open source, which allowed us to code this function. We performed our simulations in a time interval of 200,000 seconds (roughly two days).

As expected, the performance of the *dedicated* strategy is considerably lower than the *spreading* and *packing* strategies (Figure 4). This can be explained by the fact that only one flow per wavelength can be assigned, leaving significant amount of bandwidth unused. In our scenario we have used flow rates between 1 to 500 Mbps. Since one OC-24 link has a total bandwidth capacity of 1.244 Gbps, the maximum possible offloading percentage is about 40% when the maximum link capacity is in use. Obviously, this is not realistic in practice. As Figure 4 shows, the offloaded percentage lies between 20% and 30%. This number can increase up to 50% when the granularity of the flows are coarser (*e.g.*, flow rates between 1 Mbps to 1.244 Gbps). We also observed the performance when offloading only the biggest flow, as described in subsection 3.5. In this case, the simulation results show no significant differences for the *dedicated* strategy because the flows are offloaded almost in the same order.

The *spreading* and *packing* strategies perform better than the *dedicated* strategy. By allowing multiple flows per wavelength, the used capacity per wavelength increases significantly. As Figure 4 shows, both *spreading* and *packing* perform roughly between 90% and 100%. There are some moments where the *spreading* strategy show better results than the *packing* strategy, but on average the *packing* strategy (96.7%) slightly outperformed the *spreading* strategy (96.4%). The main reason for the better performing *packing* strategy is because it aims for maximizing the probability to serve a flow with a high bandwidth demand as described in section 3.4.

Unlike the *dedicated* strategy, the *packing* and *spreading* strategies perform a little less efficient when the flows are coarser (*i.e.*, higher variance in flow rate). The time a flow needs to reside at the IP level grows when the flow rate is higher, resulting in lower offload percentages. When rates vary between 1 Mbps to 1.244 Gbps, the *packing* strategy shows better performance (94.7%) in comparison to the *spreading* one (93.9%).

Looking at the 'biggest flow, highest priority' alternative, strong dropping is observed (Figure 5). This is caused by flows with a high bandwidth demand residing on the IP level due to insufficient available bandwidth on the optical level. When one of the wavelengths is able to accommodate the 'waiting' flow, a quick recovery of the offload percentage takes place. On average we can see that the *packing* strategy outperforms the *spreading* strategy when the 'biggest flow, highest priority' alternative is applied.

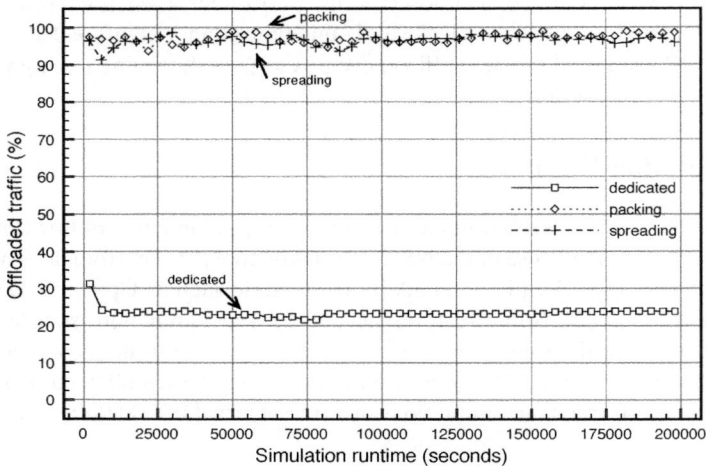

Fig. 4. Offload percentage per grooming strategy. Flow rates vary from 1 Mbps to 500 Mbps.

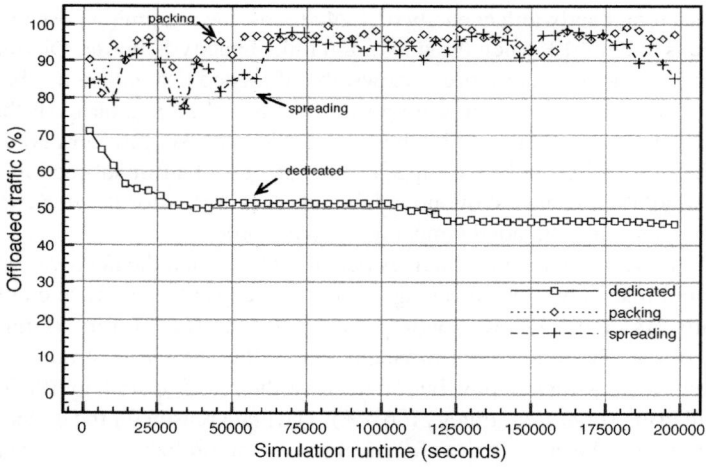

Fig. 5. Offload percentage per grooming strategy. Flow rates vary from 1 Mbps to 1.244 Gbps, using the 'biggest flow, highest priority' alternative.

4.1   Energy Consumption

By observing the energy consumption of the different strategies, we can conclude that due to the small amount of time that one or more wavelengths are idle, there is no significant difference in energy consumption among the considered strategies. We do see however a small difference in the *packing* strategy when compared with the other ones. This difference arises mainly at the start of the simulation, when the total used bandwidth has not reached the near maximum transmission rate yet. When the network load is close to maximum, all eight wavelengths are in use and, independently from the used strategy, the energy consumption is quite similar. Table 1 summarizes the percentage of time that all wavelengths were in use.

Table 1. The percentage of time with all eight wavelengths in use

Strategy	Percentage
Packing	99.82%
Spreading	99.97%
Dedicated	99.98%

5   Conclusions and Future Work

Through use of simulation we have shown the performance behavior of three different offloading strategies, *i.e.*, *dedicated*, *packing*, and *spreading*. We also observed the effects when using coarser flow rates, and we looked into an offload variant to serve flows with the biggest rate first. We conclude that, regardless the granularity of the flow rates, the *packing* strategy is superior to its *spreading* variant. At the cost of about 1%, it is possible to provide the highest priority to the biggest flow.

We also observed the energy consumption of the optical level, depending on the used strategy. Independently from the employed strategy, transmission rates are near maximum, therefore the percentage of time that one or more wavelengths are not in use is negligible. This has a direct impact on the energy consumption because the in-use wavelengths determine the difference in the percentage of consumed energy. For this reason we conclude that there is no significant difference in power consumption.

In our work we assumed that all flows have constant rates and do not change over time. As future work, it lies in the extension of this work to investigate the performance behavior when flow rates change over time. Although we concluded that there is no significant difference in power consumption, it might also be interesting to investigate if this conclusion holds when the network load is *not* near maximum capacity, increasing the chance that one or more wavelengths are not in use and therefore using less energy. At last, the use of a more complex topology, *e.g.*, where not fully utilized ingress wavelengths can be groomed into fewer egress wavelengths, may be considered in order to observe whether there is any reduction in the number of active wavelengths.

Acknowledgments. This research work has been supported by the EC IST-EMANICS Network of Excellence (#26854). Special thanks to Roeland Nuijts (SURFnet) and Cees de Laat (UvA) for their valuable contribution to this paper.

References

1. Ns-2 network simulator, http://www.isi.edu/nsnam/ns/
2. Optical WDM network simulator (OWns),
   http://www.eecs.wsu.edu/~dawn/software/owns.html
3. Bernstein, G., Rajagopalan, B., Saha, D.: Optical Network Control: Architecture, Protocols, and Standards. Addison-Wesley Longman Publishing Co., Inc., Boston (2003)
4. Chittenden, A.: Extending OWns to include protection functionality. Master's thesis, University of Pretoria (2005)
5. Drummond, A.C., da Silva, R.T.R.: IP over WDM Module for the NS-2 Simulator. In: IEEE International Conference on Communications Workshops, ICC Workshops 2008, Beijing, May 2008, pp. 207–211 (2008)
6. Dutta, R., Rouskas, G.N.: Traffic Grooming in WDM Networks: Past and Future. Tech. rep., Raleigh, NC, USA (2002)
7. Fioreze, T., Pras, A.: Using self-management for establishing light paths in optical networks: an overview. In: Proceedings of the 12th Open European Summer School, Institut für Kommunikationsnetze und Rechnersysteme, Universität Stuttgart, Stuttgart, Germany, September 2006, pp. 17–20 (2006)
8. Fioreze, T., Pras, A.: Self-management of lambda-connections in optical networks. In: Bandara, A.K., Burgess, M. (eds.) AIMS 2007. LNCS, vol. 4543, pp. 212–215. Springer, Heidelberg (2007)
9. Fioreze, T., van de Meent, R., Pras, A.: An architecture for the self-management of lambda-connections in hybrid networks. In: Pras, A., van Sinderen, M. (eds.) EUNICE 2007. LNCS, vol. 4606, pp. 141–148. Springer, Heidelberg (2007)
10. Hillier, F.S., Lieberman, G.J., Hillier, F., Lieberman, G.: Introduction to Operations Research. McGraw-Hill Science/Engineering/Math (July 2004)
11. de Laat, C., Radius, E., Wallace, S.: The rationale of the current optical networking initiatives. Future Gener. Comput. Syst. 19(6), 999–1008 (2003)
12. Mannie, E.: Generalized Multi-Protocol Label Switching (GMPLS) Architecture. RFC 3945 (Proposed Standard) (October 2004), http://www.ietf.org/rfc/rfc3945.txt
13. Sabella, R., Iovanna, P., Oriolo, G., D'Aprile, P.: Strategy for dynamic routing and grooming of data flows into lightpaths in new generation network based on the GMPLS paradigm. Photonic Network Communications 7(2), 131–144 (2004)
14. Wang, J., Cho, W., Vemuri, V.R., Mukherjee, B.: Improved approaches for cost-effective traffic grooming in wdm ring networks: Ilp formulations and single-hop and multihop connections. J. Lightwave Technol. 19(11), 1645 (2001)

MBAC: Impact of the Measurement Error on Key Performance Issues

Anne Nevin, Peder J. Emstad, and Yuming Jiang

Centre for Quantifiable Quality of Service in Communication Systems (Q2S)*,
Norwegian University of Science and Technology (NTNU), Trondheim, Norway
{anne.nevin,peder.emstad}@q2s.ntnu.no

Abstract. In Measurement Based Admission Control (MBAC), the decision of accepting or rejecting a new flow is based on measurements of the current traffic situation. Since MBAC relies on measurements, an in-depth understanding of the measurement error and how it is affected by the underlying traffic is vital for the design of a robust MBAC. In this work, we study how the measurement error impacts the admission decision, in terms of *false rejections* and *false acceptances*, and the consequence this has for the MBAC performance. A slack in bandwidth must be added to reduce the probability of false acceptance. When determining the size of this slack, the service provider is confronted with the trade-off between maximizing useful traffic and reducing useless traffic. We show how the system can be provisioned to meet a predefined performance criteria.

1 Introduction

Measurement Based Admission Control (MBAC) has for a long time been recognized as a promising solution for providing statistical Quality of Service(QoS) guarantees in packet switched networks. An MBAC does not require an *a priori* source characterization which in many cases may be difficult or impossible to attain. Instead, MBAC uses measurements to capture the behavior of existing flows and uses this information together with some coarse knowledge of a new flow, when making an admission decision for this requesting flow. Measurements are unavoidably inaccurate. This imperfection creates uncertainties which affect the MBAC decision process. The degree of uncertainty depends on flow characteristics, the length of the observation window and the flow dynamics. Flows will be accepted when they should have been rejected, *false acceptance*, and rejected when they should have been accepted, *false rejections*. For the service provider, false rejections translates into a decrease in utilization and for the end user, false admissions means that the QoS of the flow can no longer be guaranteed. Basing admissions on measurements clearly requires the understanding of the measurement error and how this impacts the performance of MBAC.

---

* "Centre for Quantifiable Quality of Service in Communication Systems, Centre of Excellence" appointed by The Research Council of Norway, funded by the Research Council, NTNU and UNINETT. http://www.q2s.ntnu.no.

F.A. Aagesen and S.J. Knapskog (Eds.): EUNICE 2010, LNCS 6164, pp. 11–20, 2010.

Consider a link where the maximum allowable mean rate is uc and the main task is to keep the average workload of the link at or below this level. An MBAC is put in place which uses measurements to find an estimate $\hat{R}$ of the average aggregate rate of accepted flows. When a new flow arrives, with mean rate ξ, it will be accepted if:

$$\hat{R} + \xi \leq uc. \tag{1}$$

With MBAC, care must be taken, since $\hat{R}$ is not the true mean rate and includes an error, which will cause *false admissions* and *false rejections*. The measurements are improved when they are taken over a longer measurement window. However, flows leaving within the window results in flawed estimates, thus the flow lifetimes set an upper limit for the window size. Given this window size, how confident can we be that this is not a false acceptance? Is this good enough? If the answer is no, the reserved bandwidth for the flows uc, must be reduced by some slack to make up for the measurement uncertainty. But how large should this slack in bandwidth be?

There is a tradeoff between rejecting too many flows thus wasting resources, and accepting too many flows resulting in QoS violations. In this work we study how the measurement errors and flow dynamics impact the performance of MBAC in terms of proper performances measures. A simple example shows how the system can be provisioned with a predefined performance criteria.

The focus in the literature has been on finding MBAC algorithms that maximize utilization while providing QoS which basically implies the determination of uc (see [1] and [2] for an overview). A deeper analytical understanding of the measurement process and its error has been sought in [3] and [4]. Our work differs significantly from previous work, in that we purely focus on the impact inaccurate measurements have on the MBAC decision process. Correlation characteristics within flows are included and we find how the uncertainty in the measurements vary with the length of the observation window. This work is based on previous work [5] and is part of a methodology and design of an analytical framework for analyzing measurement error.

The reminder of the paper is organized as follows. First the system model is introduced in Section 2. Section 3 details the rate level measurements and Section 4 introduces the flow level framework and defines the performance measures. Provisioning is discussed in Section 5 and follows up with a case study in Section 6, before the conclusion is given in Section 7.

2   System Model

Flows compete for a limited resource, a network link of capacity c, controlled by MBAC. The flow's QoS requirement can be guaranteed as long as the average aggregate rate is at or below uc, where u, $0 < u < 1$ is a tuning parameter. An optimal value for u depends on flow characteristics. In this work, u is assumed a given constant and a discussion around its optimal settings is out of scope. The MBAC works as follows: The MBAC measures the average aggregate rate

$\hat{R}$ based on observations of the aggregate rate $R(t)$ over a measurement window of size w. This measurement replaces the measurement taken in the previous window. When a new flow arrives, it will be accepted or rejected according to (1). Additional flows arriving within the same window will be rejected and lost.

Each flow i is a stationary rate process, $X_i(t)$ with mean ξ and covariance $\rho(\tau)$. A mixing of different flow types will cause increased complexity for the MBAC algorithm and also the measurement process. To simplify, only the homogenous case where flows are of the same type, will be considered. With this assumption, the *system state* N can be specified by the current number of flows. The maximum number of flows the system can handle is thus $n_{max} = uc/\xi$. It is natural to separate the timescale into the rate level where the measurements are done and the flow level where the admission decision is made. The measurements are taken by means of continuous observation and will be discussed next.

3   Rate Level: Measurements and Measurement Error

The flow rate process $X_i(t)$ is observed continuously over the window and an estimate of the mean, $\hat{X}_i$ is then: [5]:

$$\hat{X}_i = \frac{1}{w} \int_0^w X_i(t)dt \qquad (2)$$

This measured statistic varies with the window size according to [5]:

$$\zeta^2(w) = \frac{2}{w^2} \int_0^w (w - t)\rho(t)dt \qquad (3)$$

The aggregate rate process is also a stationary rate process. Conditioned on being in state n, the aggregate mean is $n\xi$. An estimate of the aggregate mean is given by:

$$\hat{R} = \sum_{i=1}^n \frac{1}{w} \int_0^w X_i(t)dt \qquad (4)$$

When the number of flows is in the order of a few tens [6] (e.g 30 flows), the aggregate rate is approximately normally distributed, thus also $\hat{R} \sim \mathcal{N}(n\xi, n\zeta^2(w))$. This assumption will be made here. The accuracy of this measurement can then be described by the $1 - \varepsilon$ confidence interval: $\hat{R} - z_{\frac{\varepsilon}{2}}\sqrt{n}\zeta(w) \leq n\xi < \hat{R} + z_{\frac{\varepsilon}{2}}\sqrt{n}\zeta(w)$, where $z_{\frac{\varepsilon}{2}}$ is the $(1 - \alpha/2)$ quantile of the normal distribution.

It is intuitive to think that in order to achieve a certain measurement accuracy all that is needed is to increase the window size. However in order for the above estimate to hold, the requirement is that no flows leave during the window, i.e. the aggregate rate process is stationary with a known distribution. Otherwise the actual estimate becomes incorrect. The flow lifetime therefore sets an upper limit for the window size. The inaccuracy of the measurement translates into uncertainty in the admission decision process done at the flow level and will be described next.

4   Flow Level and Performance Measures

Let new flows arrive following a Poisson process with parameter λ. If the flow is accepted it stays in the system for a lifetime that is negative exponentially distributed with mean $1/\mu$. A flow that is not accepted by the MBAC is lost. The *offered flow load* is the Erlang load [7] denoted by A. This is the average number of simultaneous flows if there is no blocking given by:

$$A = \lambda \cdot E(T_L). \tag{5}$$

With the assumption that $\hat{R} \sim \mathcal{N}(i\xi, i\zeta^2(w))$, if there are $N = i$ flows in the system, a new arriving flow will be accepted with a probability $q_i = P(\hat{R} + \xi \leq n_{max} \mid N = i)$. We will assume that the arrival rate is such that the probability of more than one flow arrival per window is very small. The lost traffic due to multiple arrivals within the window is thus very small and can be neglected. A flow upon arrival is promptly admitted or rejected, which justifies modeling the admission process as a loss system.

The number of flows currently accepted by the MBAC then follows a continuous time Markov chain, see Fig. 1 and the probability that there are i flows in the system is:

$$P(i) = \frac{\frac{A^i}{i!} \prod_{x=0}^{i-1} q_x}{\sum_{j=0}^{\infty} \frac{A^j}{j!} \prod_{x=0}^{j-1} q_x} \tag{6}$$

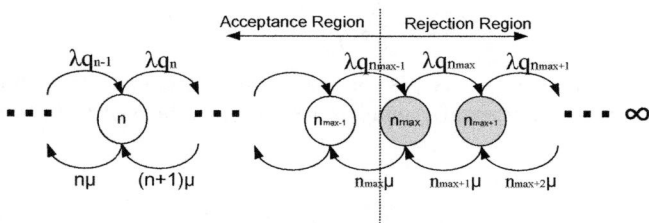

Fig. 1. State diagram of the number of sources accepted by the MBAC

In [8] a similar framework is defined, where the q_is are based on the probability that the instantaneous rate measurements are above a threshold. However, they do not specify how these measurements should be taken. Since the instantaneous rate is a random quantity without a true value, it can not be measured [9]. This is where our work deviates significantly from [8]. We study the measurement error in isolation and the q_is depend on how accurate the measurements are.

Implied by (6) and as also discussed in [8], the distribution $P(i)$ is indeed insensitive to the distribution of the flow lifetime and only depends on the expected flow lifetime.

The system state space can be divided into two regions; the *acceptance region*, $N < n_{max}$ and the rejection region, $N \geq n_{max}$ see Fig. 1.

Rejecting a flow when the system is in the acceptance region is a *False rejection* and accepting a flow when the system is in the rejection region is a *False acceptance*.

For assessing the performance and provisioning purposes, we define the following flow level performance measures:

- **Probability of False acceptance,** P_{FAcc}, is the probability that an arriving flow is accepted in the rejection region

$$P_{FAcc} = \sum_{i=n_{max}}^{\infty} P(i) \cdot q_i \qquad (7)$$

- **Probability of False rejection,** P_{FRej}, is the probability that an arriving flow is rejected in the acceptance region

$$P_{FRej} = \sum_{i=1}^{n_{max}-1} (1 - q_i)P(i) \qquad (8)$$

- **Blocking probability,** P_B, is the probability that an arriving flow is rejected.

$$P_B = P_{FRej} + P(N \geq n_{max} \cap rejection) = \sum_{i=1}^{\infty} (1 - q_i)P(i) \qquad (9)$$

- **Carried useful traffic,** A_{useful} , is the expected number of flows in the acceptance region.

$$A_{useful} = \sum_{i=0}^{n_{max}} iP(i) \qquad (10)$$

- **Carried useless traffic,** $A_{useless}$, is the expected number of flows in the rejection region.

$$A_{useless} = \sum_{i=n_{max}+1}^{\infty} iP(i) \qquad (11)$$

- **Lost Traffic,** A_{lost} , is the traffic that is blocked from the network

$$A_{lost} = AP_B \qquad (12)$$

If there are no measurement errors, the admission controller becomes *ideal*, $\hat{R} = \bar{R}$ and the distribution of flows is then as for the Erlang Loss system. This system only carries useful traffic and arriving flows will experience a blocking probability given by the Erlang B formula.

4.1   Flow Load and Window Size Limitations

In the analytical formulas used for the performance evaluation, when increasing A, it is indifferent if this is done by increasing the mean flow lifetime $1/\mu$ or increasing the arrival rate λ. In reality this is not true. Many arrivals within a measurement window will increase the blocking probability since the MBAC only admits at most one flow after a measurement update. When predicting the performance using the analytical formulas, we accept just one arrival within a window. It is assumed that the blocking probability due to multiple arrivals is negligible.

For the accepting probabilities, the assumption is that the number of flows is constant during a measurement window. If flows leave during the window, then the theoretically predicted performance will be optimistic compared to actual performance in terms of system utilization. Given a constant A, a longer measurement window (thereby reducing the measurement error) can be used for long lifetimes (infrequent arrivals) as compared to short lifetimes (frequent arrivals).

The performance measures can be directly stated upfront using the defined formulas if: 1) The lost traffic due to previous arrivals within the window can be ignored 2) The probability of a flow leaving within a measurement window is small for the actual parameter values μ and n 3) In addition we assume that the correlation at arrival points can be neglected.

5   Provisioning

The QoS provided to the flows can only be guaranteed as long as the number of flows is at or below n_{max}, thus admitting more than n_{max} flows should be avoided. If the probability of false acceptance is too high, a safeguard in terms of a slack in bandwidth can be added to make up for the measurement errors.

As in Paper [5], let the safeguard have increments of size size $l\xi, 0 < l < n_{max}$ and the refined admission control algorithm becomes:

$$\hat{R} + \xi \leq \xi n_{max} - l\xi \tag{13}$$

The critical situation arises as soon as the system reaches state n_{max}, where accepting a flow will result in the first false admission. With the condition that the system is in state n_{max} we define the *conditional performance requirement*:

$$P(FAcc \mid N = n_{max}) = P(\hat{R} + \xi \leq \xi n_{max} \mid N = n_{max}) \leq \epsilon \tag{14}$$

where ϵ is termed the *conditional performance target*.

For a given quantile and predefined window size, $P(FAcc \mid N = n_{max})$ can be kept below the target if the number of levels is [5]:

$$l + 1 = \left\lceil \frac{\sqrt{n_{max}}\zeta(w)z_\varepsilon}{\xi} \right\rceil . \tag{15}$$

Where z_ε is the ε- quantile of the standard normal distribution The resulting l can be used for provisioning the system.

Since the MBAC solely estimates the number of flows through measurements, l is independent of the system state and when there are i flows in the system, a new arriving flow will be accepted by the MBAC, with a probability $q_i = P(\hat{R} + \xi \leq n_{max} - l \mid N = i)$.

The size of l controls the probability of entering the rejection region by forcing the probability distribution towards left (Fig. 1). Obviously, if the slack is too large, the MBAC becomes too pessimistic and resources are wasted unnecessarily. The actual performance can be evaluated by means of the performance measures defined in Section 4. To the customer, the performance measures of interest are the blocking probability and the probability of false acceptance. The service provider seeks to balance the carried useless traffic and carried useful traffic.

For a given flow load the network can be provisioned to meet a desired performance target. But what flow load should be used?

The network should be dimensioned to ensure a small blocking probability under normal loads, say $P_B < 0.01$. At such low loads the probability of entering the rejection region is very small and excellent QoS can be provided to all flows. The problem arises in times of excessive demand. With an *ideal* controller, when the load increases above what is predicted the blocking probability increases to unacceptable high values. However, the QoS to the already admitted flows will not be harmed. With MBAC on the other hand, also the probability of false acceptance and useless traffic increase with increasing loads. Reviewing work in the MBAC literature, it is common practise to test the MBAC performance under a heavy flow load. For example a load corresponding to 50% blocking probability is used in [1] and an infinite load is used in [3].

We do not attempt to answer, exactly what load to use for provisioning purposes. The load must be relatively high, since the main task of MBAC is to preserve QoS to its users when the load exceeds normal values [10]. Obviously at such loads, the normal blocking probability (e.g $P_B = 0.01$) can not be met.

6   Case Study Using MMRP Source Models

In this section provisioning to fulfill some predefined performance criteria will be demonstrated with an example.

Let the flows be modeled by a two-state Markov modulated rate process (MMRP) which is a simple, yet realistic source model used to model both speech sources and video sources [11]. The MMRP process, $X(t) = rI(t)$ where $I(t)$, alternates between states $I = 0$ and $I = 1$ where r is the peak rate. The duration of the 0 and 1 states follows a negative exponential distribution with mean $1/\alpha$ and $1/\beta$ respectively.

The variance of the time average of such a source is [5]:

$$\zeta^2(w) = \frac{2r^2\alpha\beta}{w^2(\alpha + \beta)^3} \left(w - \frac{1}{\alpha + \beta}(1 - e^{-w(\alpha+\beta)}) \right) \tag{16}$$

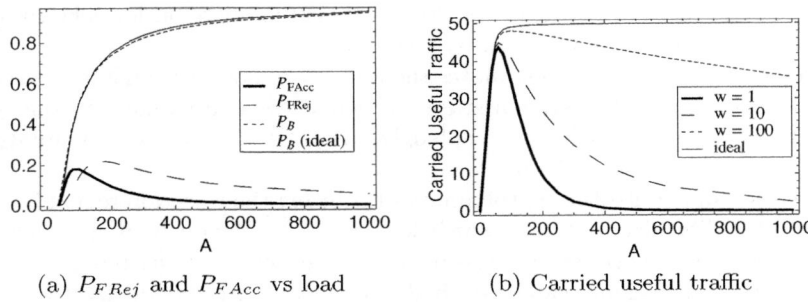

(a) P_{FRej} and P_{FAcc} vs load

(b) Carried useful traffic

Fig. 2. Probability of false acceptance as the load increase and the corresponding Carried Useful Traffic

In the following, $\alpha = \beta = r = 2$, and ξ is then 1. The flows have a QoS requirement that can only be guaranteed as long as $N \leq n_{max} = 50$.

First we shall show how the performance measures are impacted when the offered flow load, A is varied. At this stage we use no slack in the bandwidth (e.i. $l = 0$) Keeping the window size constant at $w = 1$, Fig. 2(a) shows how the performance measures P_{FAcc}, P_{FRej} and P_B are affected when the flow load increases. Low loads result in negligible false acceptance. Instead false rejections cause a slight increase in blocking probability as compared to the *ideal*. At a load of about $A = 60$, $P_{FAcc} = P_{Frej}$ and then as the load increases P_{FAcc} increases resulting in a slightly lower blocking probability as compared to the *ideal*. As the load increases towards infinity, P_{FAcc} becomes zero. The reason for this is that the system moves into the rejection region and will eventually only carry useless traffic. This is illustrated in Fig. 2(b) which shows that as the load increases the carried useful traffic approaches zero. Also shown, is that for larger window sizes, the MBAC approaches the *ideal* and the carried useful traffic falls off slower.

We are interested in balancing the tradeoff between the overall probability of false rejection and probability of false acceptance. For the user, the main concern is the probability of false acceptance and in this example the requirement is $P_{FAcc} < 0.01$. Let the offered flow load be $A = 100$ and let the window size be $w = 1$. The performance plots shown in Fig. 3(a), 3(b), 3(c), and 3(d) illustrate the tradeoff between blocking and accepting flows when $P(FAcc \mid N = n_{max})$ is varied. Consider first the performance in the light of the customer. To fulfill the requirement of $P_{FAcc} < 0.01$, Fig.3(a), shows that $P(FAcc \mid N = n_{max}) < 0.17$. At this value the blocking probability is about 5% larger than for the *ideal*.

For the service provider, the concern is false rejections and useless traffic. As $P(FAcc \mid N = n_{max})$ increases, false rejections fall off (see Fig. 3(b)) and the useless traffic increases (see Fig.3(c)). Observe in Fig.3(d) that a value $P(FAcc \mid N = n_{max}) = 0.15$ maximizes the carried useful traffic. Reducing the value and the admission controller becomes too strict due to the increase in P_{FRej}. Increasing the value passed this point on the other hand and the admission controller accepts too much useless traffic. In this case, using $P(FAcc \mid N = n_{max}) = 0.15$, also ensures that $P_{FAcc} < 0.01$.

This is an interesting fact since it shows that the carried useful traffic can be maximized by correctly tuning $P(FAcc \mid N = n_{max})$. This value of $P(FAcc \mid N = n_{max})$ can be directly mapped to a safeguard in terms of levels (15). The carried useful traffic can thus be increased by adding a safeguard. However if the safeguard is too large, the carried useful traffic will decrease because the MBAC becomes too strict.

The value of l which maximizes the carried useful traffic, will depend on the window size and the offered flow load. With a higher flow load A, the value of l which maximizes the carried useful traffic will increase. For example, the extreme load of $A = 1000$ results in $l = 9$. Decreasing the load, will have the opposite effect, and eventually as the load is reduced further adding a safeguard will only decrease utilization.

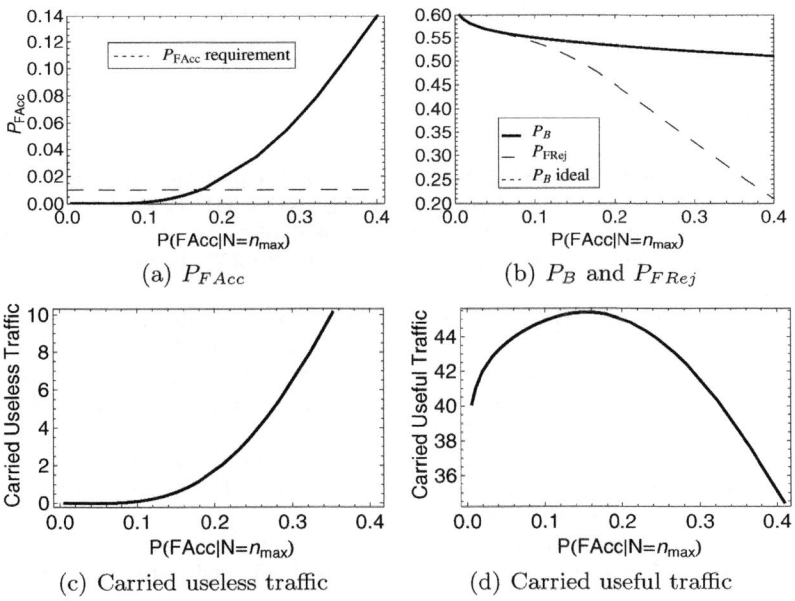

Fig. 3. The performance measures as $P(FAcc \mid N = n_{max})$ varies for $A = 100$ and $w = 10$ a)Probability of false acceptance, b) Probability of false rejection and overall blocking probability, c) Carried useless traffic and d) Carried useful traffic

Numerous simulations (with MMRP sources) to check the analytical formulas are omitted due to limited space. As long as there is no correlation between measurement windows, these simulations indicated that the formulaes are valid even when the number of aggregated flows is as low as 3.

7   Conclusion

This current work gives an in-depth understanding of how measurement uncertainties and flow dynamics impact the MBAC admission decision. An MBAC algorithm, no matter how advanced, is of little use if the measurement errors are not taken into account. These errors translate into uncertainty in the decision process. The degree of uncertainty abates with the length of the observation window. Despite the heavy reliance on measurements, there is in the literature of MBAC, surprisingly very limited work focusing on the impact of the measurement error and how it affects the admission decision.

The probability of false admissions can be reduced by adding a slack in bandwidth. However, if the slack is too large, flows are blocked unnecessarily. With some appropriate performance measures, we showed how the system can be provisioned to meet a predefined performance criteria.

In this work, we assumed that the flows were homogenous, however in a separate work we have developed the analytical tools needed to extend this work to the non-homogenous case.

References

1. Breslau, L., Jamin, S., Shenker, S.: Comments on the performance of measurement-based admission control algorithms. In: IEEE INFOCOM (2000)
2. Moore, A.W.: Measurement-based management of network resources. Technical Report, University of Cambridge, Cambridge CB3 OFD, United Kingdom (April 2002)
3. Grossglauser, M., Tse, D.N.: A time-scale decomposition approach to measurement-based admission control. IEEE/ACM Trans. Networking 11(4), 550–563 (2003)
4. Dziong, Z., Juda, M., Mason, L.G.: A framework for bandwidth management in ATM networks - aggregate equivalent bandwidth estimation approach. IEEE/ACM Trans. Networking 5(1), 134–147 (1997)
5. Nevin, A., Emstad, P., Jiang, Y., Hu, G.: Quantifying the uncertainty in measurements for mbac. In: Oliver, M., Sallent, S. (eds.) EUNICE 2009. LNCS, vol. 5733, pp. 1–10. Springer, Heidelberg (2009)
6. van de Meent, R., Mandjes, M., Pras, A.: Gaussian traffic everywhere? In: IEEE International Conference on Communications, ICC 2006, June 2006, vol. 2, pp. 573–578 (2006)
7. Iversen, V.B.: Teletraffic engineering and network planning, course Textbook (May 2006)
8. Gibbens, P.K.R.J., Kelly, F.P.: A decision-theoretic approach to call admission control in atm networks. IEEE Journal on Selected Areas in Communications (JSAC) 13(6), 1101–1114 (1995)
9. Rabinovich, S.G.: Measurement Errors and Uncertaintees, 2nd edn. Springer, New York (2005)
10. Roberts, J.W.: Internet traffic, qos and prising. Proceedings of the IEEE 92(9) (2004)
11. Schwartz, M.: Broadband integrated networks. In: Becker, P. (ed.), Prentice Hall, Englewood Cliffs (1996)

An Algorithm for Automatic Base Station Placement in Cellular Network Deployment

István Törős and Péter Fazekas

High Speed Networks Laboratory
Dept. of Telecommunications, Budapest University of Technology and Economics,
Magyar tudósok körútja 2., 1117 Budapest, Hungary
{toros,fazekasp}@hit.bme.hu
http://www.springer.com/lncs

Abstract. The optimal base station placement is a serious task in cellular wireless networks. In this paper we propose a method that automatically distributes the base stations on a studied scenario, maintaining coverage requirement and enabling the transmission of traffic demands distributed over the area. A city scenario with normal demands is examined and the advantages/disadvantages of this method are discussed. The planner and optimizing tasks are based on an iterative K-Means clustering method. Numerical results of coverage, as well as graphical results permit on radio coverage maps, are presented as base for main consequences. Furthermore, the final results are compared to another simple planning algorithm.

Keywords: cellular network planning, coverage, capacity.

1 Introduction

The radio planning of cellular wireless network is a highly investigated topic, because operators can save budget using a cost efficient planning method. The performance of a mobile network is mainly characterized by the received signal strength and signal to noise plus interference ratio at the mobile's position, hence the coverage of the network. The coverage is mainly affected by transmit power and radio propagation, hence depends on the environment of network. Gain of better coverage and increased signal to noise plus interference ratio (SINR) in a cellular network are the key objectives of mobile networking industry. Developing and using an algorithm that automatically plans the positions of base stations and provides the necessary coverage and capacity over the area with the least number of stations, is thus of utmost importance.

Our research is focused on the development and investigation of such an automatic planning algorithm. The frequency adaptation and power control are very simple in our simulation model. Any other, more complex algorithms might be used with our planning method, these would certainly result in lower number of base stations.

The key object of the planning of the new generation networks is that the map of SINR suits to user's distribution. In the position of the users with high

F.A. Aagesen and S.J. Knapskog (Eds.): EUNICE 2010, LNCS 6164, pp. 21–30, 2010.

demand should be provided with high SINR. We could achieve this criterion if the base stations are planned near the high demand users.

This new algorithm is based on a clustering method which is very popular suggestion for planning, the K-means algorithm. However, most of the methods use on K-means concentrate either on dimensioning or optimization and they require prediction of number of beginning clusters, which is not straightforward, see e.g. [1][2]. In [7], the author has used Set Cover Base Station Positioning Algorithm for BS planning. This algorithm is focussed on optimal coverage of networks. The full serving of users by placement algorithm is secondary, because the verification of SINR runs after BS placement. In [6], the authors have focussed attention on special K-means clustering with fixed number of cluster. This may produce good area coverage and have a small amount of overlap between cells. However, such a network will probably not be able to satisfy the traffic demands within each cell. In contrast, our method does both the dimensioning and optimization steps and does not require initial estimations, rather can start with an empty (in terms of number of base stations) area, with an arbitrarily placed single base station and places the necessary stations over this. However, if required, the algorithm might be used starting with an initial arbitrary network topology (location of arbitrary number of base stations) and places new base stations to fulfill coverage and capacity requirements. This is useful in the case when network deployment strategy has to be planned in order to serve increasing capacity demands in an already running network. It is important to note that the location algorithm creates the clusters based on the properties of base station which were initialized at the beginning.

2   Modelling Environment

Our goal is to investigate the capability of the new algorithm in mobile networks and evaluate coverage, SINR and quality of transmission in the network. In order to achieve our goal, we used our own developed software which simulates a mobile network, where base stations can be added easily. The model of the environment and the network implemented in the simulator is detailed in the following.

Since the communication range is an important property, the distances between nodes are displayed precisely on the graphic screen. The signal power and SINR are also shown on the screen with color spectrum changed gradually from weak to strong. Based on the signal power, we can calculate the coverage area of a given cell. The SINR parameter guides us to derive an estimated overall performance of the network.

We assume that the network to be developed by the algorithm (and the software tool implementing it) is the novel 3GPP LTE (Long Term Evolution) network. However, the method can be applied to any other radio network, even to subsequent ones (e.g. LTE Advanced). The algorithm assumes that a Radio Resource Control (RRC) mechanism is present that allocates radio resources to different cells according to arbitrary rules. In LTE, this practically means that RRC allocates Physical Resource Blocks (PRB) to cells (a PRB is 12 OFDM subcarriers transmitted in a 1 ms subframe).

2.1 Simulated Terrain

The environment of simulations is composed of three different layers. The first is the flat geographical layer which helps the speed of simulation. The second is the layer of buildings and roads. In the following, we assume that this is a large city with big houses, but other terrain types are also applicable. For the numerical results presented below, we assumed a 9 km^2 city area. The roads are generated by an algorithm that creates first the road network with confluence. The buildings will be placed near the roads with random height. The last is the layer of demands which is created according to the following scheme.

Traffic demands are given per unit area in bits per second (practically: per pixel of the map). If the position is near a road or a confluence then demand of this place will be higher, the demand is function of distance from the road. The model assumes that more traffic is demanded within buildings as well, as in reality high rate applications are used at home or in office, rather than outdoor. In the calculations presented below, the aggregate required bit rate is 900 Mb/s for the entire map.

2.2 Antenna Model

To keep the model realistic, sectorized antennas are assumed. The antenna horizontal characteristic is described by equations (1) and (2),

$$IF\ \alpha \le 90,\ then\ Power = \cos^2(\alpha) * pw. \tag{1}$$

$$IF\ \alpha > 90,\ then\ Power = 0 \tag{2}$$

where α is an angle between the main direction of sector antenna and a changing vector pointing towards the actual location under examination and pw is the transmitter gain extended by antenna gain. Hence, during the calculations, signal strength is determined (along with the path loss model) according to the direction of a given point of the map.

The vertical characteristic is described by (3).

$$Power = \cos^2(\alpha - x) * pw. \tag{3}$$

We can employ (3) in all directions where α is the vertical angle of the main direction of sector antenna and x is the vertical angle of changing vector. The BS-s are planned with the traditional layout, namely three sector antennas with 120 degrees separation between their main directions.

2.3 Propagation Models

We use COST 231 path loss model for big city environment in our simulations. This has the advantage that it can be implemented easily without expensive geographical database, yet it is accurate enough, captures major properties of propagation and used widely in cellular network planning. This model is a function of the carrier frequency f which range from 50 MHz to 2 GHz, of the

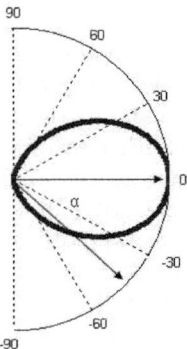

Fig. 1. Horizontal characteristic of Sector Antenna

effective height of base station h_b and the effective height of mobile h_m [5]. The attenuation is given by equation (4)

$$A_p = 46.3 + 33.9 * \log_{10} f - 13.82 * \log_{10} h_b + (44.9 - 6.55 * \log_{10} h_b)$$
$$* \log_{10} d - a(h_m) + C_m \tag{4}$$

where for large cities and $f \geq 400\text{MHz}$.

$$a(h_m) = 3.2 * (\log_{10}(11.75 * h_m))^2 - 4.97 \tag{5}$$

and $C_m = 3\text{dB}$ in large cities. Along with this model, a slow fading is also taken into account by means of a single fading margin expressed in dB.

3   Description of the Algorithm

3.1   K-means Clustering

The method is based on the K-means clustering algorithm that is briefly presented in this section. This is a dynamic clustering which attempts to directly decompose the data into disjoint clusters. We use this algorithm to cluster the demands and form sets of them (cells). The criterion function ($\rho(\text{x,y})$) which has to be minimized, is the distance between a given location of demand within the sector (x) and the position of serving base station (y) weighted by the demand of x.

The first is the assignment step. Join each demand to the closest cluster.

$$C_i^t = \{x_j : \rho(x_j, m_j) < \rho(x_j, m_i^*) \ for \ all \ i^*\} \tag{6}$$

where x_j is the location of demand, m_i is the centroid of the ith cluster (the position of serving base station). C_i^t is the closest cluster of x_j demand at the tth step. The other is update step.

$$m_i^{(t+1)} = \frac{1}{\#C_i^t} \sum_{x \subset C_i^t} x_j \tag{7}$$

where $\#C_i^t$ is the number and x is the location of demand within i^{th} cluster (C_i). This equation (7) calculates the new means to be the center point in the cluster.

3.2   Input and Output Parameters

The input parameters of simulation are the two dimension map that includes positions of roads and buildings, the distribution of traffic demands, the available bandwidth, as well as properties of base stations (power, antenna parameters). The output of the algorithm is the locations of the base stations that are needed to provide coverage and serve all traffic demands. Along with this and the information on available spectrum and traffic demands, the frequency utilization and the spectral efficiency of the network are also main outputs of the algorithm.

3.3   The Planning Algorithm

Our planning algorithm can be realized as a closed loop (Figure 2). The location algorithm waits for the input parameters, and the result of Radio Resource Control which includes allocation of radio resources (PRBs) to traffic demands. As the main contribution of this paper is the base station location algorithm (LA), the applied RRC in the subsequent analysis is supposed to be a very simple, basic operation. Namely, in the evaluations below the available band (20 MHz, 100 PRBs in LTE) is divided to three non-overlapping regions, allocated to the three sectorized cells of each base station. This simple RRC sends to location algorithm the status of service. If a sector is overloaded (the demands in its coverage cannot be served with the fixed band available), then the LA has to eliminate this problem and tries to serve the given area with more stations. This cycle will run until all demands are served.

At the beginning, the initial parameters are fed to the algorithm, practically from a database and a single base station is placed into the center of the area. In the next step, the received signal power from each antenna to every position at the map is calculated. The calculation of received signal power is based on equation (8)

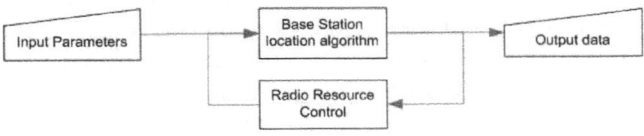

Fig. 2. Main flowchart diagram of RF planning algorithm

$$RS = TS + TAgain - PL + RAgain - C \tag{8}$$

where RS is the received signal in dBm, TS is the transmitter power in dB, $TAgain$ is the gain of transmitter antenna in dB(taking into account antenna characteristics), PL is the pathloss in dB (COST-231), $RAgain$ is the gain of receiver antenna in dB, C is the fading margin in dB. This calculation is executed for every position of the map from all antennas. After this procedure we can run the RRC mechanism. The "best server" policy is followed within the network. First of all we have to search the highest value of received signal at every position and note this value and origin of this signal (antennaID). This task is followed the calculation of other cell's interference. At last we can compute value of SINR using (9).

$$SINR = \frac{maxSignal}{Interference + Noise} \tag{9}$$

The power of thermal noise is taken to be -101 dBm in the evaluations. The algorithm assigns every location (position of demand) to the sector that will serve it. In the following step we will predict the required bandwidth (number of PRBs) for each sector. The relationship between SINR and required frequency for serving a given demand is based on the so called Alpha-Shannon Formula, suggested to be used for LTE networks (10)[4]

$$SpectralEfficiency = \alpha * \log_2(1 + 10^{\frac{SINR}{10*impfactor}}) \tag{10}$$

where α=0.75, $impfactor$=1.25, and $SINR$ is Signal Noise Interference ratio at the actual point in dB. The number of required frequency bands per sector is thus given by (11)

$$\sum_{point\ of\ sector} \frac{demand}{SpectralEfficiency} \tag{11}$$

After this step it turns out that the sector is able to serve its actual demands or not. The following steps belong to the LA. Searches the "most unserved" sector (MUS), where the absence of required frequency is the highest. If no such sector is found (all demands are served), than the algorithm will terminate. Otherwise the algorithm locates a new base station near the serving antenna of MUS in the main direction. This new BS has to have same parameters as the BS of MUS. The following step is the running of K-means clustering algorithm. The necessary ρ(x,m) metric is the distance between a position of covered demand (x) within the sector and the position of serving base station (m) weighted by the demand of x. The demands of MUS will connect with the new placed BS, because the new signals will arrive from nearer position. The update step of K-Means will shift them (Figure 3). After some cycles the moving of BSs will decrease, hence we intuitively chose the K-means clustering to run for six cycles.

The detailed flowchart diagram of the algorithm is presented in Figure 4.

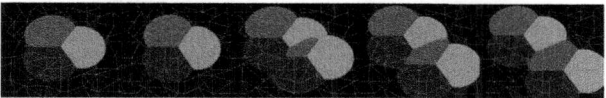

Fig. 3. Mechanism of Location Algorithm

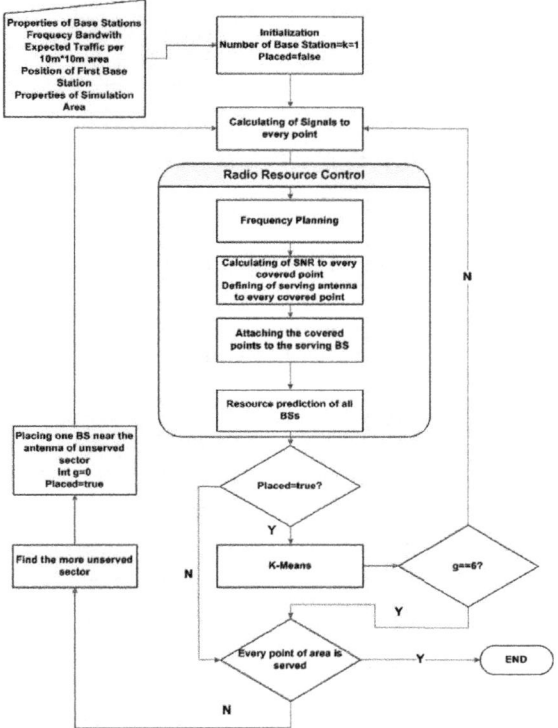

Fig. 4. Detailed flowchart diagram of RF planning algorithm

4   Numerical Results

The algorithm was used in a terrain of 9 km², which models a center of a big city. The overall traffic demand was 900 Mbps, the available frequency bandwidth is 20 MHz. The Frequency Reuse Factor is 3, therefore one sector uses 6.67 MHz frequency bandwidth.

In Figure 5 the total amount of used and unused frequency bands is presented, as the algorithm placed more and more base stations over the area. The amount of total used and unused frequency is determined by means of summing up all used (and unused) bandwidths for all cells in the network. We can see that the used total bandwidth is reaching its maximum for about 30 base stations (meaning that most of the demands are served at this point), but 10 more base

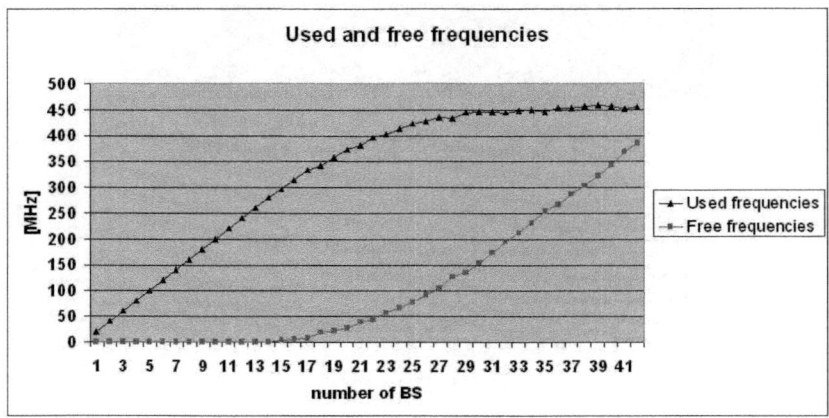

Fig. 5. Used and free frequency band during the running of planning algorithm

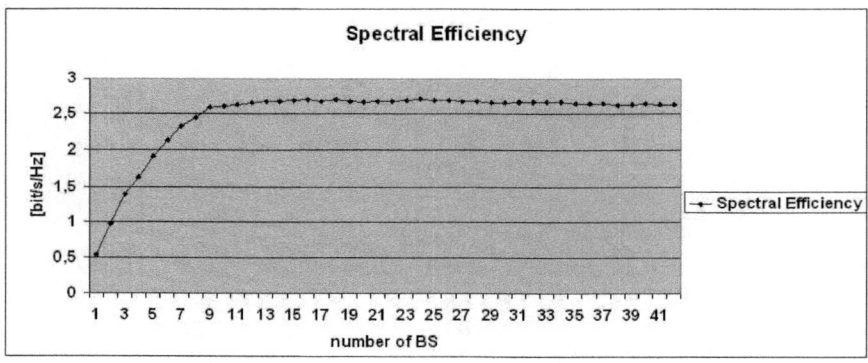

Fig. 6. Spectral efficiency during the running of planning algorithm

stations are needed to attain 100% coverage and service. This then results in the increase of unused bandwidth, meaning that the service of the last couple of percentage of demands seriously decrease efficiency. The consequences of this are that a given reasonable service requirement (in terms of coverage and capacity), say 95% (meaning that 5% of the area might remain uncovered and 5% of the demands might not be served) can save a lot in terms of required base station numbers. On the other hand, if more advanced RRC would be assumed and modelled (that allocates PRBs adaptively to demands), the efficiency of used spectrum would be higher, thus less base stations would be enough (however, the aim of this work is not to develop an optimal RRC method, but to plan the network with any given RRC).

In Figure 6, the spectral efficiency of the network is presented, as the algorithm places new base stations over the area. It reaches its peak quite early, for about

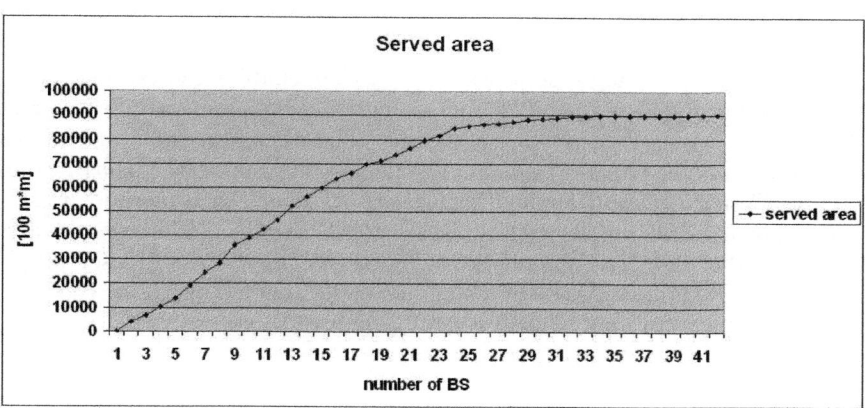

Fig. 7. Size of served area during the running of planning algorithm

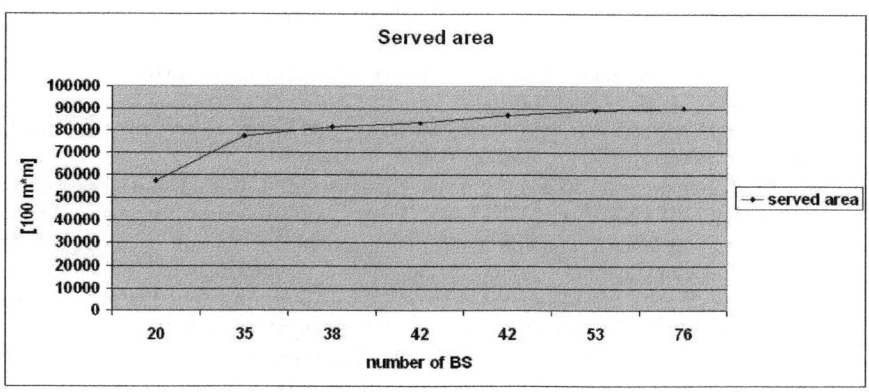

Fig. 8. Size of served area for different hexagonal BS location

13 base stations. This means that the placement of new stations will not increase spectral efficiency, however to serve all demands (and to provide more capacity), more stations are needed. A slight decrease in spectral efficiency can also be observed in case of higher numbers of base stations, that is for the same reasons that was concluded regarding Figure 5.

Figure 7 shows how the served area is increasing over the iterations of the algorithm. Again, it can be concluded, that the area of 9 km^2 is well served by around 30 stations, the rest are needed to cover negligible holes in coverage.

The algorithm proposed was compared to the results of a traditional base station placement following a regular hexagonal grid. Figure 8 shows how the served area increases as the number of base stations increase in this regular grid (We created two different hexagonal grids with 42 BSs). We can conclude that our proposed algorithm covers the area with half the number of base stations, that is a very reasonable performance increase.

5   Conclusions

In this paper a novel algorithm was shown, that enables the automatic placement and determination of the number of base stations, that can serve a cellular network area with given traffic conditions. The algorithm is based on realistic assumptions an can be used for any legacy system, with arbitrary Radio Resource Control method applied in the network. Numerical methods were presented, showing that the algorithm reaches total coverage and allows the service of all traffic demands, although it is concluded that by means of loosening coverage/capacity requirements, or by using advanced RRC algorithms, the number of necessary base stations might be significantly reduced.

References

1. Karam, O.H., Fattouh, L., Youssef, N., Abdelazim, A.E.: Employing Clustering Techniques in Planning Wireless Local Loop Communication Systems: PlanAir. In: 11th International Conference on Artificial Intelligence Applications Cairo, Egypt, February 23-26 (2005)
2. Ajay R. Mishra, Advanced Cellular Network Planning and Optimization, pp. 15–197,John Wiley & Sons Ltd. (2007)
3. Glisic, S.G.: Advanced Wireless Networks 4G, pp. 667–694. John Wiley & Sons Ltd., Chichester (2006)
4. Basit, A.: Dimensioning of LTE Network, Description of Models and Tool, Coverage and Capacity Estimation of 3GPP Long Term Evolution radio interface (2009)
5. Barclay, L.: Propagation of Radiowaves, p. 194. The Institution of Electrical Engineers, London (2003)
6. Ramamurthy, H., Karandikar, A.: B-Hive: A cell planning tool for urban wireless networks. In: 9th National Conference on Communications (2003)
7. Tutschku, K.: Demand-based Radio Network Planning of Cellular Mobile Communication Systems. In: INFOCOM 1998, pp. 1054–1061 (1998)

An Energy-Efficient FPGA-Based Packet Processing Framework[*]

Dániel Horváth[1], Imre Bertalan[1], István Moldován[1], and Tuan Anh Trinh[2]

Budapest University of Technology and Economics
[1] Inter-University Cooperative Research Centre for Telecommunications and Informatics
[2] Department of Telecommunications and Mediainformatics
Magyar Tudósok krt. 2, 1117 Budapest, Hungary
{horvathd,emerybim,moldovan,trinh}@tmit.bme.hu

Abstract. Modern packet processing hardware (e.g. IPv6-supported routers) demands high processing power, while it also should be power-efficient. In this paper we present an architecture for high-speed packet processing with a hierarchical chip-level power management that minimizes the energy consumption of the system. In particular, we present a modeling framework that provides an easy way to create new networking applications on an FPGA based board. The development environment consists of a modeling environment, where the new application is modeled in SystemC. Furthermore, our power management is modeled and tested against different traffic loads through extensive simulation analysis. Our results show that our proposed solution can help to reduce the energy consumption significantly in a wide range of traffic scenarios.

Keywords: energy management, packet processing.

1 Introduction

The future communication systems must meet several, often conflicting requirements. Such requirements are the ever-growing processing power while keeping the energy consumption down. Scalability is required without increasing the complexity. Flexibility of the software at the speed of the hardware is a requirement as well. And of course all of this at low price.

Requirements for the increased complexity of hardware design lead to a way of designing chips, where Composability, Predictability and Dependability need to be taken into account. Composability enables the development of stable software for very large systems by ensuring that parts of the software can be developed and verified separately. Predictability is the ensuring of real-time requirements on a device. Tools that calculate the timing behavior enable the correct design and scaling of systems need to be used to ensure that the real-time conditions are met. Last but not

---

[*] The research leading to these results has received funding from the ARTEMIS Joint Undertaking under grant agreement n° 100029 and from the Hungarian National Office for Research and Technology (NKTH).

F.A. Aagesen and S.J. Knapskog (Eds.): EUNICE 2010, LNCS 6164, pp. 31–40, 2010.

least, dependability relates to the behavior of the system in case of errors and problems during operation.

FPGAs are used in the prototyping of the devices but as the price of larger FPGAs become accessible, they are considered as a design alternative. For special functions that are not deployed in large quantities in the network, and which require very fast processing such as network monitoring applications, firewalls etc. an FPGA based design provides flexibility and the required performance at reasonable price.

Power consumption is also of major concern in future packet processing architectures. We address this topic with a hierarchical power management that minimizes the energy consumption at chip level.

The aim of our work is to provide a toolchain which accelerates the development process of network applications on FPGA. The toolchain will provide a set of hardware modules which builds up a variety of network devices. Key use-cases are switching, routing devices, a NAT device, a firewall/deep packet inspector, and a traffic loopback device.

For design space exploration and to validate the design, a SystemC [5] based modeling environment is used. The results of the SystemC modeling can be used to construct the final hardware models and the corresponding software. The available hardware components will also be available in generic hardware description language (Verilog/VHDL) making possible the synthesis of the hardware. The toolchain will also provide the top-level hardware model. The modules required for the generic networking applications have been selected by identifying the most important use cases.

This paper is structured as follows. Section 2 we describe the generic architecture and its power efficiency considerations are detailed in Section 3. The simulation and results obtained from it is shown in Section 4. Section 5 concludes the paper.

2   Packet Processing Architecture

The development of the new applications is also assisted by a number of available hardware component models that can be used for generic networking application development. The already available components define a generic packet processing/forwarding mechanism with extensible filtering and processing properties.

Networking applications can be composed from predefined modules and custom modules. Example predefined modules are e.g. MAC-layer input and output, switching fabric, and packet scheduler. The behavior of the network devices can be further customized by adding custom modules that apply application-specific operations at different stages of packet processing. For easy integration well defined interfaces are used.

The packet processing framework provides the background for new application development. Its extensible, modular structure is designed to allow easy integration of application-specific header operations at the ingress and egress. A method for buffering the packet payload is also provided.

2.1   Background

In the literature, we have found similar work dealing with packet processing on FPGA-based systems. Notably, we would mention the work of D. Antos et al. [1] on the design of lookup machine of a hardware router for IPv6 and IPv4 packet routing with operations are performed by FPGA, since FPGA-based implementations are an attractive option for implementing embedded applications. In this framework, part of the packet switching functionality is moved into the hardware accelerator, step by step. This allows keeping the complete functionality all the time, only increasing the overall speed of the system during the whole development process. D. Teuchert et al. [6] also dealt with FPGA based IPv6 lookup using trie-based method. Another interesting application of FPGA-based design is Gigabit Ethernet applications [2] due to the fact that FPGA-based implementations offer the possibility of changing the functionality of the platform to perform different tasks and high packet-processing rate capabilities. In particular, the authors of [2] proposed a versatile FPGA-based hardware platform for Gigabit Ethernet applications. By introducing controlled degradation to the network traffic, the authors provided an in-depth study on real-application behavior under a wide range of specific network conditions, such as file transfer, Internet telephony (VoIP) and video streaming. However, in those previous works, energy-efficiency of the system is not yet properly dealt with. As energy efficiency has become a key performance metric, techniques that can quickly and accurately obtain the energy performance of these soft processors are needed. In [3], this issue is addressed and methods for rapid energy estimation of computations on FPGA-based soft processors are provided. The authors propose a methodology based on instruction level energy profiling by analyzing the energy dissipation of various instructions with an energy estimator is built using this information. The paper deals with the estimation of energy consumption of computations on FPGA platforms, but does not propose any methods to manage and optimize it. Kennedy et al. [10] provide a frequency scaling based low power lookup method, but they optimize only the lookup function. We would also mention NetFPGA platforms that are largely in use in academic research to test for ideas and implement them on flexible hardware. Given that background, in this paper, we propose a practical energy management module on FPGA platform and validate our proposed solution through simulation analysis by using SystemC.

2.2 Packet Processing Pipeline

The packet processing pipeline is similar to the model recommended by Xilinx [9] and the model used by the Liberouter project [8]. The packet headers are processed in the pipeline while the packet payload needs to be buffered as shown on Fig. 1. In our initial work the buffering of packets is realized using the on-chip RAM referred as Block RAMs or BRAMs available in the FPGA. In the final design the packets will be buffered in an external DDR RAM. Since the memory access can be the bottleneck in our packet processing pipeline, we tried to avoid copying the packet. We have decided to use the shared-memory packet forwarding model, this way avoiding unnecessary copying of data. The model has a further advantage: even multicasting/broadcasting can be done without actually copying the data.

The arriving packets are buffered, while their header is processed in the input block. The input block applies the input filtering rules and sends the header to the forwarding module. The routing/switching module is responsible to decide on which

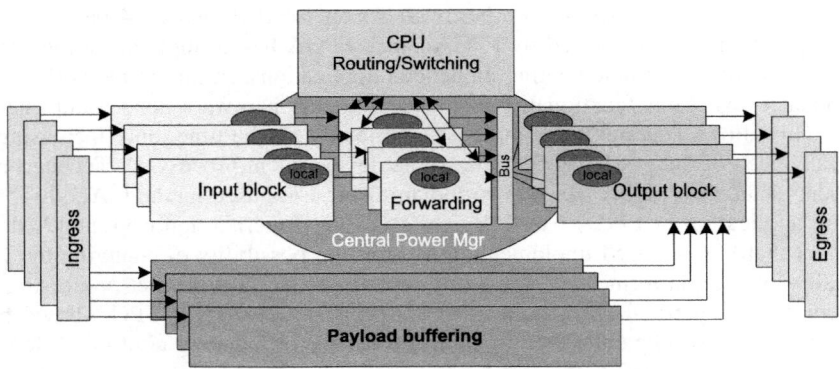

Fig. 1. Generic packet processing framework

interface the packet leave should. For flooding, broadcast and multicast multiple interfaces may be selected. For different application types, different routing/switching modules can be designed. After the decision, the packet can be output filtered, and the header will go to an output header buffer at the egress port. The packet will be assembled during transmission: when scheduled, the header will be sent from the header buffer, while the payload will be read from the payload buffer.

Each block has a local power manager, and together with the central power manager an efficient power management solution is provided.

This packet processing pipeline illustrates an example packet forwarding design. However, the design possibilities for new applications using the framework are endless; based on the existing modules monitoring, firewall, routing and switching applications can be created easily, while designing new modules any hardware-assisted packet processing application can be developed quickly. These models are generic, and planned for easy customization and implementation in other FPGA-based platforms. Furthermore, their scalability is also taken into account. For most models the scalability is ensured by design, providing cycle-accurate model of the hardware. Some of the simulation models however require high processor usage and long programming cycle. Their model is approximate.

2.3 Input and Output Blocks

The incoming frames of data are to be stored in BRAMs available on the FPGA during processing the header information. For receiving the information, we use input modules (Fig. 2.). These per-port modules get the frames from the PHY. The bitstream is deserialized and then it is forwarded both to the input buffer and header filters. The input filters perform the filtering action before reaching switch/router module. Before new frames are received, buffer reservation is checked, whether there are free slots available or not. A module called BRAM manager is responsible for the management of the BRAM buffers associated to an input.

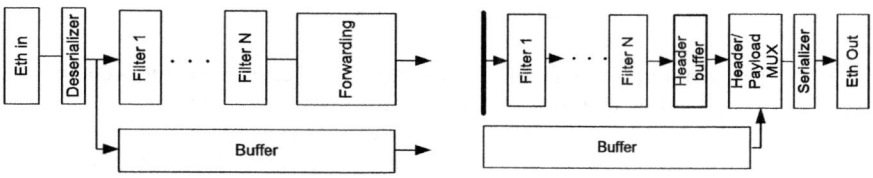

Fig. 2. Input port module   **Fig. 3.** Output port module

The output module (Fig. 3.) is responsible for sending out the routed/switched frames. To do this, the upper level modules (switches, routers) send the new packet header with payload buffer information. The ingress port and BRAM address is required to fetch the associated payload. The header information is stored in the header buffer, which is processed in a FIFO basis. This buffer plays the role of the output buffer, but since we do not want to copy the payload, only the headers are queued. The header/payload multiplexer takes the headers from the FIFO, starts sending them, and appends the payload data from the buffer indicated. The serializer is preparing the packet in a format required for the PHY.

2.4 MAC or IP Lookup Module

The MAC learning and lookup module as well as the IP lookup module is designed to be processor based. The module receives the destination MAC address to look it up, and the source MAC address to learn (Fig. 4.). The IP lookup module receives the destination IP address to look it up only. After the lookup either module returns the egress port or ports in case of multicast, broadcast or flooding. The simplest models would use a static local forwarding table; more complex models can use centralized lookup tables based on CAMs (or TCAM) and for large routing tables a CPU based lookup method can be used. The arbiter module is responsible for the access to the common bus of the output FIFOs. The switch modules send the headers and additional data to output stores. The arbiter enables them the access to the bus in a round-robin way.

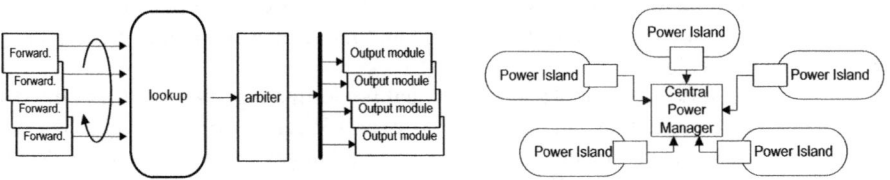

Fig. 4. Switch modules and the lookup engine   **Fig. 5.** Hierarchical power management

The fact that these modules are based on a CPU enables us to tune the duration of the lookups. This is done by controlling the frequency. Reducing frequency means power consumption reduction. In the implemented 4-port Ethernet switch, the CPU has 4 different frequency steps, resulting 4 level of performance and 4 level of power consumption.

3 Resource Management

The overall power requirement of a networking device can be reduced in several ways. The most obvious method is to switch off the parts that are not used. For example, switch off the interfaces that are not in use. Further energy can be saved by introducing power levels in the system that can be achieved either by allowing processors (or multiprocessor cores) to enter sleep mode or reducing the clock frequency and core voltage [7]. Another approach is to use energy management at the packet processing hardware level. This method requires that each module in the pipeline has an enable pin, and each module is powered only for the time period its operation is in need.

In a high-speed packet processing device the energy management should not compromise the performance. Performance degradation results in additional delays, which lead to traffic burstiness, jitter and increased end-to-end delays.

Our platform is designed to provide line-rate throughput on all interfaces. This means, that although there is buffering for the payload and header, it is not necessary to make copies of the payload, only the headers. Of course buffering may occur at the output interfaces caused by traffic aggregated from different interfaces, but the routing/switching must be designed to handle line-rate traffic from all input interfaces. In other words, while a whole packet is received on an interface, the routing/switching process should find the egress port to avoid delays on the output port, even if there are packets at other interfaces that need to be serviced with priority. In worst case scenario, 3 packets in other queues must be serviced before the round-robin scheduler services the new packet. Thus, the processor speed should be always selected according to the number of packets actually received.

The optimal energy management solution has the following requirements:

   a) power up hardware components only for the time required
   b) provide maximum energy efficiency by selecting the processing power according to the traffic conditions
   c) do not compromise the line-speed forwarding at any time

On the other hand, selecting the speed of a processor or starting/halting cores is not practical and may not be feasible on a packet arrival basis (e.g. ~700ns on a 1Gbps link vs. ~10μs change of the state of an StrongARM CPU [11]). Therefore we must predict the traffic load for longer time period and make energy management decisions on a larger time scale.

In our framework, we propose a hierarchical power management with two levels: a central power manager that is connected to several local power managers. The local power managers are responsible for the power management in their own domain, called "power island" (see Fig. 5).

The central power manager provides user-manageable functions like disabling the input/output port module and makes global decisions, while the local power managers optimize the power dissipation micro-managing the units in the power island.

3.1 Lower Level – Packet Processing Line

Each module has a chip enable pin which is connected to a local power manager. The different ports and port handling modules are connected to this local power manager, forming several power islands. The enable signal of the local power managers is collected in levels of hierarchy, and the upper level power manager is responsible of handling the enable signal of the power islands.

Within a power island the local manager is responsible of enabling the modules of the packet processing pipeline. Each module is powered only for its job, and it is powered down after completion, keeping consumption at an optimal level. The PHY layer is handled by a chip and is always on. These analogue and digital circuits are responsible for the line coding and physical requirements of the different physical layer standards. At a packet arrival they immediately inform the local managers that packet processing starts, following modules in the pipeline need to be powered up. Each module in the pipeline should be operating until the packet leaves the module and the end-of-packet signal notifies the local power manager to switch it off. The operation of the manager is optimal, as they micro-manage the pipeline to achieve maximum performance and maximum power saving.

3.2 Higher Level – Administrative and Power Control

At a higher level, two main functionalities are performed: power management and load-sensitive frequency scaling of lookup processors. The central manager presents a power management interface for the management. Power management decisions such as disabling interfaces are performed at this level. When the operator decides to disable an interface or even a subsystem, the central power manager commands all power islands related to the disabled subsystem to enter low power state.

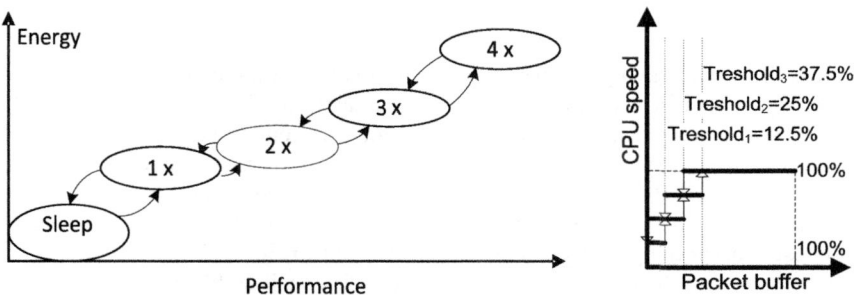

Fig. 6. Processor power management   **Fig. 7.** Buffer Thresholds

Besides the resource management interface the central manager is also optimizing the power management by managing the power of lookup/routing CPU cores. Setting processing speed or waking up/shutting down CPU cores is performed according to the offered load. In our model the CPU speed is modeled in 5 power states: the sleep mode and 1..4× frequency mode (see Fig. 6.).

The maximum performance can be achieved running at 4× speed but of course it also means higher dissipated energy. The network load is estimated according to the input buffer utilization. We have defined 3 thresholds (Fig. 7.) for buffer utilization and the CPU speed is increased or decreased whenever a threshold is crossed.

The two power management modes together provide a nearly optimal usage of available resources while keeping the system performance nearly unaffected.

4 Simulations

The modeling framework has been implemented in SystemC 2.2.0. The simulated hardware is a 4 port Ethernet switch model. The packet processing pipeline consists of a filter for own MAC address, header processing and switch module. The switch module uses a CPU based lookup where the CPU has 4 levels of speed. The switch has low buffering capabilities: each port can buffer only 4 packets. This is a realistic scenario for a Xilinx Spartan 3 FPGA without external RAM. The processing pipeline has been designed with line-speed forwarding in mind, with bottlenecks only at the egress ports, where traffic from different directions is mixed. Therefore the input queuing permits 2 headers only. The simulations were performed using traffic generators and sinks. To each port a traffic generator & sink is connected. Each traffic generator generates traffic to all other sinks, with equal probability.

In the simulation we used 100Mbit Ethernet ports. The simulated traffic was characterized by exponentially distributed inter-arrival times with expected values from 2μs to 1ms. In each simulation the packet size was fixed 64, 128, 256, 512, 1024 or 1500 bytes. During the simulations we estimated the power saving and delay for each module. As expected, the switch is capable of line speed processing, on all ports, without blocking. Header processing is performed while the packet payload is still received, and egress lookup should also be fast enough not to hold the processing pipe.

4.1 Results

To assess the power savings realized by locally managing the enable pins of different subsystems, we performed simulations with increasing traffic. To observe the effect of packet sizes, we repeated the simulations for different typical packet sizes. The obtained results are expressed as relative values compared to the "always on" scenario which is considered 100% uptime. Fig. 8. shows the uptime of the forwarding module in percentage compared to the "always on" scenario. This module needs to be "on" the longest time in the pipe as it must wait the result of the lookup. As expected, the gain decreases with the load on the network, but it is considerable even for high loads. In an extreme case the active link without traffic can save 99% on this module because it is powered off most of the time. Another example, the uptime of the destination MAC address filter is shown on Fig. 9. This module performs a simple operation; it can be "off" most of the time as it works on the headers only. The higher the packet size the higher the energy gain. At least 60% can be saved on this module.

Then we simulated the gain of managing the power levels of the switching processor. In order to keep the delay at minimum, we should only reduce the speed to a level where it is not delaying packets. The processor is entering lower power mode when the overall buffer utilization drops below a threshold. Scaling down the CPU speed comes at a price: the CPU speed cannot be changed as often as the load changes. Thus, the first packet(s) of a burst will be served at a lower speed, introducing delays for the upcoming packets.

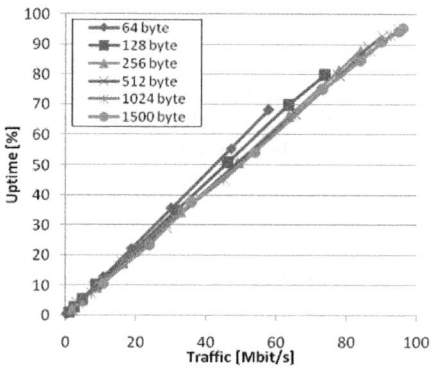

Fig. 8. Uptime of a characteristic module in the pipeline depending on packet size

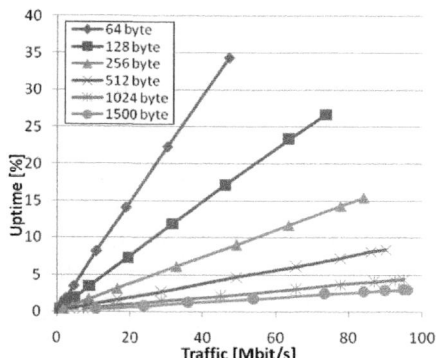

Fig. 9. Uptime of the destination MAC address filter depending on packet size

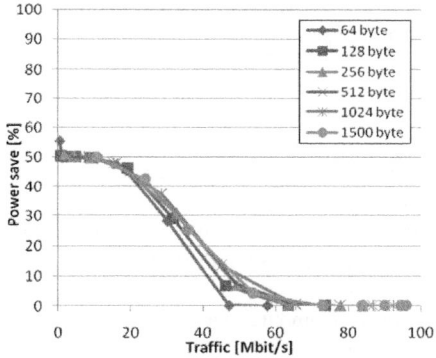

Fig. 10. Power saved with frequency tuning on CPU depending on packet size

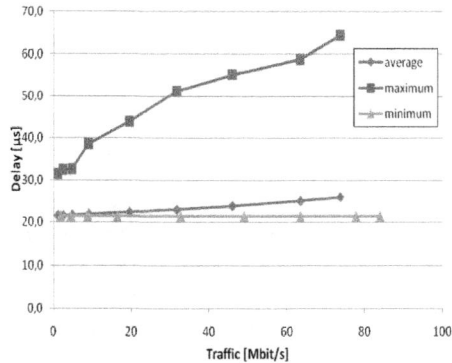

Fig. 11. Latency values for packets of 128 bytes

Again, simulations are repeated for different load levels and packet sizes. Besides the power saving, we also show the mean and maximum delays introduced.

Fig. 10. shows the power saving achieved at the lookup processor as a percentage of the always at highest performance scenario. The gain increases as the load decreases, as expected. Also, the gain increases as the packet size decreases, providing maximal gain when the overall utilization is the lowest. The gain can be as high as 50%. An example for latency at 128 byte packet-size is shown on Fig. 11, although the maximum delay is doubled, the average latency is close to the minimum, meaning that only a few packets get delayed. As the load gets higher, the CPU is running on higher performance step most of the time, handling small burst with higher speed, and delay is still not increased. As the traffic further increases, the buffers saturate, and some frames have to wait more, increasing the maximum delays.

5   Conclusions

In this paper we have presented a generic packet processing framework that can be a background for different packet processing applications like switching, routing filtering, monitoring etc. The framework is modular, which provides a scalable and composable system. For design verification a simulation environment is used, which provides cycle-accurate models for the components.

Using the simulation environment, we have designed an example switch application, with modular input and output filtering pipeline. The energy efficient operation is ensured by a hierarchical energy management. Results show that the hardware is capable of packet processing at the line rate, still achieving up to 60% gain in power for the simple modules, and in idle periods we can achieve a power saving of up to 99%, depending on traffic. For more complex modules like the forwarding module there is small gain at high loads, but again very high gain at low loads. We have also shown that with CPU scaling we can get considerable gain at a price of small increase in the traffic jitter.

Even further energy savings are possible by using dedicated hardware accelerators for specific lookup and other CPU-intensive operations.

References

1. Antos, D., Rehak, V., Korenek, J.: Hardware Router's Lookup Machine and its Formal Verification. In: ICN 2004 Conference Proceedings (2004)
2. Ciobotaru, M., Ivanovici, M., Beuran, R., Stancu, S.: Versatile FPGA-based Hardware Platform for Gigabit Ethernet Applications. In: 6th Annual Postgraduate Symposium, Liverpool, UK, June 27-28 (2005)
3. Ou, J., Prasanna, V.K.: Rapid Energy Estimation of Computations on FPGA-based Soft Processors. IEEE System-on-Chip Conference (2004)
4. Werner, M., Richling, J., Milanovic, N., Stantchev, V.: Composability Concept for Dependable Embedded Systems. In: Proceedings of the International Workshop on Dependable Embedded Systems at the 22nd Symposium on Reliable Distributed Systems (SRDS 2003), Florence, Italy (2003)
5. OSCI SystemC 2.2.0 Documentation: User's Guide, Functional Specifications, Language Reference Manual, http://www.systemc.org/
6. Teuchert, D., Hauger, S.: A Pipelined IP Address Lookup Module for 100 Gbps Line Rates and beyond. The Internet of Future, 148–157 (2009) ISBN 978-3-642-03699-6
7. Intel White Paper: Enhanced Intel SpeedStep Technology for the Intel Pentium M Processor (March 2004), ftp://download.intel.com/design/network/papers/30117401.pdf
8. Liberouter project homepage, http://www.liberouter.org/
9. Possley, N.: Traffic Management in Xilinx FPGAs, White Paper, April 10 (2006)
10. Kennedy, A., et al.: Low Power Architecture for High Speed Packet Classification. In: ANCS 2008, San Jose, CA, USA, November 6-7 (2008)
11. Iranli, A., Pedram, M.: System-level power management: An overview. In: Chen, W.-K. (ed.) The VLSI Handbook, 2nd edn., December, Taylor and Francis, Abington (December 2006)

Service Migration Protocol for NFC Links

Anders Nickelsen[1], Miquel Martin[2], and Hans-Peter Schwefel[1,3]

[1] Dept. of Electronic Systems, Aalborg University, Denmark
{an,hps}@es.aau.dk
[2] NEC Europe Ltd., Heidelberg, Germany
miquel.martin@nw.neclab.eu
[3] FTW, Vienna, Austria
schwefel@ftw.at

Abstract. In future ubiquitous communication environments, users expect to move freely while continuously interacting with the available applications through a variety of devices. Interactive applications will therefore need to support migration, which means to follow users and adapt to the changing context of use while preserving state. This paper focuses on the scenario of migration between two devices in which the actual migration procedure is executed over near-field communication (NFC) ad-hoc links. The NFC link is interesting as it gives the user the perception of trust and enables service continuity in cases where mid- or long-range wireless connectivity is unavailable. Based on an experimental performance analysis of a specific NFC platform, the paper presents a migration orchestration protocol with low overhead and low delays to be used with NFC links. Experimental results allow to conclude on the sizes of application state that can be expected to be feasible for such ad-hoc NFC migration.

Keywords: near field communication (NFC); performance measurements; service migration.

1 Introduction

Future users will be equipped with different devices and expect to be able to access personal services from any device, and even change device while using an application. The process of changing the device during application use is called *migration*. A *migratory application* supports pausing, extraction of sufficient state information to resume operation on target device, insertion of state on target device and resuming of operation. The application may rely on middleware to handle control (start,resume) and transfer of state. The middleware makes the application independent of underlying platforms and networks.

Traditional migration is conducted via high-bandwidth network links (Ethernet, WLAN or GPRS/UMTS) and a central server to coordinate the migration sequence [1]. However, using ad-hoc communication between devices and in particular the RFID based Near Field communication (NFC) can be advantageous for multiple reasons: (1) The physical closeness required to create an NFC link

F.A. Aagesen and S.J. Knapskog (Eds.): EUNICE 2010, LNCS 6164, pp. 41–50, 2010.
© IFIP International Federation for Information Processing 2010

can be interpreted as migration trigger; (2) binding the migration to the physical closeness can increase the user-trust in the migration procedure, as hijacking of sessions by remote devices can be prevented; (3) fast connection setup times and low interference probability due to its short range can be advantageous; (4) low power consumption can make it the technology of choice in migration scenarios triggered by low battery of the source device. In addition, migrating via the NFC link may be the only option in scenarios of interruption of coverage of mid- to long-range (cellular) technologies.

Migration over ad-hoc NFC links is however challenged by throughput limitations and the potential short time-windows during which the communication is possible. Our own preliminary experiments have shown that users expect migration to finish within 3-5 seconds when using NFC. The estimates are based on familiar uses of RFID and NFC, as for instance door locks or ticketing (which require a small window to complete). The consequence is that NFC may only be feasible for a small application state sizes, and we propose a protocol targeted at maximizing the feasible state size.

Transfer of active sessions is a well-studied procedure in existing research on *mobile agents* [2] [3]. The work in this paper is based on *service migration*, where service is defined as mobile interactive applications [1] [4]. The goal of service migration is to improve the user experience by a seamless migration. This is different from the goal of mobile agents, which is to autonomously achieve a goal independent of the platform. NFC has previously been studied as a part of different application types; for service discovery in Smart Spaces [5], for tracking habits in home health-care [6] or as information carrier in marketing such as in All-I-Touch [7] and [8]. Similarly, NFC performance parameters are mostly user-experience oriented; trustworthiness [9] and usability [10] [11]. Likewise, [12] and [13] look into the security implications of NFC communications.

In this paper, we propose a migration protocol optimized for resource-constrained links, such as NFC. In addition, we determine the range of state sizes that can be transferred within the available communication windows. From this investigation, we present general performance measurements, which are also applicable in other contexts than service migration. We focus on the performance implications NFC on migration. A detailed security and investigations of complex user interactions are out of the scope of this work.

We present a migration scenario in Section 2 to illustrate how use of NFC links challenges the migration architecture. Performance results of NFC usage are presented in Section 3 to obtain approximate magnitudes of boundary conditions using NFC for migration. In Section 4 the proposed protocol for using NFC in migration is described and implementation and evaluation of the protocol are presented in Section 5. Finally, Section 6 concludes the paper.

2   Background

Figure 1 illustrates a migration scenario. The user is carrying a mobile device on which a migratory client-server application is running, for instance a video

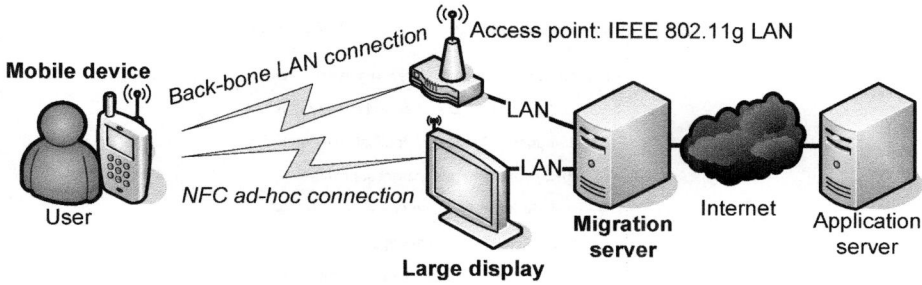

Fig. 1. A scenario of migrating the client-part of the application from the mobile device to the large display

streaming application. The client-part of the application is running on top of a middleware on the mobile device; the client-side middleware collaborates with support functions on the migration server. The client-part of the application uses special interfaces of the migration middleware such that the middleware can start, stop, pause, and resume the client-side application; furthermore, the client-side application needs to use middleware functions to store application state that must remain persistent after the migration has been executed. In case of a video streaming application, the required state information to be transferred is a URL, a time offset and potentially media relevant information regarding codecs, etc. which were received/negotiated in the beginning of the stream, for instance as a session description protocol (SDP) profile [14]. Because the video stream is not re-initialized when migrated, such information is not re-exchanged between client and server, and must thus be transferred as state. A SDP-profile is exchanged in clear-text, and its size can range from 230 bytes for a compressed video profile [15], over 860 bytes for a raw audio profile [16] and upward for more complex sessions. To avoid modification of the server-part of the application, the migration procedure is made transparent by using a mobility anchor point function (e.g. a mIP home agent [17]) in the migration server to redirect data flows during and after migration.

The client-side middleware registers devices and applications on the migration server, which orchestrates the migration through the available network connections. The users can search for suitable devices to migrate to and can trigger the migration directly from the terminal. The basic migration scenario (without ad-hoc links) from the mobile device (source) to the large display (target) includes the following activities (cf. Figure 2): 1) the user selects an application to be migrated and the large display device as target; 2) the user triggers the migration and the middleware passes the trigger event to the migration server, 3) the migration server orchestrates the migration using the client-side middleware components, which include pausing the application on the mobile device, extracting state information, transferring the state from the source device to the migration server, initializing the middleware components on the target device, transferring the state to the target (large display in Figure 1), inserting

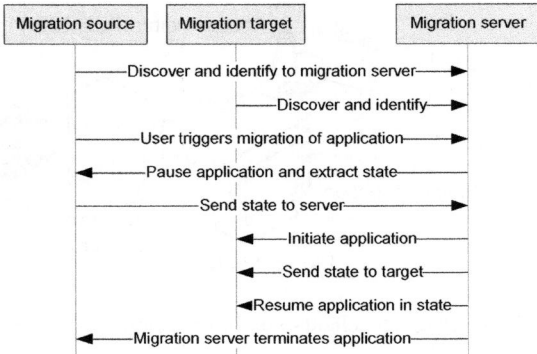

Fig. 2. Steps during a migration procedure

the state into a new application instance and resuming this application in the original state, 4) the migration server terminates the original paused application on the source device. Note that in the general case, the trigger can be generated either by the user via a manual interaction or by the migration server through an automatic decision based on general context information [18]. We focus on the scenario of user-generated triggers. See [19] for more details.

Traditional RFID communication consists of a passive RFID tag and an active RFID reader. The reader generates a radio frequency (RF) field to request a response from the tag and the tag uses the energy in the reader's RF field to respond. One NFC entity integrates both RFID tag and reader and thus allows two NFC entities to communicate peer-to-peer. The NFC specification ([20]) allows for entities to work as traditional passive RFID tag and reader for compatibility, however, only the peer-to-peer mode is considered here, in *active* mode, where all entities generate RF fields.

NFC devices must be prepared for neighbor discovery, similar to the inquiry phase of Bluetooth and the frequency scanning phase of ad-hoc mode WLAN. An NFC device can have one of two roles; *initiator* or *target*. The initiator uses its RF field to contact targets. The target only senses for initiator RF fields and does not turn on its own RF field unless requested by an initiator as part of communication. A protocol using NFC must include at least one initiator and one target to have successful communication.

3   NFC Performance Measurements

This section presents an experimental performance study of an NFC link in order to obtain ranges of several performance metrics. These ranges are important for the design of the migration protocol for resource-constrained NFC links. We investigate relevant migration parameters: neighbor discovery time, transmission delay (round-trip), and throughput. The neighbor discovery time directly

subtracts from the time budget available for orchestrating migration and transferring state. The round-trip time message delay impacts duration of exchanging
messages during orchestration. The throughput is used to estimate how much
state data can be transferred during the remaining time budget. The setup con-
sisted of two HTC 3600 smart phones running Windows Mobile 5 and with
an 'SDiD 1010' NFC dongle [21] attached. The NFC dongles were configured
in active mode with 424 kbit/s PHY data rate. A file transfer application was
deployed on both devices and one application instant acted as NFC initiator
while the other acted as NFC target. Since the maximum frame size was 186
bytes, fragmentation was implemented. All experiment where repeated 30 times
to obtain 95% confidence intervals, which is shown with each result.

Neighbor discovery time was measured by placing the devices in range and let
the initiator search for targets. The initiator blocks execution while searching for
targets. The measured interval indicates the blocking time from the moment the
application started searching until the target was discovered and the initiator
could continue. The mean discovery time was measured to be 59 ± 0.8 ms.

Transmission delay was measured by sending the smallest possible frame size
(16 bytes header, no payload) back and forth between the NFC entities. The
delay is defined as a round-trip delay, i.e. as the time from starting the send
procedure on one device until receiving the response on the same device. For
two nodes, the average delay was measured to be 106 ± 1.8 ms.

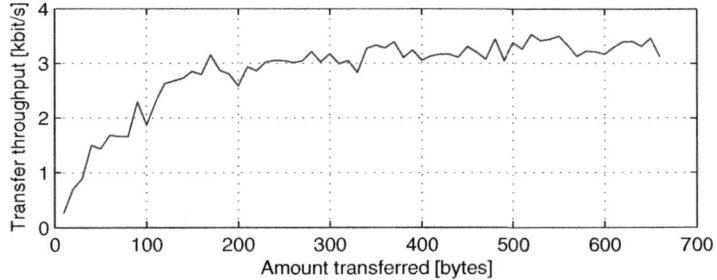

Fig. 3. Throughput measurements over NFC link

Throughput was measured by sending multiple frames continuously to mimic
a large application state object. The total size of the transferred file was varied
between 10-660 bytes and the time was measured from the first event until the
final acknowledgment was received. Results from the experiment for different
file sizes are shown in Figure 3. The figure shows that the throughput increases
rapidly with increasing file sizes on the left end and then it converges to a value
around 3.26 kbit/s for values above 300 bytes. The throughput degradation
for small file sizes is expected as the ratio between overhead and payload is
relatively high.

4   Procedure for Migration over NFC

In this section, we propose an orchestration protocol that is targeted at performing migration over a resource-constrained link such as NFC. We describe the protocol details below and provide an experimental evaluation in the subsequent section.

The driving scenario is depicted in Figure 1. Some time period before the user puts the mobile device close to the large display to activate NFC neighbor discovery, the WLAN connection disappears so that migration needs to be performed via the NFC link. The lifetime of the NFC link is limited by the window of time the user holds the mobile device close to the large display. The goal of the proposed migration protocol is to maximize the probability that the application state can be successfully transferred within the available time window. In order to maximize use of the time window, the protocol aims at starting the transfer as early as possible. When the devices have discovered each other, the state is prepared in the source device and the size of the state is exchanged at first. After that, the state is transferred. Finally, when the state has been transferred successfully, the application in the source device is terminated.

In a scenario with a back-bone connection (such as WLAN), migration decisions are made by the migration server and therefore clients spend much time waiting before transferring the state due to communication with the server. Also,

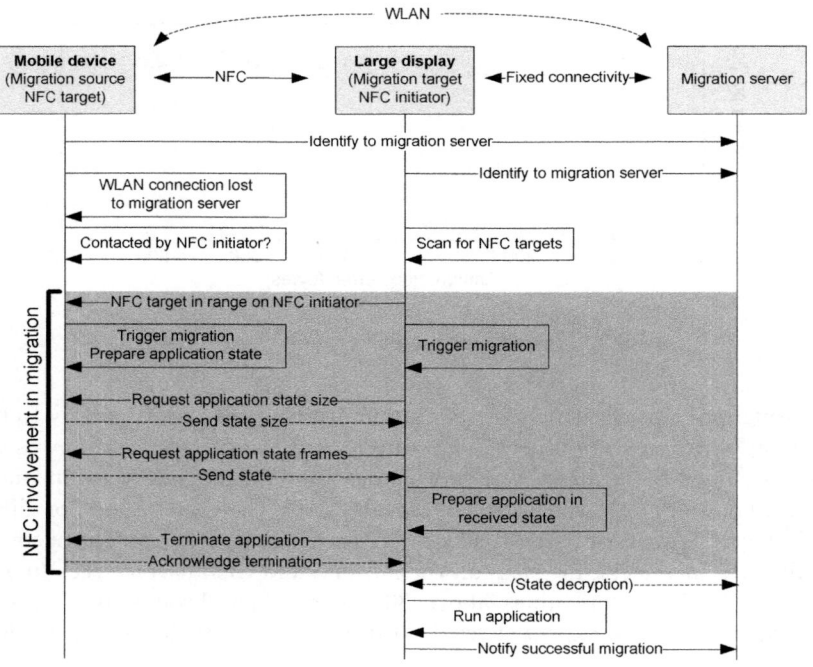

Fig. 4. Fast and low-overhead migration procedure utilizing the NFC link

when migration is controlled by the server, the transfer of the state introduces overhead as the transfer is split into an upload from the source device to the server followed by a download from the server to the target device. In our scenario where the WLAN connection has disappeared, there cannot be direct communication between source device and server. This would have to be relayed by the target device, which would be even more time consuming than normally.

In order to optimize the state transfer, delay and overhead must be reduced and interactions between source device and server cannot be performed. This requires that the decision to transfer the state needs to be made solely on the target. Moreover, the amount of signaling messages between the target and the source must be kept low due to the NFC round-trip delay. Our proposed protocol to achieve this optimization is illustrated in Figure 4 and works as follows. Both devices are assumed to have registered previously with the migration server via the WLAN connection to establish a trust relationship. The large display is configured as NFC initiator and the mobile device as NFC target. The large display actively searches for the mobile device. When the user *swipes* the mobile device close to the large display to trigger migration, the NFC initiator detects the NFC target and triggers the migration procedure on the large display.

The large display requests the application state size from the mobile device. To deliver this result, the mobile device must know which application to migrate and its state size. Several selection schemes can be employed, such as choosing the application which has the active/focused window, or the user can have selected an application to migrate manually before the swipe, e.g. when notified about the loss of connectivity. The assumption here is that the mobile device knows which application to migrate and that the application state size is known. The large display then downloads the state object from the mobile device via the NFC link. Once the download has finished, the large display sends a termination command to the mobile device, which terminates the original application. To activate the application on the large display the application must be resumed in the downloaded state. To avoid malicious stealing of applications/sessions by using migration, the state object could be encrypted and hence the large display would need to request a decryption key from the migration server or have the server decrypt the state. This way the state download can be initiated without having to establish a trust relation, which is both time consuming and would bother the user. In summary, our lean protocol relies on decentralization of decisions from the server to the migration target. The major decisions that have been decentralized are:

- Trigger the migration when 'NFC-in-range' event is detected
- Use NFC for the migration protocol and state transfer
- Use pre-defined rules or user-interaction to allow the source device to identify which application to migrate

As the role of a device in the NFC communication must be set before communication, a role selection algorithm must be in place. Selection could be based on static rules, where all devices select a certain role or have pre-assigned roles. The rule in a given scenario must ensure that roles are distributed in order to allow communication. The selection could also be based on a random-hopping scheme, where devices hop between the two roles and listen for presence of opposite roles; initiators request target responses and targets listen for initiator requests. Due to the scarceness of time in the scenario, we employ a static selection decided by the device type: Mobile devices are NFC targets and static devices are NFC initiators. The rationale is that mobile devices are typically power-constrained where as static devices are assumed to have a external power supply is in the case of the large display. As the target role requires less power than an initiator the target role is assigned to the most power-constrained device.

5   Implementation and Evaluation

To evaluate the feasibility of the lean migration protocol, it was implemented using the same NFC setup as in the performance measurements in Section 3. Here, one of the mobile phones acted as the large display. An additional goal of the evaluation was to understand how much state information can be transferred within a typical swipe. The proposed migration orchestration messages GET_STATE_SIZE, GET_STATE and TERMINATE were implemented as simple string messages and the transferred state size was varied similar to the throughput measurements. The results of one experiment for each application state size in the range between 1 and 700 bytes are shown in Figure 5.

The results for the migration duration show a linear behavior over state size S; a least-squares fit yields the relation $delay = 2.1\frac{ms}{byte} \cdot S + 386ms$ which is also depicted in Figure 5. The minimum delay at $S = 0$ is due to the required 3 message exchanges. A detailed analysis of the time stamps shows that each exchange requires 106ms plus some processing delay. The inverse of the slope of the line $\frac{1}{2.1}\frac{bytes}{ms} = 3.8kbit/s$ is in the same order of magnitude as the throughput values observed in Section 3.

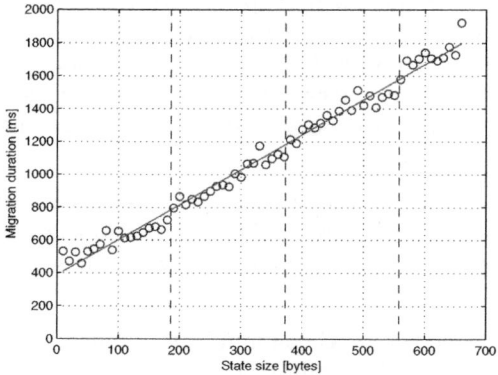

Fig. 5. Duration of a migration consisting of orchestration signaling and state transfer

The observed linear behavior can be used to estimate the limit on state sizes that are feasible to transfer in a certain time window. For instance, assuming a maximum time window of 5 seconds for the transfer, the relation yields a maximum $S=2197$. Based on this upper limit, the orchestration entity in the target can decide whether to initiate migration when only the NFC link is available.

6 Conclusion and Future Work

This paper presents a solution for migration over resource-constrained links, with the specific example of RFID based near-field communication (NFC) technology. The main challenge of migrating an application over resource-constrained links is to make best use of the limited time in which the link is available. The primary steps of migration that can be optimized time-wise are device discovery, decision to trigger and message exchange during orchestration, including transferring the actual state of the application from the source device to the target device. In a general migration scenario, all these steps are coordinated by a central migration server in the network. Through an experimental setup we have shown that in some cases, migration over ad-hoc NFC links is possible and feasible. This is achieved by shifting the primary migration steps from the server to the target device. By measuring the overhead of the optimized migration orchestration protocol and the performance of NFC it is shown that the platform can be used for migrating application state sizes up to 2000 bytes within an NFC swipe window of 5 seconds.

Future work should include investigation of how feedback to the user during migration (auditory/visual) may increase the average length of the available time window. Moreover, for simplicity reasons, we have only addressed one-way migration (from mobile device to large screen). The optimized migration protocol can be generalized to handle migration both ways by introducing a few additional handshake messages during connection establishment. The performance of such a general protocol should also be investigated.

Acknowledgments

This work was partially supported by the EU ICT FP7 project 'Open Pervasive Environments for iNteractive migratory services – OPEN', see www.ict-open.eu. The Telecommunications Research Center Vienna (FTW) is supported by the Austrian Government and by the City of Vienna within the competence center program COMET.

References

1. Paternò, F., Santoro, C., Scorcia, A.: User interface migration between mobile devices and digital tv. In: Forbrig, P., Paternò, F. (eds.) HCSE/TAMODIA 2008. LNCS, vol. 5247, pp. 287–292. Springer, Heidelberg (2008)

2. Fuggetta, A., Picco, G., Vigna, G.: Understanding code mobility. IEEE Transactions on software engineering 24(5), 342–361 (1998)
3. Carzaniga, A., Picco, G., Vigna, G.: Is code still moving around? Looking back at a Decade of Code Mobility. In: ICSE 2007, pp. 9–20. IEEE Computer Society, Los Alamitos (2007)
4. Bandelloni, R., Mori, G., Paternò, F.: Dynamic generation of web migratory interfaces. In: MOBILEHCI 2005 (2005)
5. Antoniou, Z., Varadan, S.: Intuitive mobile user interaction in smart spaces via nfc-enhanced devices. In: ICWMC 2007 (2007)
6. Iglesias, R., Parra, J., Cruces, C., de Segura, N.G.: Experiencing nfc-based touch for home healthcare. In: PETRA 2009 (2009)
7. Kneissl, F., Rottger, R., Sandner, U., Leimeister, J., Krcmar, H.: All-i-touch as combination of nfc and lifestyle. In: NFC 2009 (2009)
8. Karpischek, S., Michahelles, F., Resatsch, F., Fleisch, E.: Mobile sales assistant - an nfc-based product information system for retailers. In: NFC 2009 (2009)
9. Massoth, M., Bingel, T.: Performance of different mobile payment service concepts compared with a nfc-based solution. In: ICIW 2009 (2009)
10. Cappiello, I., Puglia, S., Vitaletti, A.: Design and initial evaluation of a ubiquitous touch-based remote grocery shopping process. In: NFC 2009 (2009)
11. Mika, H., Mikko, H., Arto, Y.-o.: Practical implementations of passive and semi-passive nfc enabled sensors. In: NFC 2009 (2009)
12. Schoo, P., Paolucci, M.: Do you talk to each poster? security and privacy for interactions with web service by means of contact free tag readings. In: NFC 2009 (2009)
13. Madlmayr, G., Langer, J., Kantner, C., Scharinger, J., Schaumuller-Bichl, I.: Risk analysis of over-the-air transactions in an nfc ecosystem. In: NFC 2009 (2009)
14. Handley, M., Jacobson, V., Perkins, C.: SDP: Session description protocol (RFC-4566). Request for Comments, IETF (2006)
15. Ulvan, M., Bestak, R.: Analysis of Session Establishment Signaling Delay in IP Multimedia Subsystem. Wireless and Mobile Networking, 44–55
16. Elkotob, M., Andersson, K.: Analysis and measurement of session setup delay and jitter in VoWLAN using composite metrics. In: MUM 2008, pp. 190–197 (2008)
17. Perkins, C., Alpert, S., Woolf, B.: Mobile IP; Design Principles and Practices. Addison-Wesley Longman, Amsterdam (1997)
18. Bauer, M., Olsen, R., Jacobsen, M., Sanchez, L., Imine, M., Prasad, N.: Context management framework for MAGNET Beyond. In: Workshop on Capturing Context and Context Aware Systems and Platforms, IST Summit (2006)
19. Martin, M., et al.: Migration Service Platform Design. Tech. Rep., ICT-OPEN EU FP7 project (2009)
20. ECMA, 340: Near Field Communication: Interface and Protocol (NFCIP-1). European Association for Standardizing Information and Communication Systems (2004), http://www.ecma.ch/ecma1/STAND/ecma-340.htm
21. http://www.sdid.com/products1010.shtml

Swarm Intelligence Heuristics for Component Deployment

Máté J. Csorba and Poul E. Heegaard

Department of Telematics,
Norwegian University of Science and Technology, N-7491 Trondheim, Norway
{Mate.Csorba,Poul.Heegaard}@item.ntnu.no

Abstract. We address the problem of efficient deployment of software services into a networked environment. Services are considered that are provided by collaborating components. The problem of obtaining efficient mappings for components to host in a network is challenged by multiple dimensions of quality of service requirements. In this paper we consider execution costs for components and communication costs for the collaborations between them. Our proposed solution to the deployment problem is a nature inspired distributed heuristic algorithm that we apply from the service provider's perspective. We present simulation results for different example scenarios and present an integer linear program to validate the results obtained by simulation of our algorithm.

1 Introduction

Implementing distributed networked software systems requires many important design decisions to be made that have strong influence on the Quality of Service (QoS) perceived by the user of the service. A major factor in satisfying QoS requirements of a software service is the configuration of the elementary building-blocks of the service and their mapping to the suitable network elements and resources that are available during execution. Moreover, it is also important for next generation software systems to be capable to (self-)adapt to foreseen and unforseen changes that appear in the execution context. This dynamism in the context of services is even increased by enabling swiftly reconfigurable hardware, allowing mobility and changes in the cardinality of users. Simply improving the QoS metrics without planing is, however, inevitably increasing the costs on the providers' side leading to multifaceted optimization problems.

We address the issue of obtaining efficient and adaptable mappings for software components of networked services as an optimization problem in a distributed environment. We model services as being built by collaborating components with several dimensions of QoS requirements including but not restricted to dependability, performance or energy saving aspects. Correspondingly, our service models are extended with costs relevant for the various dimensions of requirements in a more detailed model. Based on the service models we apply heuristics and a nature-inspired optimization method, called the Cross Entropy Ant System (CEAS) [1], [2], [3] to solve the problem of deploying service components into the network.

Distributed execution of our deployment mapping algorithm has been an important design criteria to avoid the deficiencies of existing centralized algorithms, e.g.

F.A. Aagesen and S.J. Knapskog (Eds.): EUNICE 2010, LNCS 6164, pp. 51–64, 2010.

performance bottlenecks and single points of failure. Moreover, we intend to conserve resources by eliminating the need for centralized decision-making and the required updates and synchronization mechanisms. In our earlier work [4] we selected a well-known example in the domain of task assignment problems and converted it to our context of collaborating components with execution and communication costs. In this paper we extend the initial example from [4] with two additional example system models, present an Integer Linear Program (ILP) able to solve component mapping problems with load-balancing and remote communication minimization criteria, and compare simulation results obtained by executing our algorithm on the examples presented with the optimum cost solutions given by the ILP.

Related to our work a fair number of approaches aim at improving dependability and adaptability through influencing the software architecture. QoS-aware metadata is utilized together with Service Level Agreements (SLAs) in the planning-based middleware in the MUSIC project [5]. SLAs are common means to target policy based research allocation, e.g. [6]. The SmartFrog deployment and management framework from HP Labs describes services as collections of components and applies a distributed engine comprised of daemons running on every node in a network [7]. Fuzzy learning is applied for configuration management in server environments targeting efficient resource utilization by Xu et al. in [8]. Biologically-inspired resource allocation algorithms in service distribution problems have been targeted earlier too, such as by the authors of [9]. A different approach, namely layered queuing networks are employed by Jung et al. in an off-line framework for generating optimal configurations and policies [10]. Changing the deployment mapping of applications is investigated by others as well, however due to the fact that complexity of exact solution algorithms becomes NP-hard already in case of 2-3 hosts or several QoS dimensions applicability of these methods is restricted. Heuristics, such as greedy algorithms and genetic programming are used by Malek to maximize utility of a service from the users' perspective in [11], whereas we formulate the problem from the providers' view, while still considering user requirements. The various middleware approaches can be very good candidates serving our approach as a means of instrument for deployment that is guided by our logic.

The remainder of this paper is organized as follows. First, in Sect. 2 we discuss the component deployment problem in more detail. Next, an introduction to CEAS follows in Sect. 3. In Sect. 4 an ILP is formulated for the general component deployment problem discussed in this paper. Sect. 5 presents service examples and the corresponding simulation results are evaluated. Finally, in Sect. 6 we conclude and touch upon future work.

2   The Component Deployment Problem

We define the component deployment problem as an optimization problem, where a number $|C|$ of components (labelled c_i; $i = 1, \ldots, |C|$) have to be mapped to a set N of nodes. Components can communicate via a set K of collaborations (k_j; $j = 1, \ldots, |K|$). We consider three types of requirements in the deployment problem. Components have execution costs e_i; $i = 1, \ldots, |C|$, collaborations have communication costs f_j; $j = 1, \ldots, |K|$ and some of the components can be restricted in deployment mapping. Components that are restricted to be deployed to specific nodes are called *bound* components. Accordingly, we distinguish between the three concepts of component *mappings*, meaning a set variable obtained and refreshed in every iteration of our algorithm; component *bindings*, which represent a requirement that fixes the mapping of components to a constant value; and component *deployment* that is the actual physical placement of components to nodes, including the bound components. Physical deployment of components is triggered after the mappings obtained and refreshed in every iteration by the deployment algorithm converges to a solution satisfying the requirements.

Furthermore, we consider identical nodes that are interconnected in a full-mesh and are capable of hosting components with unlimited processing demand. We observe the processing load components hosted at a node impose and target load-balancing between the nodes available in the network. By balancing the load we mean minimizing the deviation from the global average per node execution cost. Total offered execution load is obtained from the service specification by the sum $\sum_{i=1}^{|C|} e_i$, thus the global average execution cost can be obtained as

$$T = \frac{\sum_{i=1}^{|C|} e_i}{|\mathbf{N}|}. \tag{1}$$

Communication costs are considered only if a collaboration between two components happens remotely, i.e. it happens between two nodes. In other words, if two components are *colocated* (are placed onto the same node) we do consider communication between them to be free (not consuming any network bandwidth).

Our deployment logic is launched with the service model enriched with the requirements specifying the search criteria and with a resource profile of the hosting environment specifying the search space. In our view, however, the logic we develop is capable of catering for any other types of non-functional requirements too, as long as a suitable cost function can be provided for the specific QoS dimension at hand. In this paper, costs in the model are constant, independent of the utilization of underlying hardware. This limitation has been removed and explored in [12] by the authors. Furthermore, we benefit from using collaborations as design elements as they incorporate local behavior of all participants and all interactions between them. That is, a single cost value can describe communication between component instances, without having to care about the number of messages sent, individual message sizes, etc. For more information on how we use collaborations for system modelling we refer to [4] and [12].

We, then define the objective of the deployment logic as obtaining an efficient (low-cost if possible optimum) mapping of components to nodes, $\mathbf{M} : \mathbf{C} \rightarrow \mathbf{N}$, one that satisfies the requirements in reasonable time. More importantly, the actual placement of components does not change with every iteration of the algorithm but is changed only on a larger timescale, once a mapping is converged to a solid solution to avoid churn. Re-deployment, for adaptation to changes in the execution context involves migrating components, which naturally incurs additional costs. In principle migration costs can be considered as thresholds prohibiting changing the deployment mapping if the benefit is not high enough. In this work, however, we do not consider adaptation and migration costs.

The heuristics we use in our deployment logic is guided by a cost function, $F(\mathbf{M})$ that is used to evaluate the suggested mappings iteration-by-iteration. The construction of the cost function is in accordance with the requirements of the service. How we build the corresponding cost function is discussed in Sect. 3.2.

3   Component Deployment Using the Cross Entropy Ant System

3.1   The Cross Entropy Ant System

The key idea in the CEAS is to let many agents, denoted *ants*, iteratively search for the best solution according to the problem constraints and cost function defined. Each iteration consists of two phases; the *forward* ants search for a solution, which resembles the ants searching for food, and the *backward* ants that evaluate the solution and leave markings, denoted *pheromones*, that are in proportion to the quality of the solution. These pheromones are distributed at different locations in the search space and can be used by forward ants in their search for good solutions; therefore, the best solution will be approached gradually. To avoid getting stuck in premature and sub-optimal solutions, some of the forward ants will explore the state space freely ignoring the pheromone values.

The main difference between various ant-based systems is the approach taken to evaluate the solution and update the pheromones. For example, AntNet [13] uses reinforcement learning while CEAS uses the *Cross Entropy (CE) method for stochastic optimization* introduced by Rubinstein [14]. The CE method is applied in the pheromone updating process by gradually changing the probability matrix $\mathbf{p}$ according to the cost of the solutions found. The objective is to minimize the cross entropy between two consecutive probability matrices $\mathbf{p}_r$ and $\mathbf{p}_{r-1}$ for iteration r and $r-1$ respectively. For a tutorial on the method, [14] is recommended.

The CEAS has demonstrated its applicability through a variety of studies of different path management strategies [2], such as shared backup path protection, p-cycles, adaptive paths with stochastic routing, and resource search under QoS constraints. Implementation issues and trade-offs, such as management overhead imposed by additional traffic for management packets and recovery times are dealt with using a mechanism called elitism and self-tuned packet rate control. Additional reduction in the overhead is accomplished by pheromone sharing where ants with overlapping requirements cooperate in finding solutions by (partly) sharing information, see [3] for details and examples on application.

In this paper, the CEAS is applied to obtain the best deployment mapping $\mathbf{M} : \mathbf{C} \rightarrow \mathbf{N}$ of a set of components, $\mathbf{C}$, onto a set of nodes, $\mathbf{N}$. The nodes are physically connected by links used by the ants to move from node to node in search for available capacities. A given deployment at iteration r is a set $\mathbf{M}_r = \{\mathbf{m}_{n,r}\}_{n \in \mathbf{N}}$, where $\mathbf{m}_{n,r} \subseteq \mathbf{C}$ is the set of components at node n at iteration r. In the CEAS applied for routing the path is defined as a set of nodes from the source to the destination, while now we use the deployment set $\mathbf{M}_r$ instead. The cost of a deployment set is denoted $F(\mathbf{M}_r)$. Furthermore, in the original CEAS we assign the pheromone values $\tau_{ij,r}$ to interface i of node j at iteration r, while now we assign $\tau_{mn,r}$ to the component set $\{m\}$ deployed at node n at iteration r.

In CEAS applied for routing and network management, selection of the next hop is based on the *random proportional rule* presented below. In our case however, the *random proportional rule* is applied for deployment mapping. Accordingly, during the initial exploration phase, the ants randomly select the next *set of components* with uniform probability $1/|\mathbf{C}|$, where $|\mathbf{C}|$ is the number of components to be deployed, while in the normal phase the next set is selected according to the *random proportional rule* matrix $\mathbf{p} = \{p_{mn,r}\}$, where

$$p_{mn,r} = \frac{\tau_{mn,r}}{\sum_{l \in \mathbf{M}_{n,r}} \tau_{ln,r}} \qquad (2)$$

The pheromone values τ_r are updated by the backward ants as a function of the previous pheromone value τ_{r-1}, the cost of the deployment set $\mathbf{M}_r$ at iteration r, and a control variable γ_r (denoted the *temperature*) that captures the history of the cost values such that the total cost vector does not have to be stored. The CEAS function was first introduced in [1], for later extensions, details and examples see [3].

3.2   The Component Deployment Algorithm

To build a distributed cooperative algorithm, in contrast to a centralized approach, we employ the autonomous ant-like agents (denoted ants) of the CEAS method that cooperate in pursuing a common goal. Ants base their decisions solely on information available locally at a node they currently reside in. Accordingly, in our logic the information required for optimization is distributed across all participating nodes, this being a contributing factor to robustness, scalability and fault tolerance of the method.

In our deployment algorithm ants are emitted from one designated place in the network, from the so-called ant-nest. When an ant starts its random-walk over the available nodes it is assigned with the task of deploying the set of components, $\mathbf{C}$. One round-trip (not necessarily visiting every node, but arriving back at the nest eventually) of an ant is called an iteration of the algorithm, which is repeated continuously until convergence or until the algorithm is stopped. During its random-walk the ant selects the next hop to visit entirely randomly. When arriving at a node the ant's behavior depends on if the ant is an *explorer* or a *normal* ant. The difference between the two types of ants is in the method they use to select a mapping, $m_n \subset \mathbf{M}$, for a (maybe empty) subset of $\mathbf{C}$ at each node $n \in \mathbf{N}$ they visit. Normal ants decide on whether to deploy some components at a particular node based on the pheromone database at the given node, whereas explorer ants make a decision purely randomly, thus enforcing exploration of the state-space. Explorer ants can be used both initially to cover up the search space and later, after the system is in a stable state as well to discover fluctuations in the network and, thus to aid adaptation. The useful number of initial explorer iterations is depending on the given problem size (e.g. can be estimated by the number of components in a service). Normal ants, on the other hand focus entirely on optimizing the mapping ($\mathbf{M}$) iteration-by-iteration using and updating the pheromone tables.

An important building-block of the algorithm is the cost function applied to evaluate the deployment mapping obtained by one ant during its search. We denote the application of the cost function as $F(\mathbf{M})$. The function itself is built in accordance with the requirements. As discussed in Sect. 2, in this paper we consider load-balancing and remote cost minimization. Every ant doing its search for a mapping is capable of sampling load-levels at the nodes visited, thus the execution load imposed on the nodes in the network can be obtained by each ant as a list of sample values in the form of

$$\hat{l}_n = \sum_{c_i \in m_n} e_i, n \in \mathbf{N}, m \in \mathbf{M}. \tag{3}$$

The overall cost function evaluating the mapping obtained in an iteration using Eq. (1) and Eq. (3) then becomes

$$F(\mathbf{M}) = \sum_{n=1}^{\mathbf{N}} |\hat{l}_n - T| + \sum_{j=1}^{\mathbf{K}} I_j \, f_j \tag{4}$$

where

$$I_j = \begin{cases} 1, \text{ if } k_j \text{ remote} \\ 0, \text{ if } k_j \text{ internal to a node} \end{cases} \tag{5}$$

is an indicator function for evaluating collaborations between pairs of components.

Optimization of mappings is achieved by gradually modifying pheromone values aligned to sets of components. Pheromone values are organized into tables and are stored in every node participating in hosting the service being deployed. We use *bitstring* encoding to index the pheromone table, each entry representing a given combination/subset of components, i.e. a flag is assigned to every (unbound) component in the service model. Thus, the pheromone database has to accommodate $2^{|C|}$ floating point entries using this encoding. Normalizing the entries in a node an ant can obtain a probability distribution of component sets to be mapped at the particular node. Using CEAS, which is a subclass of Ant Colony Optimization (ACO) algorithms, the optimal solution emerges in a finite number of iterations once the optimum has been observed, which does in fact happen with probability close to one in ACO systems.

Indexing of the database based on component set identifiers is illustrated in the following example. Let us start with a service provided by 4 components, i.e. $|\mathbf{C}| = 4$, $\mathbf{C} = \{c_1, c_2, c_3, c_4\}$, the database size is $2^4 = 16$. If an ant needs to collect or update information e.g. regarding the deployment of the component set $\{c_2, c_4\}$, the set labelled by the bitstring $'1010'B$ has to be addressed, which is equivalent to accessing the $'1010'B = 10^{th}$ element of the pheromone table at the node the ant resides in. The lifetime of an ant, which is equivalent to one iteration of the algorithm, contains two phases. First *forward search* is conducted during which the ant looks for a deployment mapping $\mathbf{M}$ for the set $\mathbf{C}$ visiting arbitrary nodes in the network. Once a complete mapping $\mathbf{M}$ is obtained by the ant the mapping has to be evaluated using the cost function $F(\mathbf{M})$. Using F the cost of the mapping is obtained using which the ant updates the pheromone databases in the nodes visited (the forward route has been stored in the hop-list, $\mathbf{H}$) during *forward search* during the second phase of its lifetime called *backtracking*. Once an ant is finished with backtracking and arrives back to its nest a new iteration can start and a new ant will be emitted. With convergence of the (distributed) pheromone database a strong value will emerge indicating the suggested deployment mapping, while inferior combinations will evaporate. The algorithmic steps of ants' behavior are summarized briefly in Algorithm 1.

Algorithm 1. Deployment mapping of the set of components $\mathbf{C} = \{c_1, \ldots, c_{|\mathbf{C}|}\}$

---

1. Select next node to visit, n randomly and add n to the hop-list $\mathbf{H} = \mathbf{H} + \{n\}$.
2. Select a set of components $m_n \subseteq \mathbf{C}$ according to the random proportional rule (*normal* ant), Eq. (2), or in a random manner (*explorer* ant). If such a set cannot be found, goto step 1.
3. Update the ant's deployment mapping set, $\mathbf{M} = \mathbf{M} + \{m_n\}$.
4. Update the set of remaining components to be deployed, $\mathbf{C} = \mathbf{C} - m_n$.
5. If $\mathbf{C} \neq \emptyset$ then goto 1., otherwise evaluate $F(\mathbf{M})$ and update the pheromone values corresponding to the list of mappings $\{m_n\} \in \mathbf{M}$ going backwards along $\mathbf{H}$.
6. If stopping criteria is not met then initialize and emit new ant and goto 1.

In the algorithm we presented we have a trade-off between convergence speed and the quality of the obtained solution. However, during the deployment of services in a dynamic environment a pre-mature solution, which does satisfy the functional and non-functional requirements often suffices. In the next section we establish a centralized, off-line model to evaluate our deployment logic and to form a basis for cross-validation of the results obtained by simulation of the algorithm.

4   An Integer Program to Validate the Algorithm

To validate the results we obtain via simulation and using various deployment scenarios we have developed an Integer Linear Program (ILP), which we solve using regular solver software. In this ILP we take into account the two counteracting objectives we presented in Sect. 2 and define a solution variable $m_{i,j}$ that will show the optimal, i.e. lowest cost mapping of components to nodes.

We start the definition of the ILP with two parameters. First $b_{i,j}$

$$b_{i,j} = \begin{cases} 1, & \text{if component } c_i \text{ is bound to node } n_j, \\ 0, & \text{otherwise.} \end{cases} \tag{6}$$

which enables the model to fix some of the mappings to cater for bound components, if any, in the model of a given service. Second, T

$$T = \lfloor \frac{\sum_{i=1}^{|\mathbf{C}|} e_i}{|\mathbf{N}|} \rfloor \tag{7}$$

that is used to approximate the ideal load-balance among the available nodes in the network. Beside the binary solution variable showing the resulting mapping, $m_{i,j}$

$$m_{i,j} = \begin{cases} 1, & \text{if component } c_i \text{ is mapped to node } n_j, \\ 0, & \text{otherwise.} \end{cases} \tag{8}$$

we utilize two additional variables. One variable for checking whether two components that communicate via a collaboration, k_i, are colocated or not, in variable col_i.

$$col_i = \begin{cases} 0, & \text{if } c_l, c_k \in k_i \text{ and } c_l \text{ is colocated with } c_k, \\ 1, & \text{otherwise.} \end{cases} \tag{9}$$

Moreover, another variable, $Delta_j$, for calculating the deviation from the ideal load-balance among the nodes participating in hosting the components.

$$\Delta_j \geq 0, \forall n_j \in \mathbf{N} \tag{10}$$

The objective function we use in the ILP is naturally very similar to Eq. (4), however to keep the model within the linearity requirement of the ILP here we use addition.

$$min \sum_{j=1}^{|\mathbf{N}|} \Delta_j + \sum_{i=1}^{|\mathbf{K}|} f_i \cdot col_i \tag{11}$$

Having established the objective function we have to define the constraints the solutions are subjected to to obtain feasible mappings. First we stipulate that there has to be one and only one mapping for all of the components.

$$\sum_{j=1}^{|\mathbf{N}|} m_{i,j} = 1, \forall c_i \in \mathbf{C} \tag{12}$$

In addition, the ILP has to take into account that some component mappings might be restricted in the case of components explicitly bound in the model. Thus, we restrict the mapping variable $m_{i,j}$ using $b_{i,j}$.

$$m_{i,j} \geq b_{i,j}, \forall c_i \in \mathbf{C}, \forall n_j \in \mathbf{N} \tag{13}$$

Next we introduce two constraints to implicitly define the values of the variable Δ_j that we apply in the objective function. We use two constraints instead of a single one to avoid having to use absolute vales (i.e. the $abs()$ function) and thus we avoid non-linear constraints.

$$\sum_{i=1}^{|\mathbf{C}|} e_i \cdot m_{i,j} - T \leq \Delta_j, \forall n_j \in \mathbf{N} \tag{14}$$

$$T - \sum_{i=1}^{|\mathbf{C}|} e_i \cdot m_{i,j} \leq \Delta_j, \forall n_j \in \mathbf{N} \tag{15}$$

Lastly, we introduce constraints, again for implicitly building the binary variable indicating colocation of components.

$$m_{i,j} + m_{k,j} \leq (2 - col_l), k_l = (c_i, c_k) \in \mathbf{K}, \forall c_i, c_k \in \mathbf{C}, \forall n_j \in \mathbf{N} \tag{16}$$

$$m_{i,j1} + m_{k,j2} \leq 1 + col_l, k_l = (c_i, c_k) \in \mathbf{K}, \forall c_i, c_k \in \mathbf{C}, \forall n_{j1}, n_{j2} \in \mathbf{N} \tag{17}$$

Based on this definition the ILP can be executed by a solver program and mapping costs can be obtained by submitting the appropriate data describing any given scenario corresponding to our general definition of the deployment problem in Sect. 2. The optimum mapping of components to nodes will be obtained according to the objective Eq. (11) subject to Eq. (12) – (17).

In the next section we present the example models we used in our simulations and also evaluate the mapping costs obtained by finding a lower bound using the ILP presented in this section.

5   Example Scenarios

In this section we present the three example models we simulated and validated the corresponding results for.

The first example we consider has been investigated, solved and the solutions were compared to other authors work by Widell et al. in [17]. This example originates from heuristical clustering of modules and assignment of clusters to nodes [18], which problem is NP-hard. We translate the module clustering and assignment problem to execution costs and communication costs while the complexity remains NP-hard even if we only deal with a single service at a time. Fig. 1 shows the first example set of components and the collaborations between them, it comprises $|\mathbf{C}| = 10$ components interconnected by $|\mathbf{K}| = 14$ collaborations. Out of the ten components three (shaded) are bound to one of the nodes, c_2 to n_2, c_7 to n_1 and c_9 to n_3. The execution costs of components and communication costs of collaborations are shown as UML note labels.

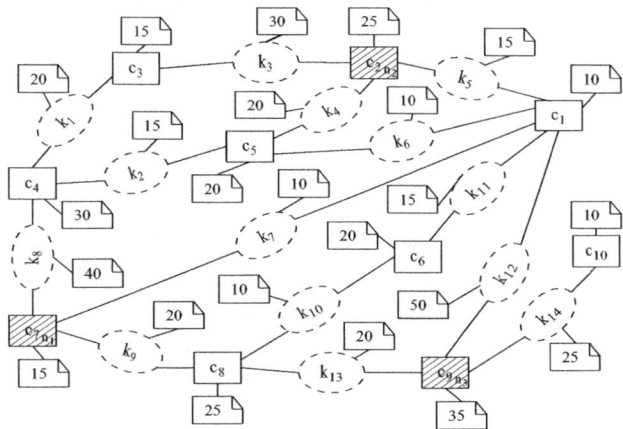

Fig. 1. Example 1, costs

The next figure (Fig. 2) shows the single optimal solution of mapping the components of the first example to three available nodes, $n_{1..3}$, i.e. it shows the optimum mapping $\mathbf{M} : \mathbf{C} \rightarrow \mathbf{N}$. In this optimum cost mapping, the deviation from the total load-balance ($T = 68$) among the nodes are $2, -8, 7$ respectively and the mapping incurs a communication cost of 100 for the remote collaborations. Thus, this mapping results in an overall cost value equal to 117. More details are discussed about this example in [4].

The second example we considered is obtained by extending the first setting into a larger service model and at the same time increasing the number of nodes available for deployment mapping (Fig. 3). The number of components has been increased to 15, out of which 5 components are bound, more collaborations have been introduced and one additional node is added to the deployment environment.

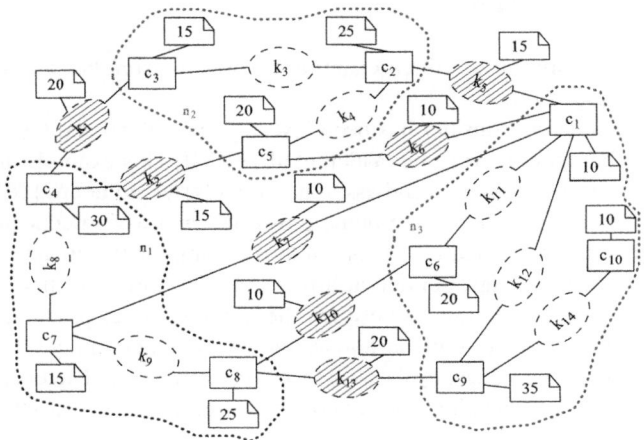

Fig. 2. Example 1, optimum mapping

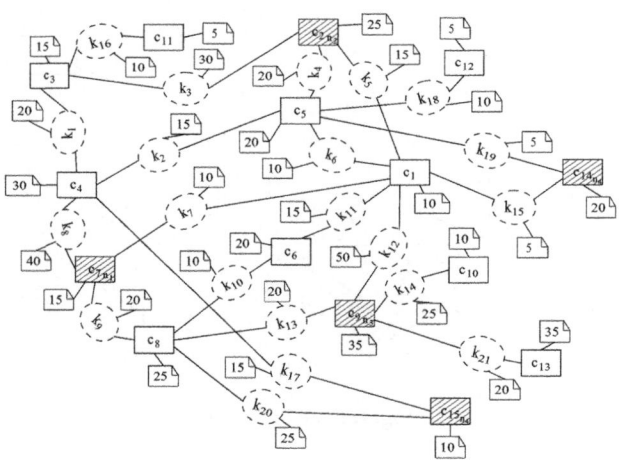

Fig. 3. Example 2, costs

This extended example is introduced to examine how our deployment algorithm behaves with the same type of problem as the computational effort needed increases or, in other words, as the solution state-space is extended.

The next figure, Fig. 4 presents the resulting mapping of components of the second service model to four equivalent nodes after a solution with optimum cost has been found.

In the third example the cardinality of **C** remains the same but the configuration is changed as well as the number of collaborations. In addition, the number of available

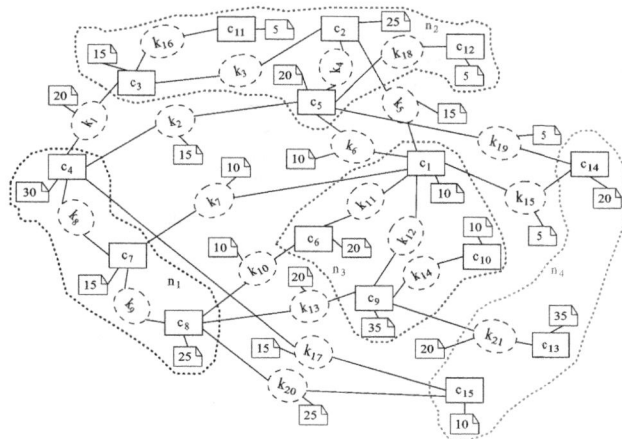

Fig. 4. Example 2, optimum mapping

nodes is increased to 6, as shown in Table 1. Due to the even more complex collaboration pattern between the components in $Example$ 3 we omit the figure showing the corresponding model and its optimal mapping to 6 nodes.

Table 1 also shows the absolute minimum cost values ($Optimum\ cost$) in the three example settings obtained by the ILP presented in Sect. 4. The presented values were generated by executing the algorithm in the simulation environment 100 times for each example model. We can see from the resulting solutions that in the first example our algorithm finds the optimum in 99 simulation runs. The somewhat larger scenario, $Example$ 2 is more difficult to solve for the algorithm, thus we have a larger deviation in the mapping costs. The average cost found is slightly above the optimum as well in this case. However, it is to be noted that the adjective $slightly$ is appropriate here as by changing the placement of a single component from the optimum configuration to a near-optimal one increases the costs not only by 1 but significantly more, this is also the reason for the increased deviation in this case. (In fact, the algorithm has generally found a variety of three different configurations, the optimum with cost 180 and two sub-optimal configurations with costs 195 and 200, this gave the average of 193 shown in Table 1.

In $Example$ 3 the algorithm managed to obtain solutions with costs closer to the absolute optimum obtained by the ILP, with less deviation at the same time. The main cause of this lies in the fact that the collaboration costs in the example are more fine grained, thus, the algorithm managed to find near-optimum solutions with only slightly

Table 1. Example scenarios

	$|C|$ (bound)	$|K|$	$|N|$	Optimum cost	Sim. avg.	Sim. stdev.
Example 1	10(3)	14	3	117	117.21	2.1
Example 2	15(5)	21	4	180	193	9.39
Example 3	15(5)	28	6	274	277.9	5.981

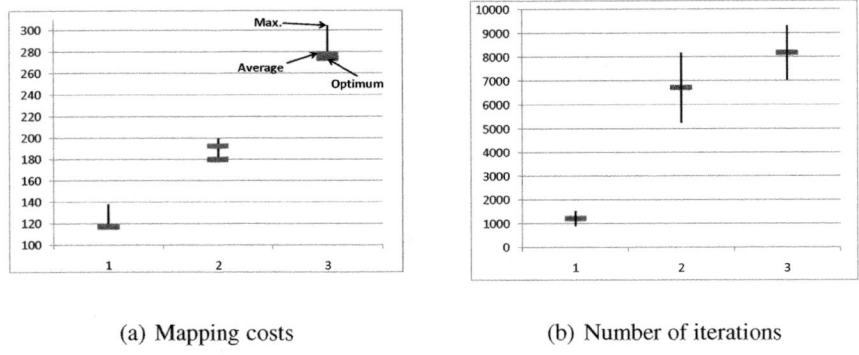

(a) Mapping costs   (b) Number of iterations

Fig. 5. Simulation results for the three examples

higher costs. The average mapping costs obtained in the 100 runs and the number of *normal-ant* iterations our CEAS-based algorithm had to perform are shown in Fig. 5.

The most significant difference for the algorithm in complexity between the original example and the extended ones lies in the size of the pheromone database needed. As the algorithm uses a binary pheromone encoding, in the first example the number of pheromone entries required is 2^7 as the number of components free to map is 7. In the larger examples, however, the pheromone database size increases to 2^{10} in each node, which results in a theoretical state-space of $4 \cdot 2^{10}$ and $6 \cdot 2^{10}$ for CEAS. Correspondingly, in the experiment we applied 2000 *explorer-ants* for the initial example, but 5000 of them in the larger examples. One iteration of a centralized global-knowledge logic, such as the ILP in Sect. 4, is not really comparable with one iteration in the distributed CEAS, which is a tour made by the ant. Nevertheless, the iterations and cuts required by the ILP while solving the three example settings are shown in Table 2. The number of required (explorer and normal) iterations in CEAS is naturally higher than what is required for the ILP with a global overview. However, we advocate that we gain more by the possibility of a completely distributed execution of our algorithm and also because of the capability of adaptation to changes in the context, once the pheromone database is built up after the initial phase. For more on the adaptation capabilities of CEAS in component deployment problems we refer to [12] and [19].

Table 2. Computational effort needed by the ILP

	Example 1	Example 2	Example 3
Simplex iterations	86	495	1075
Branch and cut nodes	0	5	33

6   Closing Remarks

We presented how the deployment of distributed collaborating software components can be supported by swarm intelligence and we have introduced our ACO-based algorithm for obtaining optimal mappings of components to execution hosts. The software components we consider are described by a model enriched with relevant costs, in this particular paper with execution and communication costs. The logic we have developed can be executed in a fully distributed manner, thus, it is free of the deficiencies most existing centralized approaches suffer from, such as performance bottlenecks or single points of failure. We have showed that our algorithm is able to obtain load-balancing among the execution hosts and minimize remote communication at the same time that constitute two contradictory objectives in the deployment mapping problem.

The two main contributions of this paper are the ILP model applicable for component deployment scenarios and the cross-validation of the results obtained by our algorithm and the ILP using the three example scenarios presented. We have used the ILP to find the optimum mappings and a lower bound for the mapping costs obtained by heuristics. It is difficult, however, to compare execution times or required iterations of the two different approaches. During service deployment in a dynamic environment we are often satisfied with pre-mature solutions as long as they adhere to all the functional and non-functional requirements. ACO-based systems are proven to be able to find the optimum at least once with a probability near one, afterwards convergence to this solution is assured within a finite number of steps. CEAS, the main cornerstone in our algorithm can be considered as a subclass of ACO algorithms.

As future work we can identify several directions, out of which we are currently focusing on extending the ILP definition to cater for requirements other than the ones discussed in this paper, such as dependability aspects. Besides, we will continue validating the behavior of our deployment algorithm with extended service models and network scenarios.

References

1. Helvik, B.E., Wittner, O.: Using the Cross Entropy Method to Guide/Govern Mobile Agent's Path Finding in Networks. In: Pierre, S., Glitho, R.H. (eds.) MATA 2001. LNCS, vol. 2164, p. 255. Springer, Heidelberg (2001)
2. Heegaard, P.E., Helvik, B.E., Wittner, O.J.: The Cross Entropy Ant System for Network Path Management. Telektronikk 104(01), 19–40 (2008)
3. Heegaard, P.E., Wittner, O.J.: Overhead Reduction in Distributed Path Management System. Computer Networks, Available online, August 13 (2009) (in press) (Accepted Manuscript)
4. Csorba, M.J., Heegaard, P.E., Herrmann, P.: Cost Efficient Deployment of Collaborating Components. In: Meier, R., Terzis, S. (eds.) DAIS 2008. LNCS, vol. 5053, pp. 253–268. Springer, Heidelberg (2008)
5. Rouvoy, R., et al.: Composing components and services using a planning-based adaptation middleware. In: Pautasso, C., Tanter, É. (eds.) SC 2008. LNCS, vol. 4954, pp. 52–67. Springer, Heidelberg (2008)
6. Ardagna, D., Trubian, M., Zhang, L.: SLA based resource allocation policies in autonomic environments. Journal of Parallel and Distributed Computing 67 (2007)
7. Sabharwal, R.: Grid Infrastructure Deployment using SmartFrog Technology. In: Proc. of the Int'l. Conf. on Networking and Services (ICNS), Santa Clara, USA (July 2006)
8. Xu, J., et al.: On the use of fuzzy modeling in virtualized data center management. In: Proc. of the Int'l. Conf. on Autonomic Computing (ICAC), Jacksonville, FL, USA (June 2007)

9. Heimfarth, T., Janacik, P.: Ant based heuristic for OS service distribution on adhoc networks. Biologically Inspired Cooperative Computing (2006)
10. Jung, G., et al.: Generating adaptation policies for multi-tier applications in consolidated server environments. In: Proc. of the 5th Int'l. Conf. on Autonomic Computing (ICAC), Chicago, IL, USA (June 2008)
11. Malek, S.: A User-Centric Framework for Improving a Distributed Software System's Deployment Architecture. In: Proc. of the doctoral track at the 14th ACM SIGSOFT Symposium on Foundation of Software Engineering, Portland, USA (2006)
12. Csorba, M.J., Heegaard, P.E., Herrmann, P.: Adaptable model-based component deployment guided by artificial ants. In: Proc. 2nd Int'l. Conf. on Autonomic Computing and Communication Systems (Autonomics). ICST/ACM, Turin (September 2008)
13. Di Caro, G., Dorigo, M.: AntNet: Distributed Stigmergetic Control for Communications Networks. Journal of Artificial Intelligence Research 9 (1998)
14. Rubinstein, R.Y.: The Cross-Entropy Method for Combinatorial and Continuous Optimization. Methodology and Computing in Applied Probability (1999)
15. Helvik, B.E., Wittner, O.: Using the Cross Entropy Method to Guide/Govern Mobile Agent's Path Finding in Networks. In: Pierre, S., Glitho, R.H. (eds.) MATA 2001. LNCS, vol. 2164, p. 255. Springer, Heidelberg (2001)
16. Wittner, O.: Emergent Behavior Based Implements for Distributed Network Management. PhD thesis, Norwegian University of Science and Technology, NTNU, Department of Telematics (2003)
17. Widell, N., Nyberg, C.: Cross Entropy based Module Allocation for Distributed Systems. In: Proc. of the 16th IASTED International Conference on Parallel and Distributed Computing and Systems, Cambridge (2004)
18. Efe, K.: Heuristic models of task assignment scheduling in distributed systems. Computer (June 1982)
19. Csorba, M.J., Meling, H., Heegaard, P.E., Herrmann, P.: Foraging for Better Deployment of Replicated Service Components. In: Senivongse, T., Oliveira, R. (eds.) DAIS 2009. LNCS, vol. 5523, pp. 87–101. Springer, Heidelberg (2009)

On Force-Based Placement of Distributed Services within a Substrate Network

Laurie Lallemand[*] and Andreas Reifert

Institute of Communication Networks and Computer Engineering (IKR),
University of Stuttgart, Pfaffenwaldring 47, D–70569 Stuttgart, Germany
{llallem,reifert}@ikr.uni-stuttgart.de

Abstract. Network Virtualization Environments have great potential for overcoming the current ossification of the Internet, fostering innovation, and allowing several concurrent architectures to run on the same physical network. This paper presents a novel physically inspired algorithm for efficiently solving the virtual network embedding problem of placing a virtual network over a substrate network. Compared to a reference heuristic, our algorithm shows lower rejection rates and improved substrate network utilization.

1   Introduction

Internet research is at a crossroads of how to develop, deploy, test, and evaluate new architectures for next generation infrastructures that address current and future challenges. Researchers already attribute the current Internet as "ossified" [TT1] or as an "impasse" [AP1] where network architects can only suggest small deployable changes due to the Internet's multi-stakeholders principle, develop new architectures on a laboratory scale only, or must fall back to overlays with their inherent limitations.

Network Virtualization Environments (NVEs) [CB1] are new architectural approaches to overcome the ossification. A large-scale physical substrate network enables the research community to conduct experiments on a similar scale. Virtualization is the key technology, which offers and isolates the physical resources in form of slices, i. e., partitions of virtual resources. Whether an NVE is only means (base for architectural research) or end (architectural principle in itself) is undecided yet. Projects like GENI [PA1] develop such an environment with the former notion in mind, but the latter not excluded. Such an architecture would also simplify the deployment of distributed services like video-conferencing, IPTV, or content-distribution.

The survey of NVEs in [CB1] lists many more related projects and research directions. One challenge identified is the virtual network embedding problem that assigns virtual networks to slices over the substrate network. Good algorithms to solve this NP-hard problem are crucial for efficient use of the substrate's resources. There exist no recommended way on how to solve this problem best

---

[*] At the time of writing, Laurie Lallemand was a student at the IKR.

F.A. Aagesen and S.J. Knapskog (Eds.): EUNICE 2010, LNCS 6164, pp. 65–75, 2010.

so far. Adaptation of graph theory heuristics of similar problems, application of simulated annealing, or use of integer linear solvers are current approaches. Our novel approach translates the problem into the physical domain of an n-body problem with different force types between the bodies and solve it there. The resulting algorithm scales polynomially with the problem's size and yields good results fast.

2   Problem Description and Related Work

The *(virtual) network embedding problem* (NEP or VNEP) consists of finding a mapping of a virtual network $G_v = (V_v, E_v)$ with components V_v and connections E_v onto a substrate network $G_s = (V_s, E_s)$ with nodes V_s and links E_s. The mapping places components onto nodes ($m_V : V_v \mapsto V_s$) and implements connections through paths in the substrate network ($m_E : E_v \mapsto \mathcal{P}(E_s)$). The sources and destinations of the paths correspond with the respective locations of components' nodes.

The substrate network offers resources for the virtual network. In our case, we consider offered capacity resources of nodes (cap_s) and offered bandwidth resources of links (bw_l). Components and connections have corresponding resource demands (cap_v and bw_e). Our optimization goal is to minimize the used resources on nodes s and links l, individually weighted with $w_{v,s}$ and $w_{e,l}$:

$$\sum_{v \in V_v} w_{v,m_V(v)} \cdot \mathrm{cap}_v + \sum_{e \in E_v} \sum_{l \in m_E(e)} w_{e,l} \cdot \mathrm{bw}_e \tag{1}$$

The mapping must not exceed the offered resources of the nodes and links:

$$\sum_{v:m_V(v)=s} \mathrm{cap}_v \leq \mathrm{cap}_s \quad \text{and} \quad \sum_{e:l \in m_E(e)} \mathrm{bw}_e \leq \mathrm{bw}_l \tag{2}$$

The virtual network is attributed with additional constraints on components and connections:

- Anchor locations $A_v \subset V_s$ of component v. Proxy components for ingress/ egress user traffic may only be positioned at certain nodes ($m_V(v) \in A_v$).
- Max. delay $\Delta t_{u,v}$ of a connection (u,v). The corresponding path must not exceed this delay ($\Delta t_{m_V(u),m_V(v)} \leq \Delta t_{u,v}$).

Figure 1 gives an example of an NEP instance. Both networks are overlaid on each other. The substrate network consists of the graph with the squared boxes. The numbered labels indicate the offered resources. The virtual network consists of the graph with the circles. The numbers indicate the resource demands. Black components are anchored to the specific nodes ($|A_v| = 1$). The remaining components can have arbitrary locations ($A_v = V_s$). The connection between components requiring capacities of 3 and 5 has an additional delay constraint.

In our approach we assume equally weighted node and link resources and precomputed shortest or equal cost multi paths, and thus can transform the

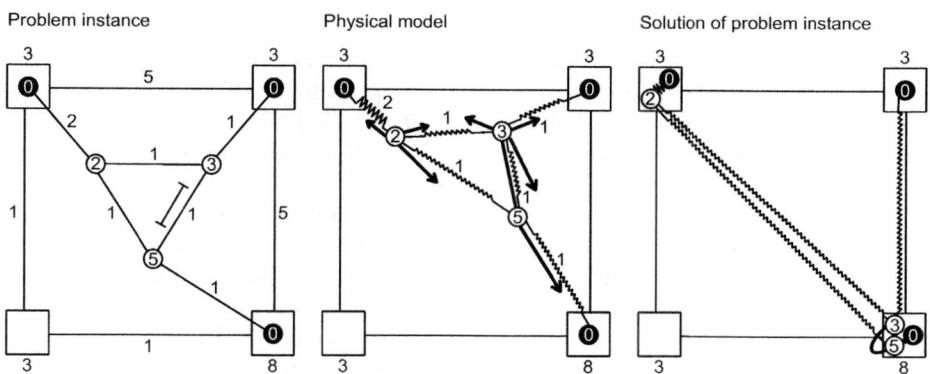

Fig. 1. Example of a network embedding problem instance (left), the corresponding physical model (middle), and its solution (right)

problem into a physical model with attracting forces between components and nodes (middle) through charges, springs, and elastic bands. The components move according to the forces until they find their final locations (right).

NEP algorithms employ various methods. [YY1] begin with a greedy component mapping and subsequently solve a multicommodity flow problem for the connections. They do not take topological information into account in the first step. [ZA1] apply a similar strategy to individual topological clusters with better results, but only consider components and connections units as resources in an unbounded environment. Others, such as [RA1], use simulated annealing. [HL1] propose an innovative distributed algorithm, but it cannot achieve competitive performance yet [CB1].

Very recent works take an integrated approach. [LK1] adapt an algorithm for finding subgraphs in the substrate topology. [CR1] model the problem as an integer linear program and approximate the solution with LP-relaxation. We take an integrated approach as well, which adapts a physical force-based model.

Graph drawing algorithms make successfully use of physical force-based models for finding layouts that intuitively visualize explicit or inherent graph properties like edge lengths, edge weights or vertex distances in the best possible way. The model in [KK1] replaces the graph with a rings-and-spring system and finds a configuration with minimal potential energy. The model in [FR1] applies attractive forces on edges' endpoints and repulsive forces on each vertex pair on a system of particles representing nodes. Our model combines both approaches.

3 Physical Model

We replace vertices with charged particles and connections with springs and elastic bands. Our model is general enough to host additional forces if further constraints arise.

3.1   Particles

For each node $s \in V_s$ with offered capacity cap_s, we introduce a negatively-charged particle $p^{-\mathrm{cap}_s}$ (anion). For each component $v \in V_v$ with capacity demand cap_v, we introduce a positively-charged particle $p^{+\mathrm{cap}_v}$ (cation). We denote the set of anions and cations as P^- and P^+. Particles are arranged within an n-dimensional Euclidean space. Currently, we use two dimensions, but our method applies to arbitrary ones. The space has an Euclidean norm $\| \cdot \|$ and defines a distance $\mathrm{d}_{p,q}$ between two particles p and q according to their positions. We overlay the space with a grid of squared cells with identical cell dimensions. The grid enables efficient lookup of short-distance neighboring particles.

The anions occupy the fixed node positions in the space. Initially, all cations with only one anchor location are bound to their anchor anion and will stay fixed there. The remaining cations are free, randomly positioned, and mobile. We denote the sets of bound and free cations with P_{bound} and P_{free} .

In this physical system, charges express the capacity resources. A $p^{+\mathrm{cap}_v}$ cation may bind with any sufficiently charged anion $q^{-\mathrm{cap}_s}$ through an reduction-oxidation reaction if $\mathrm{cap}_v \leq \mathrm{cap}_s$. After the reaction, the charge of the anion is reduced by cap_v. We will later specify the exact conditions when a cation can bind (see subsection 4.2).

$$q^{-\mathrm{cap}_s} + p^{+\mathrm{cap}_v} \; \rightarrow \; q^{-(\mathrm{cap}_s - \mathrm{cap}_v)} + p \tag{3}$$

3.2   Springs and Elastic Bands

For each connection $e \in E_v$ with positive bandwidth demand bw_e, we introduce a spring between the corresponding endpoints' cations. The static equilibrium of such a spring is the unextended state, i. e. the spring has length zero and the cations are superposed. Its spring constant is the respective bandwidth demand. Thus, components with a large bandwidth demand come closer: the connection consumes less bandwidth in the substrate network.

On a real world partially-meshed network, the Euclidean distance between two nodes is correlated with the shortest path delay between those two nodes. Using statistical estimation, we can thus model a maximum allowed Euclidean distance between two delay-restricted components p and q with a linear function on $\Delta t_{p,q}$. We introduce an elastic band between the respective cations, of length of the maximal allowed distance between the particles, denoted $\mathrm{d}_{\Delta t_{p,q}}$. The band is unstressed when the constraint is not violated. When it is violated, the band is stressed with global spring constant c.

3.3   Forces and Potential Energy

The model elements "charged particles", "springs", and "elastic bands" exert forces on the cations components in P_{free}. We can consider each individual force between each individual pair of particles. We have three types of forces FT = {cap, bw, delay} referred to as capacity-based type, bandwidth-based type, and delay-based type.

In the following, we consider one special free particle $p \in P_{\text{free}}$. We denote the force of type t that another particle q exerts on p with $\boldsymbol{f}_{p,q}^{(t)}$. The force in the electrical central field of q points to q; the force of the spring or elastic band points to q, too. The resulting force on a particle is:

$$\boldsymbol{f}_p^{(t)} = \sum_{q \in N_p^{(t)}} \boldsymbol{f}_{p,q}^{(t)} \quad , \quad \boldsymbol{F}_p = \sum_{t \in \text{FT}} \boldsymbol{f}_p^{(t)} \tag{4}$$

The magnitude $\|\boldsymbol{f}_{p,q}^{(t)}\|$ depends on the force type, the particle distance, the charges, and the spring or elastic band constants. Also, not all particle pairs are relevant. Bandwidth and delay forces only apply between spring and elastic-band-connected cations. For capacity-based forces we only concentrate on free cation-anion pairs with the possibility to bind. Thus, each type defines for each particle p a neighborhood $N_p^{(t)} \subset P$ of relevant counterparts q. Table 1 gives the magnitudes and neighborhoods for the different force types. w_{cap}, w_{bw} and w_{delay} serve to weight the relative effect of forces. Equalling w_{cap} to the area of the grid and the other weights to unity gives good results.

Capacity-based forces make compatible particles to be attracted to each other. The formula is inspired from Coulomb's Law: it is proportional to the product of charges between the two particles and diminishes quadratically with their distance. Thus, components are drawn to high capacity nodes, but only within a certain radius. Components with larger demand reach them faster. As there is no reason for two components to be spread apart, this capacity force is never repulsive: $N_p^{(\text{cap})}$ does not contain other cations. The averages of all negative and positive charges normalize the force so that its magnitude does not depend on the problem instance.

Bandwidth-based forces are spring-like interactions between any two cations that share a connection. The formula is Hook's Law, for a spring whose static equilibrium is to be completely retracted and whose spring constant is the normalized bandwidth of the shared connection.

Delay-based forces are the interactions exerted by elastic bands which gets stretched at the maximum delay distance between any two particles: it is zero until the delay distance is exceeded; then, it acts as a spring with global spring constant c, whose static equilibrium is obtained at the delay distance.

Table 1 also gives the partial potential energies that constitutes the system's potential energy differentiated by force type. The potential energy is a relative measure of the energy stored within the system. The difference in potential

Table 1. Forces magnitudes, neighborhoods and associated potential energies

$t \in$ FT	$\|\boldsymbol{f}_{p,q}^{(t)}\|$	$N_p^{(t)}$	$u_{p,q}^{(t)}$
cap	$w_{\text{cap}} \cdot \frac{\text{cap}_p \text{cap}_q}{\text{cap}_{\text{avg}+} \text{cap}_{\text{avg}-}} \cdot \frac{1}{\text{d}_{p,q}^2}$	$\{q \in A_p \mid \text{cap}_q \geq \text{cap}_p\}$	$-\text{d}_{p,q} \cdot \|\boldsymbol{f}_{p,q}^{(\text{cap})}\|$
bw	$w_{\text{bw}} \cdot \frac{\text{bw}_e}{\text{bw}_{\text{avg}}} \cdot \text{d}_{p,q}$	$\{q \in P^+ \mid (p,q) \in E_v\}$	$\frac{1}{2} \cdot \text{d}_{p,q} \cdot \|\boldsymbol{f}_{p,q}^{(\text{bw})}\|$
delay	$w_{\text{delay}} \cdot c \cdot \left(\text{d}_{p,q} - \text{d}_{\Delta t_{p,q}}\right)$	$\{q \in P^+ \mid \text{d}_{p,q} \geq \text{d}_{\Delta t_{p,q}}\}$	$\frac{1}{2} \cdot \left(\text{d}_{p,q} - \text{d}_{\Delta t_{p,q}}\right) \cdot \|\boldsymbol{f}_{p,q}^{(\text{delay})}\|$

energies between two configurations of a system indicates the work necessary to get from the first configuration to the second one. If the potential energy is minimal and the particles are not moving the system is stable.

It derives from the positions of the physical bodies and forces that are applying on them. In our case, it is calculated as the sum of the individual partial potential energies that are defined between pairs of particles exerting forces on each other. Those individual potential energies only depend on the magnitude of forces and distance between the cation and the force-inducing particles.

$$u_p^{(t)} = \sum_{q \in N_p^{(t)}} u_{p,q}^{(t)} \quad , \quad U_p = \sum_{t \in \mathrm{FT}} u_p^{(t)} \quad , \quad U = \sum_{p \in P^+} U_p \qquad (5)$$

4   Force-Based Placement Algorithm

The forces determine the dynamic behavior of our system of particles over time. We consider individual discrete time steps. At each step we calculate the movement of each particle according to the forces, determine the new positions and reset all velocities to zero.

4.1   Displacement

The linear equations of [KK1] does not apply to our model because we have other types of forces than spring-based forces. [FR1] describes a simplified model of displacement with one attractive and one repulsive force types. The displacement step is proportional to the resulting force applying on the particle and is limited by a temperature. The value of the temperature acts as a cut-off for the displacement step of each particle. A geometrically decreasing temperature series $\mathrm{T}^{(n)}$ leads to the final configuration, avoiding infinite oscillations.

We take a similar approach to displace free cations, except that our temperature is a scaling factor instead of a cut-off value and that we also use it to avoid infinite displacement due to infinite-magnitude forces. $\Delta x_p^{(n)}$ is the normalized displacement of p at time step n:

$$\Delta x_p^{(n)} = \frac{\mathrm{T}^{(n)}}{|P^+|} \cdot \frac{F_p}{\max_{q \in P} \|F_q\|} \qquad (6)$$

If one delay constraint is too much violated, i.e. if $\|f_q^{(\mathrm{delay})}\|$ exceeds the cell length of the grid, then only delay forces apply in the calculation of the next displacement steps of all particles:

$$\Delta x_p^{(n)} = \frac{\mathrm{T}^{(n)}}{|P^+|} \cdot \frac{f_p^{(\mathrm{delay})}}{\max_{q \in P} \|f_q^{(\mathrm{delay})}\|} \qquad (7)$$

Table 2. Reaction conditions

Condition	Description
Charge compatibility	$\mathrm{cap}_v \geq \mathrm{cap}_s$
Anchoring	$s \in A_v$
No delay violation	$\|\boldsymbol{f}_p^{(\mathrm{delay})}\| = 0$
Enough underlying bandwidth	Connections between all components that correspond to a cation in $P_{\mathrm{bound}} \cup \{p\}$ must be supported on links.
Proximity	$d_{p,q} \leq \mathrm{cellsize}$
Stability	Displacement due to non-electrical forces must be small (free cation slows down near a good position).

4.2 Reaction Conditions

At any time, any couple of particles that meet the requirements of Tab. 2 can react according to (3). This determines the set $P_{\mathrm{react}}^{(n)}$ of all free cation-anion pairs that can react at time step n. Each reaction actually places the component of the cation on the node of the anion. The electron transfer corresponds to a reservation of node capacity.

4.3 Stages

Due to the dominance of charge-based forces in the short range, activating all forces at once leads to mediocre results. Thus, our algorithm consists of two stages during which free particles move into equilibrium through Alg. 1. The only difference between the two stages are the forces activated, the temperature cooling schedule, and that particles can only bind in the second stage.

The first stage starts with all free particles randomly placed. Only bandwidth and delay forces are activated. Bandwidth forces leads to the equilibrium of a spring system. Delay forces ensure that no delay requirement is broken. This stage is cooled down with positive rate $\alpha_{1,\mathrm{cool}} < 1$ by means of a simple geometrically decreasing temperature ($\mathrm{T}^{(n+1)} = \alpha_{1,\mathrm{cool}} \mathrm{T}^{(n)}$).

The second stage starts from the end position of the first stage. All forces are activated and cations/components are drawn to anions/nodes. At the end of each time step, the couple of particles that meet the reaction conditions of Tab. 2 and whose particles are closer to each other react. Because the electrical state of anions change, cations have to reorganize after each reaction. Hence we need a more aggressive temperature cooling schedule that cools down very fast (rate $\alpha_{2,\mathrm{cool}} < \alpha_{1,\mathrm{cool}}$) and reheats with factor $\alpha_{2,\mathrm{heat}} > 1$ after one cation has bound or no cation has found any suitable anion during the current cooling round:

$$\mathrm{T}^{(n+1)} = \begin{cases} \alpha_{2,\mathrm{cool}} \mathrm{T}^{(n)} & \text{if } \mathrm{T}^{(n)} > \mathrm{thresh} \wedge |P_{\mathrm{free}}^{(n+1)}| \geq |P_{\mathrm{free}}^{(n)}| \\ \alpha_{2,\mathrm{heat}} \mathrm{T}^{(n)} & \text{otherwise} \end{cases} \qquad (8)$$

Algorithm 1. Displacement algorithm for both stages

Input: P, the set of all particles
Input: $x_p^{(0)}$, the initial position of particles
Input: $T^{(n)}$, a series of temperatures
Input: $C^{(n)}$, a series of stopping conditions

```
 1: n ← 0
 2: repeat
 3:     n ← n + 1
 4:     for p ∈ P do {calculate forces and potential energy}
 5:         for t ∈ FT do
 6:             Calculate f_p^(t) according to (4)
 7:         end for
 8:         Calculate u_p^(t) according to (5)
 9:     end for
10:     if not C^(n) is met then {displace particles}
11:         for p ∈ P_free^(n) do
12:             Calculate Δx_p^(n) according to (6) or (7)
13:             x_p^(n) ← x_p^(n-1) + Δx_p^(n)
14:         end for
15:     end if
16:     if P_react^(n) ≠ ∅ then {bind particles}
17:         Choose random (p,q) ∈ arg min_{(p,q)∈P_react^(n)} {d_{p,q}}
18:         Bind p to q according to (3)
19:     end if
20: until C^(n) is met
```

At the end of the second stage, remaining cations, if any, are classified by charge in decreasing order and greedily forced to react with the nearest anion for which all reactions conditions except the last two are fulfilled. If some cations are left after this process, the placement fails.

Each stage must stop once all forces are equilibrated. When this happens, the system is near a stable state of low potential energy, which means that the current position of particles is a satisfying compromise regarding initial requirements. In [KK1], the algorithm stops when the potential energy is under a not further specified threshold. In [FR1], it stops after an experimentally found fixed number of moves. Unfortunately, both thresholds are problem-specific in our model.

Thus, we stop a stage when the magnitude of the averaged resulting force for the last moves is low for most particles. This condition has the advantage of treating particles that are not moving anymore and particles that are infinitely oscillating equally. Sometimes, particles have not enough time to arrange themselves so that this first condition is met. To avoid loosing time while the system does not evolve anymore, we also end a stage when the average variation of potential energy for the recent last moves is low enough. $C^{(n)}$ aggregates all those conditions into one stopping criterion.

5   Evaluation

Our evaluations compare the performances of our *force-based* algorithm (FB) in different scenarios with the greedy *best-node-first* heuristic (BNF) of [YY1] and, for small instances only, with the *optimal* solution (OPT). For OPT we developed a mixed integer linear program formulation and solved it with SCIP [Ko1], with almost prohibitive running times for the generated instances already.

5.1   Evaluation Scenarios

We based our evaluations on Monte-Carlo simulations with 100 or 500 problem instances per diagram value. Our networks are undirected graphs.

For each problem instance, we generate random substrate network graphs G_s of 10 or 50 nodes according to the Waxman model [Wa1] with parameters $\alpha = 0.1$ and $\beta = 0.9$. Node capacities are uniformly distributed integers within [10, 50]. Links have infinite available bandwidth. Our virtual network G_v is an Erdös-Rényi [ER1] random graph with a variable number of non-anchored components from 2 to 80 and a 25% connection probability between two components. Component capacities are uniformly distributed integers within [10, 25] or [10, 50]. Each of those non-anchored components has a 30% chance of being connected to a proxy, modeled by an additional component anchored to a single node.

5.2   Comparison

Both diagrams of Fig. 2 display the results of our experiments, in particular rejection rates, as light grey markers with associated logarithmic axis on the right, and bandwidth consumption, as blacks markers with associated axis on the left. OPT results appear as upwards triangles, FB results as diamonds and BNF as downwards triangles. The capacity consumption is always the same regardless the algorithm, and therefore not included.

Our first scenario concerns small problem instances (Fig. 2, left). We independently compare FB and BNF with the optimal solution and thus only can use 100 problem instances per diagram value. Independent means that for bandwidth

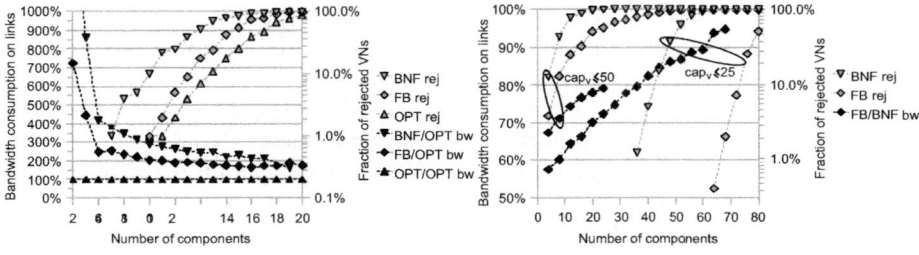

Fig. 2. Comparison of rejection rates (grey) and bandwidth consumption (black) of our force-based algorithm (FB) with a best-node-first heuristic (BNF) and optimal solutions (OPT)

consumption we consider non-rejected instances of any algorithm irrespective of any other. In our case, for FB we always consider at least the same instances as for BNF. Each substrate network has 10 nodes. There are 2 to 20 components with required capacities in the interval between 10 and 25.

With increasing virtual network size, we observe a growth in the rejection rates, the rejection rate of FB being much lower than that of BNF. The rejection rate of FB is actually closer to OPT than to BNF. Rejection is low – below 10% – for at most 8 components in BNF, for at most 11 components in FB, and for at most 12 components in OPT. In this area, FB's rejection rate is below one fifth of BNF's one.

With increasing virtual network size, the averaged normalized bandwidth consumption with respect to OPT decreases. The main contributing factor for this decrease is the reduced probability of finding really bad placements. For this reason, meaningful bandwidth consumption results concerns only areas with a low rejection rate. The bandwidth consumption of FB is 2 to 4 times higher than the optimum value, and at most 75% that of BNF in independent comparison. Further investigation revealed that if we compare only problem instances that have been accepted by both BNF and FB directly, FB bandwidth consumption even drops below 40% of that of BNF.

Our second scenario deals with large problem instances and compares FB and BNF directly (Fig. 2, right). We could not run OPT anymore in reasonable time. We use 500 problem instances per diagram value. Each substrate network has 50 nodes. There are 4 to 80 components with required capacities between 10 and cap_v with $cap_v = 25$ or 50.

Again, FB rejects significantly less problems than BNF. For the small capacities configuration, the rejection rate of FB is below 10% as long as the number of components is below 72. BNF already exceed this threshold with more than 40. For the high capacity configuration, the limit is 4 components for FB, while BNF always rejects more than 10% problems. Concerning bandwidth consumption, the bandwidth consumption of FB is always less than 80% that of BNF in low rejection areas for both components' capacities configurations.

6   Conclusion

Network Virtualization Environments are a promising approach for large-scale Internet network architecture experimentation and can also form the base of a future Internet architecture on its own. One challenge there is solving the NP-hard network-embedding problem. We transposed such problem instances into a physical system with charged particles, springs and elastic bands, and designed a force-based placement algorithm inspired from graph layout algorithms. However, our model introduces more attractive force types. Therefore, our algorithm runs in two stages instead of just one. Our approach currently supports capacity and bandwidth offers/demands as well as delay constraints. It can be easily extended through additional force types.

Monte-Carlo simulations indicate that our algorithm improves the rejection rate and the bandwidth consumption over a BNF heuristic, while still being scalable to large problem instances in terms of running time. In particular our force-based algorithm rejects less instances of hard problems than the heuristic, and significantly less instances of average difficulty. In the latter case, average bandwidth consumption obtained by the force-based algorithm is at least 20% lower than the independent results of the BNF heuristic. In direct comparison, it is even at least 60% lower. In small simulation scenarios where we could obtain the optimum solutions, our algorithm consumes twice the optimal bandwidth.

References

[AP1]   Anderson, T., Peterson, L., Shenker, S., Turner, J.: Overcoming the Internet Impasse through Virtualization. IEEE Computer 38(4), 34–41 (2005)
[CB1]   Chowdhury, N.M.M.K., Boutaba, R.: A survey of network virtualization. Computer Networks 54(5), 862–876 (2010)
[CR1]   Chowdhury, N.M.M.K., Rahman, M.R., Boutaba, R.: Virtual Network Embedding with Coordinated Node and Link Mapping. IEEE Conference on Computer Communications (INFOCOM 2009), April 19-25, pp. 783–791(2009)
[ER1]   Erdös, P., Rényi, A.: On Random Graphs. Publicationes Mathematicae Debrecen 6, 290–297 (1959)
[FR1]   Fruchterman, T.M.J., Reingold, E.M.: Graph drawing by force-directed placement. Softw. Pract. Exper. 21(11), 1129–1164 (1991)
[HL1]   Houidi, I., Louati, W., Zeghlache, D.: A Distributed Virtual Network Mapping Algorithm. In: IEEE Conference on Computer Communications (ICC 2008), May 19-23, pp.5634–5640 (2008)
[KK1]   Kamada, T., Kawai, S.: An Algorithm for Drawing General Undirected Graphs. Inf. Process. Lett. (1), 7–15 (April 1989)
[Ko1]   Konrad-Zuse-Zentrum für Informationstechnik Berlin: SCIP—Solving Constraint Integer Programs, http://scip.zib.de/
[LK1]   Lischka, J., Karl, H.: A Virtual Network Mapping Algorithm based on Subgraph Isomorphism Detection. In: Proceedings of the 1st ACM workshop on Virtualized infrastructure systems and architectures (VISA 2009), August 17, pp. 81–88 (2009)
[LT1]   Lu, J., Turner, J.: Efficient Mapping of Virtual Networks onto a Shared Substrate (2006)
[PA1]   Peterson, L., Anderson, T., Blumenthal, D., Casey, D., Clark, D., Estrin, D., Evans, J., Raychaudhuri, D., Reiter, M., Rexford, J., Shenker, S., Wroclawski, J.: GENI Design Principles. IEEE Computer 39(9), 102–105 (2006)
[RA1]   Ricci, R., Alfeld, C., Lepreau, J.: A Solver for the Network Testbed Mapping Problem. ACM SIGCOMM Computer Communications Review 33(2), 65–81 (2003)
[TT1]   Turner, J.S., Taylor, D.E.: Diversifying the Internet. In: IEEE Global Telecommunications Conference (GLOBECOM 2005), November 28-December 2, pp. 755–760 (2005)
[Wa1]   Waxman, B.M.: Routing of Multipoint Connections. IEEE Journal on Selected Areas in Communications 6(9), 1617–1622 (1988)
[YY1]   Yu, M., Yi, Y., Rexford, J., Chiang, M.: Rethinking Virtual Network Embedding: Substrate Support for Path Splitting and Migration. ACM SIGCOMM Computer Communication Review 38(2), 17–29 (2008)
[ZA1]   Zhu, Y., Ammar, M.H.: Algorithms for Assigning Substrate Network Resources to Virtual Network Components. In: INFOCOM 25th IEEE International Conference on Computer Communications, Proceedings, pp. 1–12 (2006)

Enabling P2P Gaming with Network Coding

Balázs Lajtha, Gergely Biczók, and Róbert Szabó

High Speed Networks Laboratory
Dept. of Telecommunications and Media Informatics
Budapest University of Technology and Economics
{lajtha.balazs,biczok,szabo}@tmit.bme.hu

Abstract. The popularity of online multiplayer games is ever-growing. Traditionally, networked games have relied on the client-server model for information sharing among players, putting a tremendous burden on the server and creates a single point of failure. Recently, there have been efforts to employ the peer-to-peer paradigm for gaming purposes, however, latency-sensitive action games still pose a formidable challenge. The main contribution of this paper is the design of a novel peer-to-peer gaming framework based on random linear network coding. We briefly evaluate the performance of the proposed framework; our initial results suggest a significant reduction in network latency that comes at the expense of a small data traffic overhead. Although further evaluation is clearly needed, we believe that our approach can be the foundation of a truly peer-to-peer communication architecture for networked games.

1 Introduction

Playing computer games is the favorite pastime of hundreds of millions of people. The population of players is very diverse, men and women, children and grandparents, construction workers and university professors all have a tendency to use their computer for entertainment purposes. In the past decade, online games gained popularity; specifically, multiplayer online games are the most successful, such as Call of Duty (a first-person shooter) or World of Warcraft (a role-playing game) [1]. Designing an architecture which meets the strict requirements of online gaming, offers a good quality of experience for the users, and proves to be efficient under the unpredictable conditions of the current Internet is a challenging task.

Traditionally, networked games have relied on the client-server architecture: players' home computers act as clients, with (one or) multiple servers situated in the higher tiers of the network. As an alternative, some games allow for hosting a server at sufficiently equipped home computers, making multiplayer gaming possible in a local network (LAN) setting. Either way, massively multiplayer games can put a tremendous burden on a single server, both from the traffic and computational load viewpoint. This can result in inefficient network resource utilization and also creates a single point of failure. The notion of a peer-to-peer gaming architecture emerges naturally, however, only a small number of role-playing games use this concept: action games are more sensitive to network delays, as lags may render the gameplay experience unsatisfactory.

On the other hand, the recently proposed network coding principle [2] could open new avenues for packet switched networks. While network coding has been proven to

F.A. Aagesen and S.J. Knapskog (Eds.): EUNICE 2010, LNCS 6164, pp. 76–86, 2010.

be effective in a wireless environment [3] and also in core networks, it's usefulness in traditional P2P applications like file-sharing and video streaming has been widely debated [4] [5] [6]. From a networking point of view, multiplayer gaming shares similarities with both content distribution applications, and as such, it could potentially benefit from network coding.

In this paper we present a practical network coding approach to multiplayer gaming over a peer-to-peer overlay. We replace the standard unicast forwarding mechanism with a random network coding mechanism. Since home computers, especially ones used for gaming, are powerful, coding and decoding packets on-the-fly is feasible. Our initial performance evaluation shows encouraging results: average latency of player state updates is reduced in a wide range of scenarios. Specifically, network coding outperforms unicast by more than 30% when participating peers are heterogeneous in terms of access bandwidth. The cost for this improvement is a potential overhead regarding generated data traffic; however, this extra traffic is proportionally marginal in a number of scenarios, and its absolute volume is always low. We argue that applying random network coding in this context is an extremely promising research direction, as it has the potential to be the foundation of a truly peer-to-peer architecture for networked games.

The remainder of this paper is structured as follows. Section 2 gives an overview on network coding techniques and online games. Our main idea, the application of network coding for peer-to-peer online gaming is presented in Section 3. We evaluate the performance of the proposed method in Section 4. Finally, Section 5 identifies open issues and concludes the paper.

2   Related Work

Network coding. Network coding was proposed to be used for improving packet-switched network throughput by Ahlswede et al. in [2]. In a classical routed approach packets are transmitted without change from data sources to destinations. Meanwhile, in a system implementing network coding, nodes are allowed to modify packets, encode and re-encode packets on the path, as long as the receivers are able to extract the content of the original packets. Network coding usually doesn't use compression, but achieves a larger throughput in multicast or broadcast scenarios with the optimal usage of network topology.

A typical example for the exploitation of network topology is the "butterfly" (see Fig. 1) . In this network a source (1) sends two flows at the rate of one packet per time slot to both node 6 and 7. Each edge has a throughput of one packet per time slot. With no network coding, four packets have to share the minimum cut of three edges ($2 \rightarrow 6$, $4 \rightarrow 5$ and $3 \rightarrow 7$). The problem can only be solved by increasing the capacity of link $4 \rightarrow 5$. On the other hand, application of network coding (e.g., using bitwise $a + b$) at node 4 and 5 allows bottleneck link $4 \rightarrow 5$ to carry the combined information of the two packets. With information decoded at nodes 6 and 7 (e.g., using bitwise $(a+b)-a$ and vice versa), network coding achieves a throughput of 4 packets per time slot.

In [2] it was proven that with network coding the information rate from a source to a set of nodes can reach the minimum of the individual max-flow bounds. In [7] a constructive proof was given that the theoretical maximum information rate can be

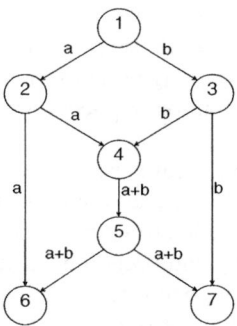

Fig. 1. Network coding in the butterfly topology

achieved by linear network coding. In linear network codes flows are broken into vectors over a field. Each node in the network can linearly combine vectors to create a new packet. Giving a method for constructing optimal network codes using only linear transformations opened the way to the application of network coding when topology is available and changes are seldom.

In dynamically changing networks, advertising network changes and rebuilding the coding infrastructure creates a large overhead. Instead of a static coding scheme, random linear network coding (RLNC) has been proposed in [8]. Random codes differ from traditional linear codes in that linear combinations are generated by each node on the fly. Transmitted packets contain the combination of packets and the coefficients associated with each source vector present in the packet. Authors showed that the performance of RLNC exponentially approaches that of linear network coding. By doing so, RLNC can provide a solution as good as any network coding scheme making use of the network topology. RLNC makes the application of network coding possible in ad hoc networks and networks with a high churn rate, with no central authority or strong distributed computing required to design and maintain network coding schemes. Benefits of applying RLNC in P2P applications were studied in [5] for file transfer applications. An implementation of a P2P media system using RLNC was proposed in [6].

Chiu et al. [4] showed that applying network coding alone at peers in an overlay network will not result in improved throughput, which questions the effectiveness of network coding in a file-sharing scenario. However, our focus is on peer-to-peer gaming, where reducing network latency is a first order concern, while bandwidth consumption is not critical.

Networked games. Games can be seen as discrete interactive simulation: the game model is evaluated and changed periodically in a game loop. Users interact with the game model through avatars. The avatar control mechanism is not event based, user input is scanned at a specific point of the ever-running game loop and processed based on the state of the avatar. In present networked multiplayer games each player maintains a local copy of the game model. Input generated by all players has to be available to each player in order to keep these copies synchronized. This is achieved through periodical player state update messages.

The client-server architecture has the benefit of a central authority that maintains the game model and broadcasts both player state and game object state updates. Servers can implement security features, e.g., filtering out malicious players based on their in-game or network activity. A central server can also optimize bandwidth consumption and eliminate some cheats by sending player state updates only from those players and objects that may be visible to a certain player. Servers can be operated by the game's publisher or a company dedicated to host online games, but this is not always necessary: in a number of games any player can act as a server. The disadvantages of both approaches are obvious: dedicating resources to host multiplayer games is expensive, as servers must be localized to maintain low response times, and game load is concentrated to the evening hours [9]. On the other hand, making one of the players take up the server's role will consume the player's local resources and reduce her and (due to limited bandwidth) other players' gaming experience. Moreover, this way a single point of failure exists, hence a connectivity problem at the server can prevent all the participants from playing.

As a response to the above-defined problems, some massively multiplayer online games (MMOG, mostly role-playing and adventure) are migrating to P2P networks. MMOGs are played by thousands of players contributing to the same huge and detailed virtual environment. This virtual world is permanent, with players joining and leaving. In these games player state updates can be less frequent and the emphasis is on dispatching game object state updates to all players. To make such games easier to scale, P2P overlay based games were proposed in [10], focusing on the partitioning of the game-space to optimize communication by separating users into groups. The proposed techniques are not fast enough for first person view games such as action games, shooters (FPS) and simulators. Authors of [11] proposed Colyseus, a framework which utilizes efficient object location and speculative prefetching besides game-state partitioning to deal with latency requirements. Another relevant piece of work is the Donnybrook system [12]: it uses a sophisticated method to estimate which objects and other players are important to a given player, thereby reducing the frequency of state updates. Additionally, updates are disseminated via overlay multicast. While the achievements of these works are valuable, real-world games requiring high responsiveness are still server-based.

In this paper we concentrate on the efficient communication of frequent player state updates (synchronization). We believe that reducing network traffic and latency through limiting recipient lists, clustering based on game state and speculative prefetching are important ingredients of a peer-to-peer gaming framework, however, an efficient dissemination scheme for frequent in-game updates is essential for a practical system. The nature of such kind of traffic is multicast, therefore we expect that overlay-based network coding could significantly help in such a scenario [13].

3   Overlay Network Coding for P2P Gaming

In a multiplayer gaming scenario, latency of player state updates comes from two factors, the one-way trip time of the packet and available bandwidth. Both are dependent of factors like home networking equipment, cross-traffic generated by the end user,

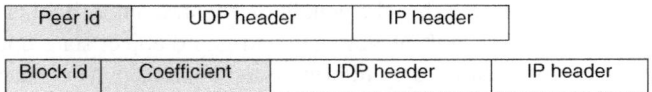

Fig. 2. Packet headers: unicast vs. RLNC

overall network load, geographical distance and traffic shaping equipment used by network providers.

If a server-based solution is deployed, each node uploads only a single packet to the server, and downloads the packets of every other peer from the server. Bandwidth limitations are usually present only at the server, especially when the server connects to the Internet through an asymmetric home access link such as DSL or cable. In a server-based scenario, latency due to trip time depends only on Internet factors such as BGP policies, load balancing mechanisms, congestion and link availability.

Peer-to-peer gaming architectures use an overlay network to disseminate player state updates among peers. There are two basic characteristics of the overlay which determine overall latency: its topology and the forwarding mechanism used. From the topology standpoint, overlays can be fully or partially connected. We restrict our investigations to fully connected overlays for the sake of simplicity.

In a full mesh overlay the simplest packet forwarding strategy is unicast, when each peer sends its update packet directly to every other peer. We propose an alternative forwarding mechanism, network coding. The main difference brought by network coding is the possibility of combining available packets before forwarding them. While optimal bandwidth consumption and computational overhead can be achieved with fixed linear coding schemes, this approach is not suitable for network gaming over an unreliable medium. With a fixed coding scheme packets are expected to be present at given nodes at a given time. In peer-to-peer gaming packet generation is indeed synchronized, but packet transport times vary in a wide range. Hence buffers have to be introduced to ensure that all packets required for decoding are present as the coding scheme requires it. This is not desirable when the goal is to reduce overall latency. Random linear network coding can work with asynchronous packet flows, as the coding scheme doesn't specify which packets have to be combined, packets contributing to the outgoing packet are chosen from the available packets randomly. The price of this freedom is a larger packet header, containing the coefficients of the original packets that are combined to get the current one (see Fig. 2). Moreover, decoding is more computationally intensive, as a new linear equation system has to be solved for every block to be decoded.

The basics of random network coding as implemented in our framework are shown in Fig. 3. Equations and variables correspond to a single game step, hence no indexes will be introduced for the time. At each game step, an update message is generated by each peer. Each message corresponding to a single network coding packet. Assume that a virtual array A is composed of these n messages. This array is split into k blocks of rows, denoted by A_i. Having smaller number of rows in a block reduces the computation needs of the decoding process. Packets received from neighbors contain the block number i of the message it was generated from, the coefficient c used and the encoded message b. Received messages for each block are stored at peer j along with the local

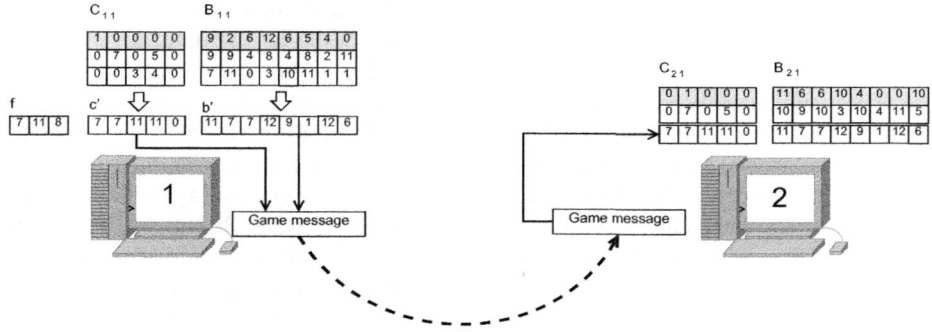

Fig. 3. Random network coding in action (modulo class: 13)

message in the working arrays B_{ji} (local message is stored in B_{ji} if message of peer j is part of block i). Received c coefficients for these messages are stored in the arrays C_{ji} with the local message having a coefficient of 1. At the beginning of the game step, arrays B_{ji} and C_{ji} are empty except the local message of each peer stored in one of the B_{ji} arrays and the associated coefficient in C_{ji}. Random linear network coding guaranties that the equation

$$B_{ji} = C_{ji} \times A_i \qquad (1)$$

(obviously true at the beginning) will always hold. This is ensured through the coding mechanism.

A given outgoing message of peer j is created from a chosen block i. The message to be sent is generated with the help of a random vector f of the length of the number of rows in B_{ji}. The message consisting of c' and b' is constructed as follows:

$$c' = f \times C_{ji} \quad \text{and} \quad b' = f \times B_{ji}. \qquad (2)$$

Note, that

$$b' = f \times B_{ji} = f \times C_{ji} \times A_i = c' \times A_i. \qquad (3)$$

When peer l receives the message, c' and b' will be appended to the arrays B_{l_i} and C_{l_i}:

$$(B_{l_i})' = B_{l_i}|b' \quad \text{and} \quad (C_{l_i})' = C_{l_i}|c'. \qquad (4)$$

Hence if

$$B_{l_i} = C_{l_i} \times A_i \qquad (5)$$

then

$$(B_{l_i})' = (C_{l_i})' \times A_i. \qquad (6)$$

In order to retrieve the original messages (rows of A_i), the above equation has to be solved for all k blocks at each peer j. Inverting C_{ji} requires at least $\lceil \frac{n}{k} \rceil$ linearly independent messages received from each block. It has been shown in [14] that the probability of selecting linearly dependent combinations becomes negligible even for a small code field size. This result was achieved in a streaming application where network coding was applied at the source where an entire block of data was available.

In peer-to-peer gaming, data are available distributed between all peers, and network coding is only applied during message forwarding. In the beginning of a given game round, each peer has access only to a limited number of messages and peers in small proximity will usually have knowledge of the same subset of messages. This in turn may lead to sending messages that are not independent from those the destination peer has already received. These irrelevant messages may constitute a considerable traffic load, but may not affect overall delivery latency. Our first network coding method, *RLNC*, is the basic implementation of a linear network coding system. RLNC operates on the full mesh, every peer sends a message periodically to each neighbor. The packet is a random linear combination of already received messages. The second method, *RLNC+*, implements an additional mechanism for reducing unnecessary transmissions: each peer maintains a message array which contains messages sent to and received from its neighbors. Subsequent outgoing messages are sent only to neighbors that could be interested in the content of the packet, this way, unnecessary packet sending could be spared.

4   Performance Evaluation

To provide an initial performance assessment we developed a JAVA-based simulation tool. The environment consists of the configuration generator, responsible for creating, saving and loading network topology, the simulator and the log processing pipeline. Our network core simulator provides delayed packet delivery on point to point links. For the proof-of-concept evaluation, peers were connected directly to this core, and no computation overhead (decoding) was taken into account.

Network model. Our simulator was created to handle network traffic generated by player state updates, and didn't considered game object changes. We also presumed that addresses of the players are available, and players do not leave or join during a match as this is the case in most match-based multiplayer games. We placed our players in the Internet, as in LAN games connections are more reliable, bandwidth is almost always sufficient and the impact of network latency on gameplay can be neglected.

Simulations were run over networks consisting of 10, 20, 30, 40 and 50 peers. The traffic generated by the user interactions was uniform in time as player updates are not event but state based. We assumed UDP as the transport protocol of choice, since real-time traffic does not tolerate the added latency of the feedback loop in TCP. A typical player state update messages size ranges from 10 to 100 bytes, and is sent to each other player 10-60 times per second [15]. In our experiments we used a 10 Hz update frequency and player state update message of 50 bytes, totaling 78 bytes with IP and UDP headers. When using random linear network coding, the message size increases with the coefficient. This depends on the block size, and ranges from 5 bytes for 10 peers to 25 bytes for 50 peers, as we used a single block for all peers. For coding we used a fixed field size of 13 (modulo class).

We measured a round-trip time of 15-87 ms towards different European servers [16]. Based on these results we used one-way peer-to-peer latency values of 5 to 40 ms uniformly distributed with a jitter of 5 ms. Since typical home broadband connections are asymmetric, we assumed that upload bandwidth would determine the effectiveness of different forwarding mechanisms (downlink is assumed to be unlimited). Based on

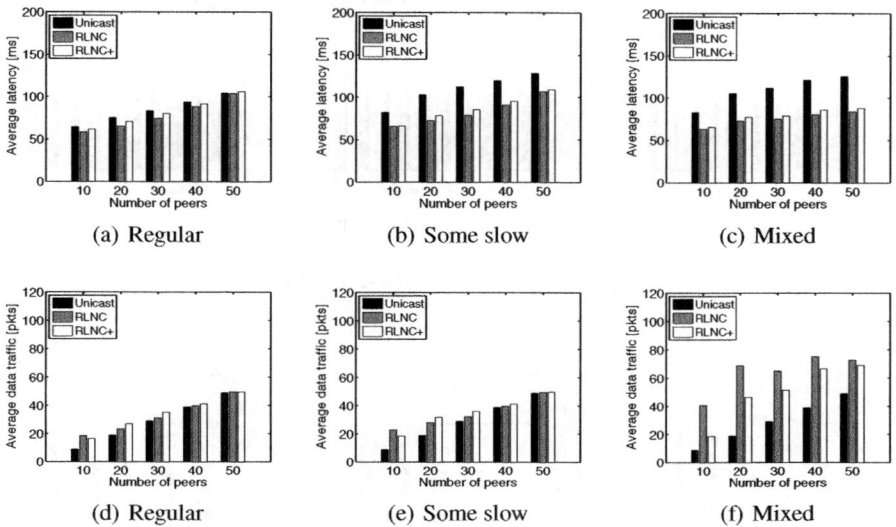

Fig. 4. Average latency (upper) and data traffic generated (lower) for different peer distributions

bandwidth characteristics three peer classes were used: slow peers had an upload band-width just enough to send out a single packet to each other peer (62.4 kbit/s to 312 kbit/s depending on the number of participating peers), the regular peer had 624 kbit/s and fast peers were granted a 8 Mbit/s uplink. We used four different scenarios in terms of participating peer distribution: regular (regular peers only), some slow (90%regular and 10% slow), some fast (90% regular and 10% fast), and mixed (80% regular, 10% slow, 10%fast).

Experimental results. When processing the results, our main concern was the aver-age latency. We defined latency as the time elapsed from the beginning of the game round until the respective peer becomes aware of all other players' states. In the unicast scenario this happens when a packet from every peer is received. In the network cod-ing scenario received packets were scanned for a solution to the linear equation created from the received coefficients after the reception of each packet. If the equation becomes fully determined all player update packets can be decoded. Other key performance indi-cators were the maximum latency in a single round and the number of packets received before decoding could be achieved. Average and maximum latency measures quality of experience, while the generated data traffic measures network load. Note, that all latency, maximum latency and traffic figures are averaged over 100 game rounds.

Average latency values for different participating peer populations can be seen in Fig. 4(a)-4(c). In a network with regular peers, network coding (RLNC) performs slightly better then traditional unicast mechanism, even with its larger packet size which reduces the packet sending rate. This is due to it's flooding-like behavior: packets originating from one source will find the shortest paths, while in the unicast solution the direct link between two peers may have a larger delay. When some slower peers are also present in

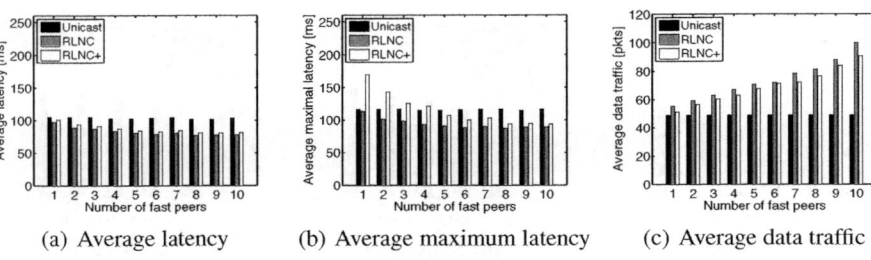

(a) Average latency   (b) Average maximum latency   (c) Average data traffic

Fig. 5. Impact of fast peers (50 peers, some fast)

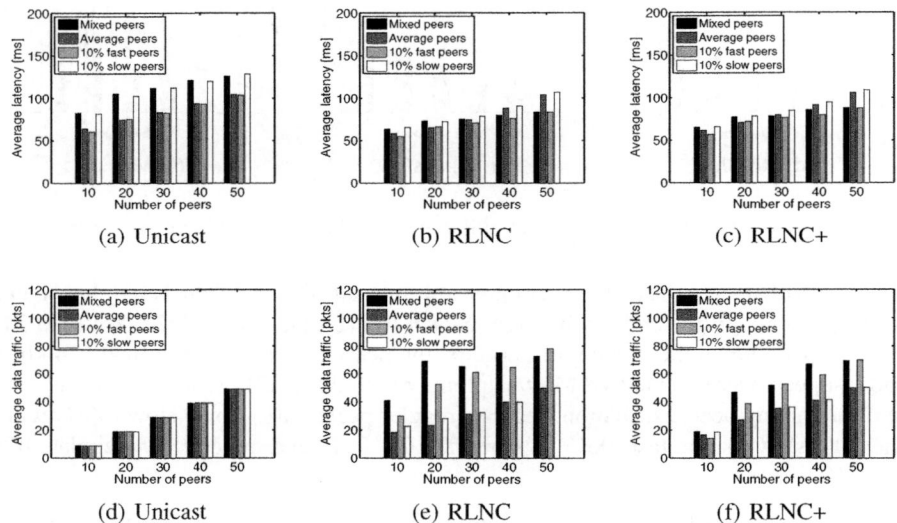

(a) Unicast   (b) RLNC   (c) RLNC+

(d) Unicast   (e) RLNC   (f) RLNC+

Fig. 6. Average latency (upper) and data traffic generated (lower) for different forwarding mechanisms

the overlay, network coding shows greater improvement over the traditional forwarding method. On the top of that, the heterogeneity of the mixed scenario (slow, regular and fast peers are also present in the system) induces the highest gain for RLNC among all scenarios. Improvement in this setting can be more than 30%.

Note, that network coding with redundancy protection (RLNC+) performs comparably to RLNC in terms of average latency. However, its added benefit becomes visible in Fig. 4(e)-4(f), where generated data traffic is shown. In the first two scenarios traffic generated by network coding mechanisms is only slightly more than that of unicast. On the other hand, RLNC produces much more traffic in the mixed peer setting, with RLNC+ reducing this overhead significantly.

The next series of experiments studied the impact of fast peers on the system. We used a network of 50 peers, with 1 to 10 fast peers beside the regular ones. Fig. 5 shows how the latency benefit of network coding increases with the ratio of fast peers. The amount of generated data traffic gives an explanation for this behavior: the more peers with high bandwidth connections are present, the higher redundancy can be achieved in the system. This redundancy provides improved latency figures. Note, that RLNC+ produces considerably high maximum latency values, when only a few fast peers are present. This shows that restricting the scope of recipients is more beneficial when peer heterogeneity is higher.

Latency and data traffic results grouped by forwarding mechanisms are shown in Fig. 6. It can be observed that network coding improves the average latency in every scenario. It is important to emphasize that 150 ms is barely tolerable [17], while 100 ms is acceptable for online shooter and action games; our results suggest that this limit can be met even in case of a large number of participants. Moreover, the overhead of network coding compared to unicast is almost non-existent if only regular or some slow peers are playing; in more heterogeneous scenarios, there is some reasonable overhead (with RLNC+ generating slightly less traffic). Note, that this overhead might be recognizable percentage-wise, but the absolute volume of extra data traffic remains very low (20 extra packets per peer in a 50 peer scenario).

5   Conclusion and Future Work

The recently proposed technique of network coding has been shown to boost network capacity compared to the traditional store-and-forward mechanism in a variety of scenarios. Here, we have presented a very different use-case for the same method: reducing network latency. Our promising results in this area shows a clear need for further understanding of the possible benefits and side-effects of network coding.

In this work, we have introduced a practical framework for peer-to-peer networked gaming based on random linear network coding. We have shown that our proposed method outperforms traditional unicast in terms of average network latency in a wide range of scenarios. Furthermore, the absolute traffic overhead has been shown to be low for all settings analyzed.

Our initial performance evaluation indicates that the proposed method is worthy of further research. One important area is the specific codes used. Also, more simulations are needed to investigate the impact of more complex topologies, packet loss and computational overhead. Moreover, a prototype implementation and testbed measurements with a real game are essential to fully understand the behavior of our system.

References

1. Jones, D.C.: The monster that is – World of Warcraft,
   http://www.bme.eu.com/news/world-of-warcraft (accessed April 15, 2010)
2. Ahlswede, R., Cai, N., Li, S.-Y.R., Yeung, R.W.: Network Information Flow. IEEE Transactions on Information Theory (2000)
3. Katti, S., Rahul, H., Hu, W., Katabi, D., Medard, M., Crowcroft, J.: XORs in the air: practical wireless network coding. In: Proc. of ACM SIGCOMM 2006 (2006)
4. Chiu, D.M., Yeung, R.W., Huang, J., Fant, B.: Can Network Coding Help in P2P Networks? In: Proc. of WiOpt 2006 (2006)

5. Wu, Y., Hut, Y.C., Li, J., Chou, P.A.: The Delay Region for P2P File Transfer. In: Proc. of ISIT 2009 (2009)
6. Wu, C., Li, B., Li, Z.: Dynamic Bandwidth Auctions in Multioverlay P2P Streaming with Network Coding. IEEE Transactions on Parallel and Distributed Systems (2009)
7. Li, S.-Y.R., Yeung, R.W., Cai, N.: Linear Network Coding. IEEE Transactions on Information Theory (2003)
8. Ho, T., Medard, M., Koetter, R., Karger, D.R., Effros, M., Shi, J., Leong, B.: Random Linear Network Coding Approach to Multicast. IEEE Transactions on Information Theory (2006)
9. Online Gamers Research, http://www.video-games-survey.com/online_gamers.htm (accessed April 15, 2010)
10. Knutsson, B., Lu, H., Xu, W., Hopkins, B.: Peer-to-Peer Support for Massively Multiplayer Games. In: Proc. of INFOCOM 2004 (2004)
11. Bharambe, A., Pang, J., Seshan, S.: Colyseus: A Distributed Architecture for Online Multiplayer Games. In: Proc. of NSDI 2006 (2006)
12. Bharambe, A., Douceur, J.R., Lorch, J.R., Moscibroda, T., Pang, J., Seshan, S., Zhuang, X.: Donnybrook: Enabling large-scale, high-speed, peer-to-peer games. In: Proc. of ACM SIGCOMM 2008 (2008)
13. Zhu, Y., Li, B., Guo, J.: Multicast with Network Coding in Application-Layer Overlay Networks. In: IEEE JSAC (2004)
14. Wu, Y., Chou, P.A., Jain, K.: A Comparison of Network Coding and Tree Packing. In: Proc. of IEEE ISIT (2004)
15. Feng, W., Chang, F., Feng, W., Walpole, J.: Provisioning Online Games: A Traffic Analysis of a Busy Counter-Strike Server. In: Proc. of IMW 2002 (2002)
16. Speedtest, http://www.speedtest.net (accessed April 15, 2010)
17. Armitage, G.: An experimental estimation of latency sensitivity in multiplayer Quake 3. In: Proc. of ICON 2003 (2003)

A Virtual File System Interface
for Computational Grids

Abdulrahman Azab and Hein Meling

Dept. of Electrical Engineering and Computer Science
University of Stavanger, N-4036 Stavanger, Norway

Abstract. Developing applications for distributed computation is becoming increasingly popular with the advent of grid computing. However,
developing applications for the various grid middleware environments require attaining intimate knowledge of specific development approaches,
languages and frameworks. This makes it challenging for scientists and
domain specialists to take advantage of grid frameworks. In this paper, we
propose a different approach for scientists to gain programmatic access
to the grid of their choice. The principle idea is to provide an abstraction layer by means of a *virtual file system* through which the grid can
be accessed using well-known and standardized system level operations
available from virtually all programming languages and operating systems. By abstracting away low-level grid details, domain scientists can
more easily gain access to high-performance computing resources without learning the specifics of the grid middleware being used. We have
implemented such a virtual file system on HIMAN, a peer-to-peer grid
middleware platform. Our initial experimental evaluation shows that the
virtual file system only cause a negligible overhead during task execution.

1 Introduction

Grid computing is an appealing concept to support the execution of computationally intensive tasks; grid computing generally refers to the coordination of
the collective processing power of scattered computing nodes to accommodate
such large computations. Peer-to-peer grid computing extends this idea, enabling
all participants to provide and/or consume computational power in a decentralized fashion. Although appealing at first sight, grid computing frameworks are
complex pieces of software and thus typically come with some obstacles limiting
adoption to expert programmers capable of and willing to learn to program using the grid middleware application programming interfaces (APIs). Therefore,
a major challenge faced by developers of grid computing frameworks is to expand the base of grid application developers to include non-expert programmers,
e.g. domain specialists using their own domain-specific tools. One approach is to
devise transparency mechanisms to hide the complexities of the grid computing
system. In this paper, we propose the design of a programming interface for a
peer-to-peer grid middleware through a *virtual file system* (VFS) [5, 11, 22]. The
rationale behind this idea is that the file system interface is well-known, and is

F.A. Aagesen and S.J. Knapskog (Eds.): EUNICE 2010, LNCS 6164, pp. 87–96, 2010.

accessible from just about any domain specific tool or even simple file-based commands. The approach also lends itself to devising a file system organization exposing varying degrees of complexity to its users. For example, complex management operations can be performed using one part of the file system hierarchy, whereas domain specialists can submit tasks through another part of the file system limiting the exposure to a most essential part of the system. The main difference between this approach and the existing approches is that it uses the most familier interface between the user and the OS, i.e. FS, which is a data management interface to provide interaction with a computation management environment, i.e. the Grid.

We have implemented such a VFS layer above a middleware core based on our HIMAN platform [1, 3]. The VFS layer sits between the end user/programmer and the client program enabling them to assign a task to be executed on the grid simply by calling the write() function of the virtual file system on a specific virtual path. Similarly, when task execution has completed, the user can easily obtain the results by invoking the read() function on another virtual path. To evaluate the overhead imposed by the VFS layer, we compared a simple parallel matrix multiplication task running with and without the VFS layer. The results confirm our intuition that the VFS layer only adds a negligible computational overhead.

The rest of the paper is organized as follows: Section 2 gives an overview of the related efforts. Section 3 presents a general overview of grid middleware architectures. Section 4 describes the new added virtual file system layer and how it interacts with the other components in the middleware. Section 5 presents and discusses the results obtained from the performed experiments. Section 6 presents conclusions and future work plans.

2   Related Work

The concept of virtual file systems was first introduced in the Plan 9 operating systems [22] to support a common access style to all types of IO devices, be it disks or network interfaces. This idea has since been adopted and included in the Linux kernel [11], and support is also available on many other platforms, including Mac OS X [18] and Windows [5, 8]. There are a wide range of applications of such a virtual file system interface, including sshfs, procfs, and YouTubeFS.

In the context of grid computing, providing access to a (distributed) file system architecture is both useful and commonplace. Transmitting the data necessary for computation is made easier with a distributed file system, and a lot of research effort has gone into data management for the grid, e.g. BAD-FS [4], and finding efficient protocols for data transfer, e.g. GridFTP [14]. BAD-FS [4] is different from many other distributed file systems in that it gives explicit control to the application for certain policies concerning caching, consistency and replication. A scheduler is provided with BAD-FS that takes these policies into account for different workloads. The Globus XIO [2] provides a file system-based API for heterogeneous data transfer protocols aimed for the Grid. This allows applications to take advantage of newer (potentially) more efficient data transfer protocols in the future, assuming they too provide the XIO APIs. These file systems tackle the problem of interacting with the data management component and file transfer protocols of a data grid environment. However, none of them try to solve the generic interaction with the computational grid middleware for task submission, results collection in a suitable manner.

Some work has been carried out in the trend of using VFS to provide transparent interaction with remote target systems in a distributed environment. UEM [27] is a unified execution model for cloud and cluster computing which provides VFS interface for the execution, monitoring, and control of remote processes. This is carried out by presenting each process in the system as a virtual directory. Within each process' directory, virtual files are serving as interfaces for information retrieval and control of the remote process. The structure of nodes within the distributed system is displayed as a hierarchical directory structure in away that each node directory includes a number of sub-directories one for each remotely called process within the node. The drawback of this approach is that the user has to specify at which node should the process be allocated. This can be simple in case of simple distributed structure of the system. But as the complexity of the system structure increases, the complexity of the VFS structure will also increase, which will in turn reduce the transparency of the interface. Parrot [25] is an interposition agent [15] which provides seamless integration between standard Unix applications and remote storage systems. This integration makes a remote storage system appear as a file system to a Unix application. One drawback is that the user has to specify the location of the remote machine. Besides, all commands has to be included in a Parrot command, which decreases the provided transparency. Many other systems [10, 17, 19, 20, 24] also provide an intermediate layer between local user interfaces and remote applications for providing transparent access to the target applications. All of these systems require that the user has to know how to deal with the front end application which is system specific.

Our proposed VFS layer provides an interface which is well known to all computer users and locates itself as the front end application. The user does not have to provide any information about the remote machine or to be aware of the structure of the distributed system. It also hides the complexity of the process management, i.e. scheduling and fault-tolerance, from the user. To our knowledge, most of the existing computational grid middleware environments [6, 7, 9, 13, 16, 17, 21, 23] require a special language level API for interaction with the scheduler and to collect the results.

3   HIMAN Overview

The VFS layer has been implemented on HIMAN [3], a pure peer-to-peer grid middleware that support serial and parallel computational tasks. HIMAN has three main components: (i) a *worker* component responsible for task execution, (ii) a *client* component responsible for task submission and execution monitoring, and (iii) a *broker* responsible for the task allocation. All nodes in the grid have all three components. The client and broker of a submitting node is then responsible for task submission, allocation, and execution monitoring in a decentralized scheduling fasion. Any node can submit tasks, and can also serve as executor for other tasks at the same time. Since HIMAN enables the execution of both serial and parallel tasks [1], the current GUI interface requires the user to specify many configuration parameters such as: locations of the code and data files, number of subtasks, execution environment, and scheduling policy. This can be too complex for the non-expert users.

4   Virtualizing Grid Access

Existing grid computing platforms [6, 12, 13, 26] require that the user configure a wide range of parameters, have sufficient background in distributed computing, and proficient knowledge of the grid middleware APIs to be able to execute tasks on the grid. This might be appropriate for expert users, but for regular users, a simple and familiar interface is desirable. By far the most ubiquitous programming interface available on computer systems is the file system interface. Thus, our proposed VFS layer placed above our HIMAN grid computing middleware exposes such an interface to the application programmer, making it easy to interact with the middleware using a well-known interface. Moreover, it enables users to submit tasks using the `write()` function, and monitor and collect results from the computation using the `read()` function of the VFS layer.

4.1   The VFS Layer

Our VFS layer is implemented as an additional interface layer above the client using the Callback file system library [5]. Callback is similar to the FUSE library available on Linux and Mac OS X [11, 18], and enable building virtual file systems in user space. The Callback library contains a user mode API to communicate with user applications, and a kernel mode file system driver for communicating with the Windows kernel. The role of the Callback library in our VFS layer is to enable users to provide input files and collect result files using simple file system commands. In order to make the proposed VFS applicable and easy to implement in other grid systems which are based on different platforms, the communication between the VFS layer and the client is performed by means of simple UDP text-based commands. In order to implement VFS on another grid middleware, a small UDP communication module has to be added to the client.

The installation of the VFS layer on a grid client machine is very simple. The user simply runs an MSI package, which installs the VFS layer as a Windows service. The service is configured to startup automatically when the user logs on, or manually by executing a `net start` command. Once the service is started, the VFS drive will be mounted. The VFS layer will communicate with the grid client upon the execution of specific file system commands. The virtual drive will be unmounted automatically when the user logs off, or it can be unmounted manually by executing a `net stop` command.

The VFS layer is composed of two main modules: Task Submission Module (TSM) and Result Collection Module (RCM).

Task Submission Module. TSM contains two procedures. The first is the virtual volume procedure, and will be called when the Callback file system is mounted, and creates a virtual volume visible in Windows Explorer. This new volume will be used by the user to provide the input files. The second is the task submission procedure and is responsible for forwarding the input files to the client in order to start the task submission process. Task submission requests are handled in the `CbFsFsWrite()` function of the Callback user mode library, so that it is triggered only when a `write()` command for a specific file is executed on the virtual drive. The default case is when the program file is copied to the virtual drive. This can be changed by the user to determine which `write()` command will trigger the TSM. The task submission procedure depicted in Fig. 1 involves the following steps:

1. The user executes a `write()` command, through the file system interface (e.g. Windows Explorer), to write the input files into the virtual volume.
2. The file system interface forwards the command to the Windows kernel.
3. The kernel forwards the command to the Callback file system driver.
4. The file system driver responds by calling the `CbFsFsWrite()` function in the Callback user mode library.
5. This function invokes the task submission procedure in the TSM.
6. The task submission procedure responds by presenting the input files in the file list of the virtual volume, so that they can be accessed with the file system interface, and forwarding the input files to the client.
7. The client submits the task to the grid.
8. During the execution, the user can monitor the execution process (i.e. the progress) by executing a `read()` command on a specific text file on the virtual drive.

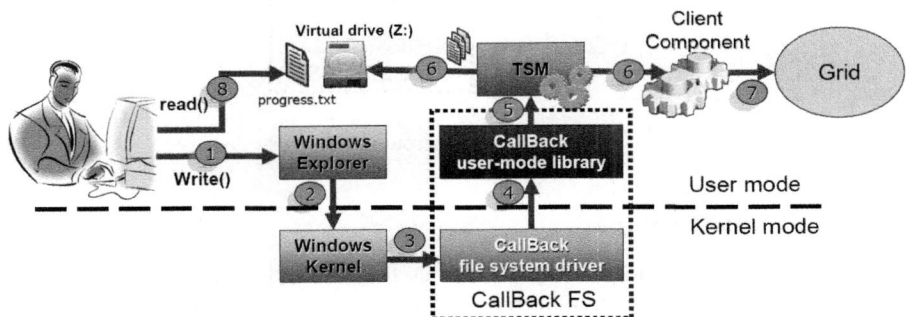

Fig. 1. Task submission procedure

Result Collection Module. The objectives of the result collection module are twofold: (i) provide a file system interface through which the user can collect the result files from task execution, and (ii) signal the completion of task execution to the user. The result collection procedure depicted in Fig. 2 involves the following steps:

1. When task execution is completed, the results are sent to the client process on the submitting node [3].
2. The client sends the collected results to the RCM. The result files are then stored in a temporary physical directory.
3. The RCM responds by creating a new virtual directory (e.g. `Results`) in the virtual drive mapping the temporary physical directory. Creation of the `Results` directory indicates to the user that task execution has complete.
4. The user can collect the result files by executing a `move` command on the files in the `Results` directory to move the result files to a physical path.
5. Windows Explorer forwards the command to the kernel.
6. The kernel forwards the command to the Callback file system driver.
7. The file system driver responds by calling the `CbFsRenameOrMoveEvent()` event procedure in the Callback user mode library.
8. The `CbFsRenameOrMoveEvent()` procedure responds by moving the result files to the given physical path, and removing them from the file list of the `Results` directory of the virtual drive.

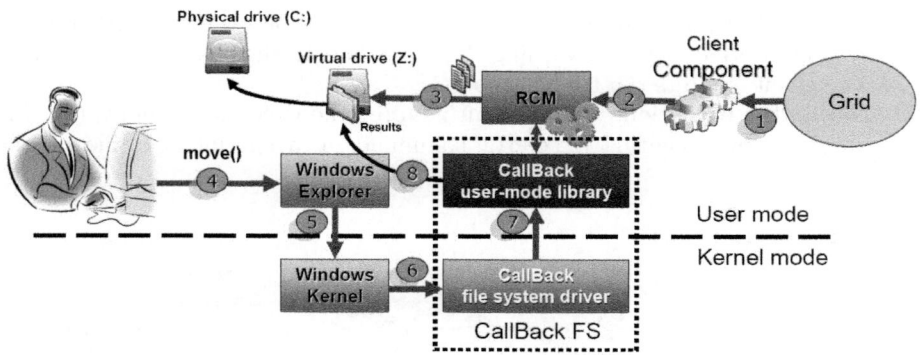

Fig. 2. Result collection procedure

4.2   A Simple Example

In this section we provide a simple example describing how to use the VFS layer for executing a parallel task of matrix multiplication. To accomplish this, the user must provide as input: an executable code file (e.g. `code.dll`), and the necessary data input files (e.g. `input?.data`). The number of input files correspond to the number of parallel subtasks that will be generated by the HIMAN middleware during execution of the task.

For illustration, the DOS command prompt is used (interactively) below to issue file system commands to interact with the grid middleware. The virtual drive is mounted on drive Z:, and the input files are stored in C:\Input\. The following steps describe the whole execution procedure, including results collection.

1. The user writes the data files to the virtual drive by typing:

   copy C:\Input*.data Z:

2. The user writes the code file to the virtual drive by typing:

   copy C:\Input\code.dll Z:

   As explained in Section 4.1, this command will trigger the TSM to invoke the client to begin task execution, and to a create virtual file progress.txt on drive Z: for monitoring progress.

3. During the execution, the user can monitor the execution progress for the subtasks by typing:

   type progress.txt

   ```
   Subtask 1. Worker:193.227.50.201 Progress: 10%
   Subtask 2. Worker:74.225.70.20   Progress: 15%
   Subtask 3. Worker:87.27.40.100   Progress: 20%
   Subtask 4. Worker:80.50.96.119   Progress: 6%
   Subtask 5. Worker:211.54.88.200  Progress: 12%
   ```

4. The user can repeat the above command to keep up with the execution progress. The user need not consider management issues such as scheduling, fault tolerance, and connectivity. These issues are seamlessly handled by the HIMAN grid middleware [3]. In case of a worker failure, the worker address will be changed for the associated subtask in the progress file.

5. When the execution of one or more subtasks is completed, this will be revealed in the progress.txt file as follows:

   ```
   Subtask 1. COMPLETED
   Subtask 2. Worker:74.225.70.20 Progress: 90%
   Subtask 3. Worker:87.27.40.100 Progress: 88%
   Subtask 4. COMPLETED
   Subtask 5. COMPLETED
   ```

6. Upon the completion of all subtasks, the Results virtual directory will appear, enabling the user to move all the virtual files to a physical directory as follows:

   move Z:\Results*.* C:\Results

5 Initial Performance Evaluation

In order to demonstrate the applicability of our VFS layer, we have performed a simple experimental evaluation to reveal the overhead caused by the VFS layer compared to interacting directly with the HIMAN middleware (through a GUI interface).

Two experiments were performed using a classic parallel matrix multiplication task for multiplying two square matrices of size: a) 1500×1500 and, b) 2100×2100. In both cases, different number of workers (i.e. parallel subtasks) varying from one to six were used. The implemented benchmark is to calculate the task submission time (TST) for both cases: direct submission (using the GUI), and submission throgh the VFS interface. In case of direct submission, TST interval starts when the user submits the task after setting all required configuration options. In case of VFS, the TST starts from the first step in the TSM, Section 4.1. TST ends when all subtasks have been transfered to workers. off course in case of direct submission, the user will take some time setting the configuration options, but as long as this time depends on the user it is not included in the TST. The Result collection phase is not considered since in both cases the result files are copied to a physical path and the user has to collect them. The results are shown in Fig. 3.

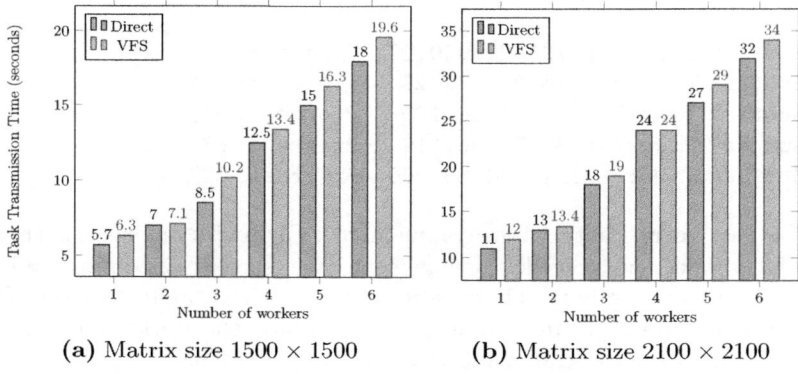

(a) Matrix size 1500×1500 (b) Matrix size 2100×2100

Fig. 3. Task transmission time against the number of workers using Direct and VFS interface submissions

In Fig. 3, it is clear that the computational overhead which is represented by the increase in the task transmission time in case of using the VFS interface instead of the built in client GUI, is nearly negligible. Since the transmission of the input files to workers is done in a serial fashion, the overhead is due to the time taken by the VFS to build the virtual files in memory and to communicate with the client for transmitting each of input files. The reason behind implementing Serial transmission in HIMAN is to avoid the large traffic in the submitting node which will be resulted from sending many data files at the same time, especially in case or large files.

The two approaches, direct and VFS, have identical CPU overhead simply because in both cases all pre-processing procedures are carried out by the grid client. The memory overhead is also identical since in both cases, the task input files are loaded into memory in order to be processed by the grid client upon the execution.

6 Conclusions and Future Work

Given the complexity of grid computing middleware, providing an easy to use interface for task submission is a significant challenge. In this paper, we have proposed a new technique for accessing computational grid middleware through a virtual file system interface. The proposed technique has been implemented on the HIMAN middleware, enabling non-expert users to submit compute tasks to the grid. Our initial evaluation indicate that the overhead imposed by the VFS interface over the direct interaction approach is negligible.

Our current implementation support serial and simple parallel task execution; in future work we will extend the VFS layer with support for complex parallel task execution. Moreover, we also plan to design a VFS-based grid portal that supports multiple grid computing frameworks, e.g. Globus and Condor. To accomplish this we will leverage the FUSE [11] framework on Linux to provide cross platform support for programming languages and runtime environments that can run on multiple operating systems. We will also add support for multi-user scenarios taking security and scalability issues into account.

References

1. El-Desoky, A.E., Ali, H.A., Azab, A.A.: A pure peer-to-peer desktop grid framework with efficient fault tolerance. In: ICCES 2007, Cairo, Egypt, pp. 346–352 (2007)
2. Allcock, W., Bresnahan, J., Kettimuthu, R., Link, J.: The globus extensible input/output system (xio): A protocol independent io system for the grid. In: IPDPS 2005: Proceedings of the 19th IEEE International Parallel and Distributed Processing Symposium (IPDPS 2005) - Workshop 4, Washington, DC, USA, pp. 179.1. IEEE Computer Society Press, Los Alamitos (2005)
3. Azab, A.: HIMAN: A Pure Peer-to-Peer Computational Grid Framework. Lambert Academic Publishing (2010)
4. Bent, J., Thain, D., Arpaci-Dusseau, A.C., Arpaci-Dusseau, R.H., Livny, M.: Explicit control in a batch-aware distributed file system (March 2004)
5. Callback File System, http://www.eldos.com/cbfs/
6. Condor project, http://www.cs.wisc.edu/condor/
7. D4Science: Distributed collaboratories Infrastructure on Grid ENabled Technology 4 Science, http://www.d4science.eu/
8. Dokan file system, http://dokan-dev.net/en/
9. EGEE: Enabling Grids for E-Science in Europe, http://public.eu-egee.org/
10. Ernst, M., Fuhrmann, P., Gasthuber, M., Mkrtchyan, T., dCache, C.W.: A distributed storage data caching system. In: Computing in High Energy Physics, Beijing, China (2001)
11. Filesystem in Userspace, http://fuse.sourceforge.net/

12. GLite: Light weight middleware for grid computing,
    http://glite.web.cern.ch/glite/
13. The Globus toolkit, http://www.globus.org/toolkit/
14. GridFTP, http://globus.org/toolkit/data/gridftp/
15. Jones, M.B.: Interposition agents: Transparently interposing user code at the system interface. In: fourteenth ACM symposium on Operating systems principles, Asheville, North Carolina, United States, pp. 80–93 (1994)
16. Lederer, H., Pringle, G.J., Girou, D., Hermanns, M.-A., Erbacci, G.: Deisa: Extreme computing in an advanced supercomputing environment (2007)
17. Leech, M., Ganis, M., Lee, Y., Kuris, R., Koblas, D., Jones, L.: Socks protocol version 5 (1996)
18. MacFUSE, http://code.google.com/p/macfuse/
19. Miller, B., Callaghan, M., Cargille, J., Hollingsworth, J., Irvin, R.B., Karavanic, K., Kunchithapadam, K., Newhall, T.: The paradyn parallel performance measurement tools. IEEE Computer 28(11), 37–46 (1995)
20. Narasimhan, P., Moser, L.E., Melliar-Smith, P.M.: Exploiting the internet inter-orb protocol interface to provide corba with fault tolerance. In: Usenix Conference on Object-Oriented Technologies and Systems, Portland, Oregon (1997)
21. NorduGrid: Nordic Testbed for Wide Area Computing and Data Handling, http://www.nordugrid.org/
22. Pike, R., Presotto, D., Thompson, K., Trickey, H., Winterbottom, P.: The use of name spaces in Plan 9. SIGOPS Oper. Syst. Rev. 27(2), 72–76 (1993)
23. G.R.: Grid3: An application grid laboratory for science. In: Computing in High Energy Physics and Nuclear Physics, p. 18. Interlaken, Switzerland (2004)
24. Son, S., Livny, M.: Recovering internet symmetry in distributed computing. In: CC-Grid, Los Alamitos, CA, IEEE Computer Society, Los Alamitos (2003)
25. Thain, D., Livny, M.: Parrot: Transparent user-level middleware for data-intensive computing. Scalable Computing: Practice and Experience 6(3), 9–18 (2005)
26. UNICORE: Uniform Interface to Computing Resources, http://www.unicore.eu
27. van Hensbergen, E.V., Evans, N.P., Stanley-Marbell, P.: A unified execution model for cloud computing. In: Large Scale Distributed Systems and Middleware (LADIS 2009), October 2009, ACM, New York (2009)

Labeled VoIP Data-Set
for Intrusion Detection Evaluation

Mohamed Nassar, Radu State, and Olivier Festor

INRIA Research Center, Nancy - Grand Est
615, rue du jardin botanique, 54602
Villers-Lès-Nancy, France
FirstName.LastName@loria.fr

Abstract. VoIP has become a major application of multimedia communications over IP. Many initiatives around the world focus on the detection of attacks against VoIP services and infrastructures. Because of the lack of a common labeled data-set similarly to what is available in TCP/IP network-based intrusion detection, their results can not be compared. VoIP providers are not able to contribute their data because of user privacy agreements. In this paper, we propose a framework for customizing and generating VoIP traffic within controlled environments. We provide a labeled data-set generated in two types of SIP networks. Our data-set is composed of signaling and other protocol traces, call detail records and server logs. By this contribution we aim to enable the works on VoIP anomaly and intrusion detection to become comparable through its application to common datasets.

1 Introduction

Voice over IP (VoIP) has become a major paradigm for providing flexible telecommunication services while reducing operational and maintenance costs. The large-scale deployment of VoIP has been leveraged by the high-speed broadband access to the Internet and the standardization of dedicated protocols. The term is often extended to cover other IP multimedia communications in general and convergent networks. VoIP services are much more open if compared to PSTN networks. A typical VoIP service is composed by three main parts: the user premises, the VoIP infrastructure for signaling and media transfer, and a number of supporting services (e.g. DNS, TFTP). From a technical point of view, a SIP-based VoIP service is similar to an email service more than a conventional telecommunication service. Hence, VoIP suffers from the same threats of the TCP/IP networks and services. In fact, VoIP faces multiple security issues including vulnerabilities inherited from the IP layer and specific application threats. The attacks against VoIP can typically be classified into four main categories [1]: (a) service disruption and annoyance such as flooding and SPIT (Spam over Internet Telephony), (b) eavesdropping and traffic analysis, (c) masquerading and impersonation such as spoofing, (d) unauthorized access and fraud such as Vishing (Voice over IP

F.A. Aagesen and S.J. Knapskog (Eds.): EUNICE 2010, LNCS 6164, pp. 97–106, 2010.

phishing). These attacks can have significant consequences on the telephony service, such as the impossibility for a client of making an urgent phone call.

The research community started to investigate the best ways of protection, detection and response for the VoIP services [2]. Researchers argue that intrusion detection is necessary to struggle against VoIP fraudsters. The main difficulty remains to obtain real-world traces of attack incidents or even normal traffic. Labeled data-sets are necessary to evaluate the accuracy results of the proposed techniques which are often based on learning. It is also hard to obtain these data from VoIP providers which are constrained by user privacy agreements.

In this context, we propose a framework to annotate and customize VoIP traffic. The normal traffic is generated by profiled emulated users based on a social model. The attack traffic is based on currently available VoIP assessment tools. We deploy and configure two illustrative VoIP networks in a local test-bed and we generate a labeled data-set that can be used to compare different detection algorithms. The data-set is composed of signaling and other protocol traces comprising call detail records and server logs. Thus, correlation approaches can also be applied to the data.

The rest of the paper is composed as follows: In Section 2 we expose the related works on VoIP intrusion detection. In section 3 we describe the data types. Section 4 presents the traffic generation model and the generation of attacks. Section 5 describes the test-bed and the labeled data-set. Finally Section 6 concludes the paper and discusses the future work.

2   VoIP Intrusion Detection

An Intrusion Detection System (IDS) is a second line of defense behind protection systems like authentication, access control and data encryption. Intrusion detection refers to the process of monitoring computer networks and systems to track violations of the applied security policies. VoIP intrusion detection consists on automating the analysis of specific VoIP data sources. Many works have focused on the Session Initiation Protocol (SIP [3]) as the de-facto signaling protocol for the Internet and the Next Generation Networks (NGN).

Basically, SIP allows two communicating parties to set up, modify and terminate a call. Text-based with heritage from HTTP and SMTP, SIP is a request-response transaction-based protocol. A SIP Dialog is composed of one or more transactions. The SIP addressing scheme is based on URIs (Uniform Resource Identifier) in the form of sip:user@domain.

Niccolini et. al. [4] propose a network-based IDS to be deployed at the entry point of the VoIP network. This system is composed of two stages of detection: a rule-based stage and a behavioral-based stage. It implements a preprocessing logic of SIP into the SNORT IDS. SCIDIVE [5] is a knowledge-based IDS based on a rule matching engine. This system performs two kinds of detection: stateful and cross-protocol. The stateful detection focuses on multiple incoming packets belonging to the same protocol. The cross-protocol detection scheme spans packets from several protocols, e.g. a pattern in a SIP packet is followed by another pattern in a RTP packet. SCIDIVE has to be deployed at multiple points of the VoIP network (clients, servers, proxies) in order to correlate events at a global level. Sengar et. al. [6] propose specification-based detection on an extended finite state machine representing the interaction between the different protocols. By tracking deviations from the interacting protocol state machine, a series of SIP, RTP and cross-protocol attacks are detected. The VoIP Defender [7] is a highly-scalable architecture building a generic framework for supporting the detection algorithms.

To struggle against flooding, Chen [8] modifies the original state machines of SIP transactions and applies detection thresholds on the error statistics generated by those modified machines. Similarly in [9], the authors apply thresholds at three different scopes: transaction level, sender level and global level based on a modified User-Agent Server (UAS) state machine. SIP DNS flooding (flooding by SIP requests containing irresolvable domain names) are considered by Zhang et. al. [10] who propose practical recommendations in order to mitigate their effects. Sengar et. al. [11] propose a VoIP Flood Detection System (vFDS). The vFDS measures the difference between two probability distributions (one at training and one at testing) using the Hellinger Distance. In [12], the authors propose a general methodology for profiling SIP traffic at several levels: server host, functional entity (registrar, proxy) and individual user. They propose an entropy-based algorithm detecting SPIT and flooding attacks. As noticed, the learning plays a major role in most of the cited approaches.

Many of the cited contributions have similar approaches although they can not be compared because they are evaluated on different and often unavailable data-sets. A common data-set for evaluating VoIP intrusion detection is missing. We aim by this contribution to provide a totally repeatable test-bed for generating and customizing VoIP traffic as well as a labeled data-set composed of normal, different kind of attacks and mixed normal/attack traces.

3   Data Types

Monitoring distributed applications such as VoIP is challenging because in many cases monitoring only one source of data is not sufficient to reveal the complete picture of some malicious activities. Therefore we have decided to provide three types of data sources: the network traffic, the call detail records and the server logs. Each of these types is important for a group of detection approaches. Moreover, correlation of patterns spanning multiple data sources may reveal more evidence about certain anomalies. Next, we highlight the importance of each type:

- The network traffic -especially the protocols that are essential to the normal operation of VoIP (e.g. SIP, DNS)- is the typical data source for network-based anomaly detection. In [13], we defined 38 statistics (or probes) on the SIP traffic and showed their efficiency for detecting a series of flooding and SPIT attacks.
- The Call Detail Records (CDR) are particularly important for fraud detection and SPIT mitigation. The call history helps building individual user profiles and peer groups in order to detect fraudulent calls.

– The log and statistics of VoIP servers are the typical data source for host-based intrusion detection. For example, the Opensips SIP proxy[1] provides six groups of probes: core, memory, stateless statistics, transaction statistics, user location and registration. These probes can be monitored online in order to reveal abnormal processing and memory loads.

4   Traffic Generation Model

Several investigations use tools like Sipp[2] for SIP traffic generation. Sipp generates traffic by repeating call flow scenarios described in custom XML, hence it is convenient for benchmarking and assessing stress conditions at SIP servers. However Sipp -in its current state- is not able to emulate real SIP traffic. In fact, Sipp doesn't represent a SIP user-agent state machine and is more oriented towards the transaction layer than the call layer. Otherwise, our unit of traffic generation is the VoIPBot (the source code is available at [14]) which we have initially designed as an attack tool [15]. The new version of the VoIPBot supports emulation of a normal user profile. Its design is based on four main components: the protocols stack, the communication agent, the encryption engine and the SIP state machine as shown in Figure 1.

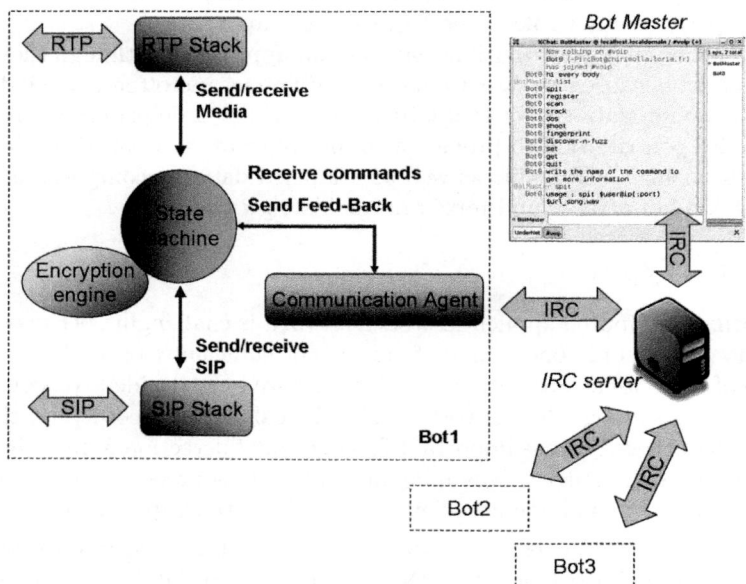

Fig. 1. Using VoIP Bots for simulating real traffic

---

[1] http://www.opensips.org
[2] http://sipp.sourceforge.net

The stacks of the different protocols provide the bot with an application interface to use the different signaling and media transfer protocols. We use the Jain-SIP stack for sending and receiving, manufacturing and parsing SIP messages. We use the Java Media Framework (JMF) stack for coding and decoding, compressing and expanding, encapsulation and demultiplexing of RTP media flows. The communication agent is based on the Internet Relay Chat (IRC) in order to exchange information and commands with the bot manager. In fact the bot connects to a pre-defined room on the manager IRC server at bootstrap. This scheme allows us to manage several groups of bots in an easy and efficient way. The encryption engine enables the bot to create digest authentication from credentials when authentication is required within the registration process. The SIP state machine drives the operations of the bot with respect to the commands issued by the manager. It defines the bot behavior upon occurrence of SIP events (i.e. receiving a SIP request `RequestEvent` or a SIP response `ResponseEvent`) and `TimeOut` events. The transition from a state to another is constrained by predicates on a set of global and local variables. The bot is configured to generate normal traffic based on the following parameters:

- The SIP port and IP: Currently, we are launching several bots from the same machine/IP. Each bot has a SIP port based on its identifier (e.g. Port 5072 is reserved to Bot12). Therefore, any SIP user-agent in the traces must be identified by the IP/port(SIP) couple rather that only the IP. The virtualization techniques can be further used in order to provision an IP for each bot.
- The SIP registrar: The bot registers to one or more SIP registrars using a user-name and a password in order to initiate and receive calls. For convenience, both the username and the password of a bot are set to its identifier.
- The probabilistic distribution of emitting calls: Traditionally in PSTN models the call arrival follows a Poisson distribution. The VoIP calls follow the same distribution [16]. We setup the bots to make calls with respect to a Poisson distribution by picking one destination form a phonebook. The mean number of calls per unit of time (λ) has to be defined. The bots have different seeds (based on their identifiers) for their pseudo-random generators. Each bot sleeps for a random boot-time before starting the emission of calls.
- The probabilistic law of call holding time: Currently the bot sets the duration of a call with respect to an exponential distribution such as in PSTN models. The mean of the call duration ($1/\lambda$) has to be defined. If the callee doesn't hang-off during the chosen call duration, the bot sends a BYE and closes the session. However, authors of [16] argued that in VoIP the call holding times follow rather heavy-tailed distributions and suggest the generalized Pareto distribution for modeling them. Moreover, the authors showed that the aggregated VoIP traffic has fractal characteristics and suggested the fractional Gaussian noise model for the aggregated VoIP traffic. We intend to adopt these findings in our future model.
- The social model: The bot belongs to one or more social cliques. The bot is socially connected with another bot if both belong to the same clique. P_{social}

(e.g. 0.8) represents the probability that a bot calls a destination which it is socially connected with. $1 - P_{social}$ (e.g. 0.2) is the probability that a bot calls a destination which it is not socially connected with. We provide a script (within the VoIPBot bundle) that assigns random social cliques to a group of bots based on a given size interval (e.g. each clique has between 3 and 7 bots).

- The behavior upon receiving a call: Using statistics from traditional telephony, about 70% of calls are answered in roughly 8.5 seconds while unanswered calls ring for 38 seconds [17]. For instance, the bot takes the call and responds with BUSY only if it is already in-call. The time of ringing follows uniform distributions over pre-defined intervals (e.g. uniformly distributed from 5 to 10 seconds).

These parameters are configured through the IRC interface or at boot-time.

The generation of attacks is based on available VoIP assessment tools. The VoIPBot can also emulate a "malicious" behavior by committing attack actions commanded by the manager. The VoIPBot supports scan, spam a wav file as given by a URL, crack a password and register with, download and send exploits, fingerprinting VoIP devices, shoot crafted SIP messages and flooding.

Inviteflood, Sipscan and Spitter/asterisk[3] are used to generate flooding, scan and SPIT attacks. Inviteflood floods the target with INVITE messages. It doesn't acknowledge the target's response causing many retransmissions and memory overload. It allows to set the SIP URI of the destination and the rate of the attack, hence making possible different scenarios (e.g. invalid SIP user-name, invalid domain, etc.). Sipscan supports three types of scanning using REGISTER, OPTIONS and INVITE messages. It doesn't support the control of the scan intensity which makes its effect similar to a flooding. Spitter uses the Asterisk IP PBX as a platform to launch SPIT calls based on an Asterisk call file. The Asterisk call file format allows configuring calls towards different destinations with different parameters. It is possible to configure a different sender for each SPIT call which makes detection hard if based only on the caller's SIP URI.

5   Test-Bed and Labeled Data-Set

Physically, the test-bed is composed of four machines (Intel Pentium 4 CPU 3.40GHz and 2G RAM memory running Ubuntu 9.10) connected through a hub (10 Mb/s). The machines are assigned static private IP addresses. A router plays the role of DNS server and Internet gateway for the test-bed. We set-up two illustrative VoIP networks based on available open source servers. The first network uses Asterisk: an open source VoIP PBX[4]. The second one is based on a SIP proxy called Opensips (former SIP Express Router).

---

[3] http://www.hackingvoip.com/sec_tools.html
[4] http://www.asterisk.org

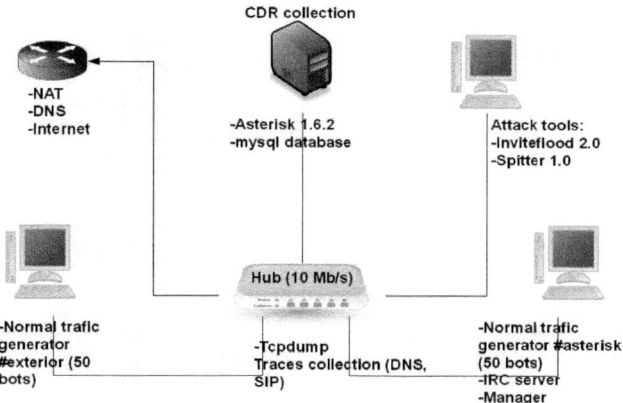

Fig. 2. Asterisk illustrative network

5.1   Asterisk Network

Asterisk is growing fast typically for Small Office/Home Office environments (SOHO). Regarding the SIP terminology, Asterisk can be considered as a registrar server, location server, and back-to-back user agent. Our illustrative network based on Asterisk is shown in Figure 2. We have two groups of VoIPBots emulating SIP users: the first group registers to Asterisk while the second group represents external users. Routing calls to and from external users is fixed at the Asterisk dial-plan. The first group connects to the #asterisk channel at the central IRC server and the second group connects to the #exterior channel. The manager issues commands to one group of bots by writing to the corresponding channel. It is possible to command each bot aside by opening a private channel with it.

5.2   Opensips+MediaProxy+Radius Network

Opensips is a high performance SIP proxy since it is able to manage a high number of registered users and thousands of calls per second without the need of a highly sophisticated hardware. Opensips supports the proxy, registrar and redirect SIP functionalities. We configure a MySQL back-end to Opensips for consistent storage. MediaProxy[5] provides NAT traversal capabilities for media streams. It is composed of one media-dispatcher and one or more media-relays. The media-dispatcher runs on the same host as Opensips and communicates with its mediaproxy module through a Unix domain socket. The media-relay connects to the dispatcher using TLS. FreeRadius[6] works in tandem with Opensips and Mediaproxy to manage the call accounting.

---

[5] http://mediaproxy.ag-projects.com

[6] http://freeradius.org

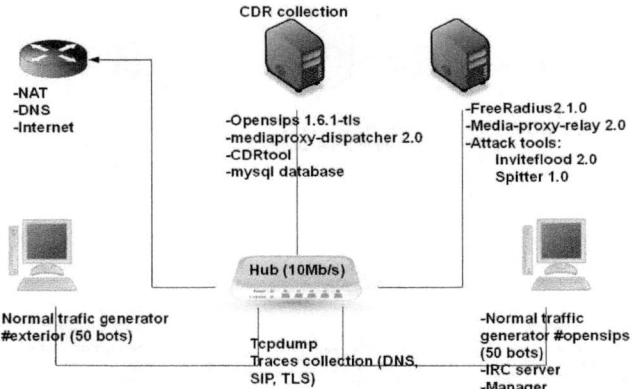

Fig. 3. Opensips illustrative network

We setup this triplet to serve two groups of bots: the first group is directly registered to the Opensips server and the second group represents the external extensions. Routing calls to and from external users is fixed at the Opensips routing logic. The Call Detail Records are collected using the CDRTool[7] which is an open source web application that provides accounting and tracing calls for Opensips. The CDRTool provides the SIP trace for each call based on the siptrace module of Opensips. It provides media information of each call based on the MediaProxy accounting.

5.3   Labeled Traces

We configure a set of normal and attack scenarios over the two aforementioned networks. The configuration choices are constrained by ensuring a relatively short time for each experiment (between 1 and 30 minutes) and a supportable overload for machines. In particular, we deploy a relatively low number of users (100 bots) and compensate this by a high number of calls per user and small call durations. The resulting traces of all experiments are available at http://webloria.loria.fr/~nassar/voip-dataset/. Each experiment is represented by a folder containing at least 4 files: a file containing the network trace in the tcpdump format, a text file representing the server log, a CSV file containing the CDRs and a meta-description file. The meta-description contains information about the experiment scenario (the number of bots, their repartition, the social model, the configuration of each group of bots, the network topology, etc.) and information about the attacks (if any) such as type, source and intensity.

The attack data are self-labeled in the sense that they carry a particular URI representing the attack tool. For example, in all the SIP messages and calls generated by Inviteflood and Spitter, we set the From-URI to start by "iflood" and

---

[7] http://cdrtool.ag-projects.com

Table 1. Labeled traces

Platform	$\lambda_{Poisson}$ (bot·hour)	Attack tool	Attack intensity	Trace duration	Number of calls
Asterisk	10	None	None	6 min	110
	10	Inviteflood	100 I/s	2 min	118
	10	Spitter	10 C.C.	2 min	82
Opensips	100	None	None	10 min	905
	100	Inviteflood	1000 I/s	2 min	1050
	100	Spitter	no-limit	2 min	452

"spitter" respectively. Table 1 shows a subset of the traces. The flooding intensity is measured by Invite/s while the SPIT intensity is measured by concurrent calls. A SPIT with no-limit intensity means that all the destinations are called in the same time.

6   Conclusion

In this paper, we have presented a unique experience of customizing and generating a labeled VoIP data-set in a controlled environment. We aimed by this contribution to make the recent works on VoIP attacks detection comparable. We provide SIP traffic traces, call detail records and server logs for a set of normal and attack scenarios.

In the future, our intention is to expand the test-bed by using virtualization techniques. We will extend the generation models in order to support more call statistical distributions, more attack types and more complex scenarios (e.g. with NAT traversal, inter-domain routing). The social model has to be refined based on the results of the social networks research. We also aim to evaluate the different proposed intrusion detection methodologies based on the data-set.

References

1. VoIPSA: VoIP security and privacy threat taxonomy. Public Release 1.0 (October 2005), http://www.voipsa.org/Activities/VOIPSA_Threat_Taxonomy_0.1.pdf
2. Reynolds, B., Ghosal, D.: Secure IP telephony using multi-layered protection. In: Proceedings of The 10th Annual Network and Distributed System Security Symposium, San Diego, CA, USA (Feburary 2003)
3. Rosenberg, J., Schulzrinne, H., Camarillo, G., Johnston, A., Peterson, J., Sparks, R., Handley, M., Schooler, E.: RFC3261: SIP: Session initiation protocol (2002)
4. Niccolini, S., Garroppo, R., Giordano, S., Risi, G., Ventura, S.: SIP intrusion detection and prevention: recommendations and prototype implementation. In: 1st IEEE Workshop on VoIP Management and Security, pp. 47–52 (April 2006)
5. Wu, Y., Bagchi, S., Garg, S., Singh, N., Tsai, T.K.: SCIDIVE: A stateful and cross protocol intrusion detection architecture for Voice-over-IP environments. In: International Conference on Dependable Systems and Networks (DSN 2004), June 2004, pp. 433–442. IEEE Computer Society, Los Alamitos (2004)

6. Sengar, H., Wijesekera, D., Wang, H., Jajodia, S.: VoIP intrusion detection through interacting protocol state machines. In: Proceedings of the 38th IEEE International Conference on Dependable Systems and Networks (DSN 2006). IEEE Computer Society, Los Alamitos (2006)

7. Fiedler, J., Kupka, T., Ehlert, S., Magedanz, T., Sisalem, D.: VoIP defender: Highly scalable SIP-based security architecture. In: Proceedings of the 1st international conference on Principles, systems and applications of IP telecommunications (IPT-Comm 2007). ACM, New York (2007)

8. Chen, E.Y.: Detecting DoS attacks on SIP systems. In: Proceedings of 1st IEEE Workshop on VoIP Management and Security, San Diego, CA, USA, April 2006, pp. 53–58 (2006)

9. Ehlert, S., Wang, C., Magedanz, T., Sisalem, D.: Specification-based denial-of-service detection for SIP Voice-over-IP networks. In: The Third International Conference on Internet Monitoring and Protection (ICIMP), pp. 59–66. IEEE Computer Society, Los Alamitos (2008)

10. Zhang, G., Ehlert, S., Magedanz, T., Sisalem, D.: Denial of service attack and prevention on SIP VoIP infrastructures using DNS flooding. In: Proceedings of the 1st international conference on Principles, systems and applications of IP telecommunications (IPTComm 2007), pp. 57–66. ACM, New York (2007)

11. Sengar, H., Wang, H., Wijesekera, D., Jajodia, S.: Detecting VoIP floods using the Hellinger distance. IEEE Trans. Parallel Distrib. Syst. 19(6), 794–805 (2008)

12. Kang, H., Zhang, Z., Ranjan, S., Nucci, A.: SIP-based VoIP traffic behavior profiling and its applications. In: Proceedings of the 3rd annual ACM workshop on Mining network data (MineNet 2007), pp. 39–44. ACM, New York (2007)

13. Nassar, M., State, R., Festor, O.: Monitoring SIP traffic using support vector machines. In: Lippmann, R., Kirda, E., Trachtenberg, A. (eds.) RAID 2008. LNCS, vol. 5230, pp. 311–330. Springer, Heidelberg (2008)

14. Nassar, M., State, R., Festor, O.: The VoIP Bot project,
http://gforge.inria.fr/projects/voipbot/

15. Nassar, M., State, R., Festor, O.: VoIP malware: Attack tool & attack scenarios. In: Proceedings of the IEEE International Conference on Communications, Communication and Information Systems Security Symposium (ICC 2009, CISS). IEEE, Los Alamitos (2009)

16. Dang, T.D., Sonkoly, B., Molnar, S.: Fractal analysis and modeling of VoIP traffic. In: Proceedings of Networks 2004, pp. 217–222 (2004)

17. Duffy, F., Mercer, R.: A study of network performance and customer behavior during-direct-distance-dialing call attempts in the USA. Bell System Technical Journal 57, 1–33 (1978)

Document Provenance in the Cloud:
Constraints and Challenges

Mohamed Amin Sakka[1,2], Bruno Defude[1], and Jorge Tellez[2]

[1] Novapost, Novapost R&D, 13, Boulevard de Rochechouart 75009 Paris-France
{amin.sakka,jorge.tellez}@novapost.fr
[2] TELECOM& Management SudParis, CNRS UMR Samovar, 9, Rue Charles
Fourrier 91011 Evry cedex-France
{mohamed_amin.sakka,bruno.defude}@it-sudparis.eu

Abstract. The amounts of digital information are growing in size and
complexity. With the emergence of distributed services over internet and
the booming of electronic exchanges, the need to identify information origins and its lifecycle history becomes essential. Essential because it's the
only factor ensuring information integrity and probative value. That's
why in different areas like government, commerce, medicine and science,
tracking data origins is essential and can serve for informational, quality,
forensics, regulatory compliance, rights protection and intellectual property purposes. Managing information provenance is a complex task and it
has been extensively treated in databases, file system and scientific workflows. However, provenance in the cloud is a more challenging task due
to specific problems related to the cloud added to the traditional ones.

1 Introduction

With the advent of cloud computing and the emergence of web 2.0 technologies, the traditional computing paradigm is shifting new challenges. This new
paradigm brings issues related to the trustworthiness of data as well as computations performed in third-party clouds. Certainly, the digital form of electronic
data has many advantages of being easily accessible, accurate and more useful.
However, the mobile and modifiable nature of digital entities requires specific
metadata and techniques ensuring information probative value. More specifically, in a cloud-computing scenario, most of the data will reside in data clouds,
while the applications will run in the cloud. In this case, users require additional
insurances regarding the confidentiality, privacy and integrity of information.
Required metadata is provenance and can be defined as information about the
origin, context or history of the data. Depending on the domain, provenance is
useful for many purposes and its systems can support different types of usage.
According to studies on information provenance [9,5,7], it can serve for audit
trail and justification, regulatory compliance and forensics, replication recipes,
attribution and copyright, as well as data quality and informational purposes.
Despite it seems obvious, information provenance issues introduces hard challenges. Provenance is not obvious because it overtakes widely recording the whole

F.A. Aagesen and S.J. Knapskog (Eds.): EUNICE 2010, LNCS 6164, pp. 107–117, 2010.

data history [14]. Provenance is listed as a hard problem in some information science recent surveys like the UK Computing Society's Grand Challenges in Computing [10] and the InfoSec Council's Hard Problems List. Provenance is challenging because it's difficult to collect and is often incomplete, it comes from untrusted and unreliable sources, it's heterogeneous and non-portable and it requires confidentiality and privacy insurance. In this paper, we present information provenance challenges in the cloud and how it can serve to enhance information integrity. Especially, we focus on services for electronic documents and more precisely storage and archival services.

Traditionally organizations and companies buy storage systems for each of its sites or acquire equipment for its data center locations. Generally, this represents enormous costs especially when we take into account the cost of managing, supporting, and maintaining these systems. For these reasons, outsourcing storage and archival to external clouds is a good alternative to reduce costs. It permits also to discard the hard task of locally managing and maintaining storage systems according to legal compliance and security recommendations [11]. However, this leverages many issues about probative value and implies an immediate need to information provenance and lifecycle.

This paper is organized as follows. Starting from a case study, section 2 illustrates provenance needs and constraints for documentary clouds. The third part presents provenance as a cloud challenge. Section 4 presents provenance management approaches, some related works and their lacks. In the last section we conclude and present our future works.

2   Introducing Provenance Constraints

2.1   Context: A Bank Record Case Study

We are going to present a case study illustrating information probative value and provenance challenges for electronic documents. This case study comes from Novapost (www.novapost.fr) which is a French company specialized on providing collect, distribution and archiving services for electronic documents (especially human resources documents like pay slips, or personal confidential documents like bank records). Novapost aims to consolidate the existing services by providing dematerialization and long term archiving solutions allowing to improve archive management (formats migration, information integrity, document lifecycle management, interoperability and reversibility issues), ensure compliance with regulations relative to the probative value of electronic documents, ensure continuity and reliability in the management of information lifecycle and its provenance and guarantee seamless integration with customers through SaaS (Software as a Service) solution provided in partnership with IBM and based on innovative technology and highly secure infrastructure.

The case study scenario is the following: a customer of Novapost services (a bank) sends its customers bank reports to Novapost in AFP[1] format. Novapost

---

[1] AFP is a document format originally defined by IBM to drive its printers and support the typical form printing on laser printers.

documents service (provided as a cloud service) applies a specific processing depending on the received document type (splitting, templating, copying ...), then generates individual bank reports in PDF/A [22] format for archiving, interacts with external trust authority (signature service composed by a signature server and a timestamp server), communicates with IBM cloud for legal archiving to archive them in IBM's datacenters. The archived files are accessible to bank customers via web portal or a widget plugged on the bank website.

2.2 Provenance Constraints

As we have mentioned before, the heart of Novaposts solutions is document lifecycle management and documentary workflows requiring high probative value of the processed documents. The purpose of the R&D department in Novapost is to provide solutions to address the problems confronting dematerialization actors. We address the issues of ensuring electronic documents probative value, their lifecycle traceability as well as guaranteeing the readability and the sustainability of their processing tracks for long term.

In France, the probative value of electronic documents is subject to compliance with the conditions requested by the Article 1316-1 of the Civil Code mandating to identify the emitter persons/entities and to provide an integrity insurance through the whole document lifecycle. To ensure documents probative value and to perpetuate the informational capital of electronic archives, we need to keep a plausible trustworthy history of their provenance. This explains the strong requirements for traceability mandated by record management standards [15,17,19].

If we consider the standard lifecycle of a document, it includes steps of generation, additional processing, transfer and archiving. This means that document's technical integrity could change while its legal integrity is the same. At present, traceability management solutions for informational flows, especially documents are based on an accumulation of log files, coming from heterogeneous systems and multiple remote legal entities. Document provenance information's are distilled in multiple files that we need to cross to retrace its history. Add to that the sustainability of the responsible entities for the conservation of these trace elements is also problematic. So, provenance in the cloud is considered as a transverse problem, at the crossroad between different types of constraints. According to our study, provenance constraints can be categorized into legal, business and technical constraints (cf. Fig 1):

 - Legal constraints: they are relative to the regulatory compliance about conservation duration of personal information as well as compliance with legislation governing information privacy like CNIL recommendation in France (www.cnil.fr) or the CDT in USA (www.cdt.org). Add to that the recent international law and the principle of territoriality which is not always in harmony with the distributed nature of the clouds over countries. For our

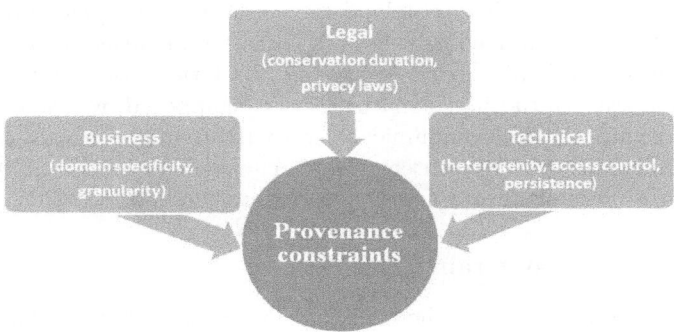

Fig. 1. Provenance constraints

case study for example, track elements are distributed across multiple systems and managed by multiple legal entities. We don't know what is the legal entity responsible for keeping these tracks neither where should they be kept and for how long.

– Business constraints: first of all, the word track is not generic, it depends on the business. For example, a track element in the bank system is different from a track element in the archival system. Also, the granularity of provenance depends on the context and the specificity of each cloud. Business constraints involves the definition of accessibility rules and profiles to provenance since confidentiality requirements imposes hiding the participation of some elements or some actors. These constraints are relevant to the definition of the suitable provenance collection and management policy to alleviate semantic interoperability issues.

– Technical constraints: they are relevant to provenance collection and management. It involves interoperability at the syntactic level between provenance records generated by heterogeneous systems, security issues including provenance integrity, access control and privacy. Performance, reliability and scalability constraints are also important because we are in a cloud context mandating from a provenance service to be scalable and available. Add to that trust and reliability about track elements provided by external systems/services during their creation or their transfer. In our case study, the identification and the crossing of log files about the processing of a document (known that document identifiers are not the same across different systems) is problematic and may rise probative value loss. Even if we can cross these elements, the heterogeneity of formats and data structures affects their readability. The other issue is guaranteeing confidentiality and privacy: an administrator of the archival system should not be allowed to access the contents of customer's bank records. Inversely, a customer should not access to the details about the systems that generates its bank record.

3   Challenges for Provenance in the Cloud

As illustrated by the case study, provenance becomes more and more important and challenging. The need for a provenance technology provided as an end-to-end service to guarantee information integrity and probative value is accentuated. Principally, the challenges can be divided into two categories. The first category includes known provenance challenges. It involves object identification and coupling with provenance, provenance reliability and confidentiality. It contains also the level of genericity of such a service. The second category contains new challenges which occur only in a cloud context and that are essentially relevant to performance, availability and scalability. Clouds are characterized by their ability to scale dynamically on demand. They can scale up and down according to the workload. So, how can we define a provenance service in the cloud and what are its characteristics? Would it be possible to use existent cloud computing frameworks like Hadoop (http://hadoop.apache.org) or more high level frameworks like Chukwa (http://hadoop.apache.org/chukwa) or Pig (http://hadoop.apache.org/pig) to meet high scalability and fault tolerance characteristics?

3.1   Identification and Mapping Digital Objects with Their Provenance

The first question is how to identify objects flowing between different clouds? The problem is that every cloud has its own policy to identify objects and that these policies are generally confidential. The challenge is that the technical integrity of a numeric object can change whereas its legal integrity remains the same. For these reason, traditional identification techniques based on hash computing are not suitable and cannot provide a unique object identifier. The second question is how to keep the link between a digital object and its provenance information along the different lifecycle phases and how to ensure provenance persistence? Persistence is the ability to provide an object provenance even if the object was removed. This data independent persistence is mandatory because in many cases where the ancestor object is removed, the provenance graph will be disconnected and the descendents objects provenance will be undetermined. So how provenance-data mapping should be managed? Would it be possible to consider provenance as an intrinsic metadata of each object or it will be better to consider the document and its provenance separately and just keep a descriptor referencing the object and its provenance.

3.2   Trust and Reliability

In the cloud, information crosses the traditional boundaries and passes through untrusted clouds. Generally, clouds are untrusted since the guarantee provided regarding the origin and the transformations are minimal, unclear or unreliable. So provenance can be forged or tampered. Having a tempered or low quality provenance can be worse than having no provenance at all. To ensure trustworthy provenance, we need to provide the mandated security elements [15,19]:

- Integrity: the assurance that provenance data is not tampered. Any forgery or tampering in provenance records will be detected.
- Availability: an auditor can check the integrity and the correctness of provenance information.
- Confidentiality: provenance information should not be public, only authorized authorities can access to it. This means that systems must have the ability to restrict views of data and its provenance according to different confidentiality levels.
- Atomicity: at storage time, both provenance and data should be stored or neither should be stored.
- Consistency: at retrieval time, data returned should be consistent with provenance.

3.3   Heterogeneity and Granularity

As illustrated by the use case, provenance is generated and managed by different entities according to different policies. It is often heterogeneous and not interoperable. Depending on the context, provenance needs and capabilities differ from one cloud to another and can't be exactly the same. The problem is that there is no portable and standard plausible one format fits all provenance that can be wired into general-purpose systems. Provenance should be tracked at different levels of granularity by the use of a flexible granularity approach. Such an approach should be based on a standardized provenance format that can be extended to a coarse-grained format or reduced to a fine-grained format. In this scope, a standard format for provenance called Open Provenance Model (OPM) [16,23] was proposed to the scientific community. This model standardizes provenance syntax and proposes a vocabulary for provenance. However, it presents security lacks because traditional access control techniques don't apply to provenance graphs and that the adaptation of existing access control models to provenance is insufficient [3]. That's why we think that the current OPM model is not sufficient for covering provenance in the cloud needs, especially security and privacy needs.

3.4   Cloud Constraints: Extensibility, Availability and Scalability

The cloud is designed to be scalable and available on demand. Existing provenance solutions in other environment (workflow, database...) dont consider the availability or scalability in their design. In addition, cloud is not extensible and couldn't be modified or extended. Furthermore, the availability of the provenance service has to match the high ability of the cloud. If it's not the case, the overall availability is reduced to the limited availability of the provenance service. Regarding scalability, the use of a database can make provenance queriable but not scalable. This limitation is due to updates synchronization between clients. These updates can cause a distributed service lock introducing a distributed deadlock or a scalability bottleneck due to a single lock. In this case, the need is to communicate with a cloud database because the use of a parallel database is hard to maintain and is in contradiction with clouds model. Storing the provenance in a separate service introduces the issue of coordinating updates between object store (documents) service and the database service .

4   Provenance Approaches and Related Works

4.1   Provenance Approaches

To address the aforementioned challenges, we present in this section approaches for provenance management. We have identified three types of approaches for provenance collection and management:

1. Centralized approach (cf. Fig 2): it consists on the definition of a centralized policy for collecting and managing provenance from end-to-end. In this case, a unique trusted authority provides a provenance service. This approach permits to alleviate data structure heterogeneity by using a unique standardized provenance format. However, we have to imagine that this authority can access external clouds through specific API. This authority should provide strong authentication techniques permitting to its clients to trust this centralized service and to communicate without revealing critical information.
2. Hybrid approach (with import/export feature): this approach allows the import and the export of provenance between a centralized provenance service and external clouds (cf. Fig 3). The import feature allows retrieving the requested information without an additional management effort because it's supposed that the imported information is directly exploitable. In the same way, the centralized authority can directly export exploitable provenance to external services requesting it. Within this approach, the provenance authority should define a management policy to alleviate syntaxic and semantic interoperability issues.
3. Federated approach (cf. Fig 4): in this approach provenance management responsibilities are delegated to each participant cloud which should handle it autonomously. Provenance is distilled through multiple files that we need to

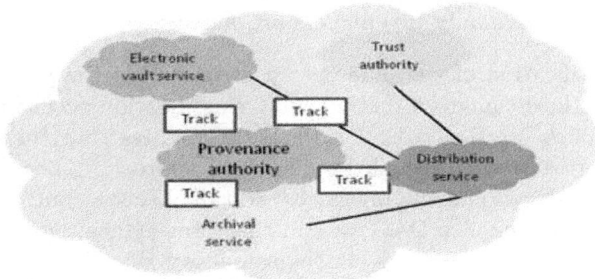

Fig. 2. Managing provenance within a centralized approach

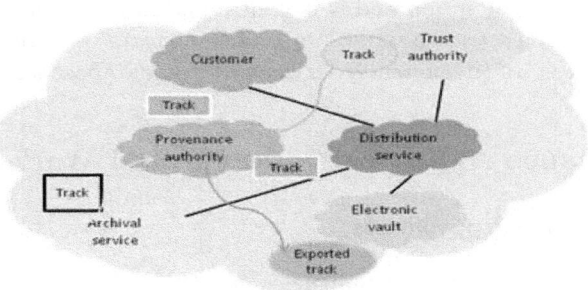

Fig. 3. Managing provenance within a hybrid approach

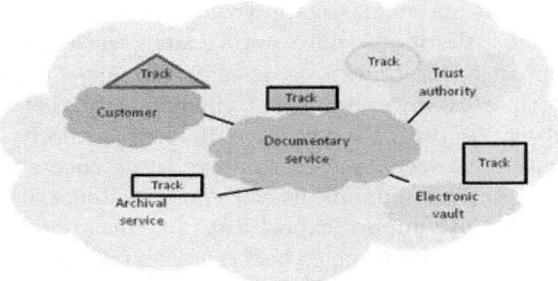

Fig. 4. Managing provenance within a federated approach

cross to produce the provenance chain. Within this approach, limits are heterogeneity between provenance formats coming from different heterogeneous entities implying semantic and syntactic interoperability issues. To have a probative value, it is necessary that provenance elements can be folded and be consistent at both ends of a transaction. Since one of the parties dont meet these conditions, the whole provenance chain becomes suspicious. More importantly, the accuracy of information implies a need for corroborating provenance elements between two clouds. This limits evidence character to transactions where only two players are concerned.

After the identification of the different constraints and the presentation of the approaches, we have achieved the following classification according to the management approach (cf. Table 1). This table illustrates that the challenge level about trust and reliability is high for the three approaches because provenance access control is always challenging since traditional access control techniques are not relevant for provenance DAG [3] and that provenance tempering is possible even within centralized approach. In the same way, managing cloud constraints has always the level (H) in the three approaches.

Table 1. Classification of provenance constraints

Constraint/ Approach	Identification Mapping	Trust / reliability	Heterogeneity / granularity	Cloud specific constraints
Centralized	L	H	L	H
Hybrid	H	H	M	H
Federated	H	H	H	H

Challenge level: - High: H
                    - Medium: M
                    - Low: L

4.2   Related Works

Previous researches on provenance tackles principally the problem from one of two sides: database and files management or workflow and grid environments. Works on database and files provenance [2,4,21] focused on data- driven business processes and relies on the ability to trace information flows through SQL manipulations or by increasing file headers and tagging them [1,20]. On the other side, workflow provenance research [8,9] tackles scientific processing (like scientific workflows in collaborative environments for biologists, earth scientists...) and deals with less transparent but predefined process executions. Many other works on provenance addresses specific problems like provenance in software development known as versionning systems which manages distributed intra-files provenance but are not able to manage inter-files provenance. We also mention recent work on securing provenance from tempering based on cryptographic mechanisms [12].

The particularity of the provenance needs in our work is the superposition of a set of traditional provenance constraints with cloud specific constraints. Most of these works assumes the ability to alter the underlying system components, and dont propose a native provenance infrastructure addressing the problem as an end-to end property corresponding to all the aforementioned constraints.

5   Conclusion

In this paper we have presented a pragmatic case study illustrating the need for provenance to ensure electronic documents probative value. It allowed us to illustrate and analyze the challenges and to categorize the inherent constraints motivating our research on information provenance.

Provenance is becoming more and more important as information is shared across networks and exceeds internal boundaries. Actually, the most used medium for information sharing, the web does not provide this feature. But in the cloud, which is not yet a mature technology, provenance issues can be addressed from now.

Provenance in the cloud is an open research problem and our future work will be focused on provenance deployment as a service in the cloud. We aim to propose a centralized provenance service model for the cloud. This model should be generic and extensible. It should provide an abstraction of application domain and accept proprietary extensions to cover the specifics of heterogeneous business processes. In this model, we aim to address security, privacy and access control issues. Regarding security issues, we want to introduce semantics to enhance provenance confidentiality. Regarding privacy, we wish to develop techniques assuring information usefulness in respect with privacy needs without any leakage. This requires the definition of new access-control techniques and data access schemas for provenance graphs. In the future, we think that the challenge for provenance systems is to introduce remembrance which is a kind of an RFID for all types of digital objects. This concept was introduced in [13], it consists in augmenting all data objects with persistent memory and to consider that provenance is an intrinsic property of data objects. Remembrance can be very interesting in cloud computing scenarios and can immediately show that a piece of data is originated from an untrusted object or application. It exceeds traditional versioning and snapshots systems and tries to give a complete end-to-end history as an atomic property of any digital object. To be useful and exploitable, remembrance should be coupled with strong privacy and access control techniques.

References

1. Agrawal, P., Benjelloun, O., Sarma, A.D., Hayworth, C., Shubha, U., Nabar, C.U., Sugihara, T., Widom, J.: ULDBs: Databases with Uncertainty and Lineage. In: Trio: A System for Data, Uncertainty, and Lineage. VLDB 2006, pp. 1151–1154 (2006)
2. Bhagwat, D., Chiticariu, L., Tan, W.C., Vijayvargiya, G.: An Annotation Management System for Relational Databases. In: VLDB, pp. 900–911 (2004)
3. Braun, U., Shinnar, A., Seltzer, M.: Securing provenance. In: Third USENIX Workshop on Hot Topics in Security (HotSec) (July 2008)
4. Buneman, P., Khanna, S., Tan, W.C.: Why and Where: A Characterization of Data Provenance. In: Van den Bussche, J., Vianu, V. (eds.) ICDT 2001. LNCS, vol. 1973, pp. 316–330. Springer, Heidelberg (2000)
5. Cameron, G.: Provenance and Pragmatics. In: Workshop on Data Provenance and Annotation (2003)
6. Cheney, J., Chong, S., Foster, N., Seltzer, M., Vansummeren, S.: Provenance: A Future History. In: International Conference on Object Oriented Programming, Systems, Languages and Applications, pp. 957–964 (2009)
7. Da Silva, P.P., McGuinness, D.L., McCool, R.: Knowledge Provenance Infrastructure. IEEE Data Engineering Bulletin 26, 26–32 (2003)
8. Davidson, S., Cohen-Boulakia, S., Eyal, A., Ludascher, B., McPhillips, T., Bowers, S., Freire, J.: Provenance in Scientific Workflow Systems. IEEE Data Engineering Bulletin 32, 44–50 (2007)
9. Goble, C.: Position Statement: Musings on Provenance, Workow and Semantic Web Annotations for Bioinformatics. In: Workshop on Data Derivation and Provenance (2002)
10. Grand challenges in computing research conference, UK Computing Society (2008), http://www.ukcrc.org.uk/press/news/challenge08/gccr08final.cfm

11. Hasan, R., Yurcik., W., Myagmar, S.: The Evolution of Storage Service Providers: Techniques and Challenges to Outsourcing Storage. In: Proceedings of the 2005 ACM workshop on Storage Security and Survivability (2005)
12. Hassan, R., Sion, R., Winslett, M.: Preventing History Forgery with Secure Provenance. ACM Transactions on Storage (2009)
13. Hassan, R., Sion, R., Winslett, M.: Remembrance: The Unbearable Sentience of Being Digital. In: Fourth Biennial Conference on Innovative Data Systems Research (2009)
14. INFOSEC Research Council (IRC) Hard problem list. Technical report (November 2005), http://www.cyber.st.dhs.gov/docs/IRC_Hard_Problem_List.pdf
15. ISO 14721:2003. Space data and information transfer systems - Open Archival Information System Reference model (OAIS), http://www.iso.org
16. Miles, S., Groth, P.T., Munroe, S., Jiang, S., Assandri, T., Moreau, L.: Extracting causal graphs from an open provenance data model. Concurrency and Computation: Practice and Experience 20(5), 577–586 (2008)
17. MoReq2 specifications. Model Requirements for the management of electronic records Update and Extension (2008), http://www.moreq2.eu
18. Muniswamy-Reddy, K.K., Holland, D.A., Braun, U., Seltzer, M.: Provenance-aware storage systems. In: USENIX Annual Technical Conference, General Track, pp. 43–56 (2006)
19. NF Z42-013. Electronic archival storage-Specifications relative to the design and operation of information processing systems in view of ensuring the storage and integrity of the recording stored in these systems, http://www.boutique.afnor.org
20. Sar, C., Cao, P.: Lineage file system. Technical Report (January 2005), http://crypto.stanford.edu/~cao/lineage
21. Simmhan, Y., Plale, B., Gannon, D.: A survey of data provenance in e-science. SIGMOD Record (Special section on scientific workflows) 34(3), 31–36 (2005)
22. ISO Standard for using PDF format for the long-term archiving of electronic documents ISO-19005-1 - Document management - Electronic document file format for long-term preservation - Part 1: Use of PDF 1.4 (PDF/A-1), http://www.pdfa.org
23. Moreau, L., Plale, B., Miles, S., Goble, C., Missier, P., Barga, R., Simmhan, Y., Futrelle, J., McGrath, R.E., Myers, J., Paulson, P., Bowers, S., Ludaescher, B., Kwasnikowska, N., Van den Bussche, J., Ellkvist, T., Freire, J., Groth, P.: The Open Provenance Model (v1.01) specifications. Future Generation Computer Systems (2009)

Wireless Handoff Optimization: A Comparison of IEEE 802.11r and HOKEY

Kashif Nizam Khan and Jinat Rehana

Norwegian University of Science and Technology, NTNU
Helsinki University of Technology, TKK
{kashifni,rehana}@stud.ntnu.no

Abstract. IEEE 802.11 or Wi-Fi has long been the most widely deployed technology for wireless broadband Internet access, yet it is increasingly facing competition from other technologies such as packet-switched cellular data. End user expectations and demands have grown towards a more mobile and agile network. At one end, users demand more and more mobility and on the other end, they expect a good QoS which is sufficient to meet the needs of VoIP and streaming video. However, as the 4G technologies start knocking at doors, 802.11 is being questioned for its mobility and QoS (Quality of Service). Unnecessary handoffs and re-authentication during handoffs result in higher latencies. Recent research shows that if the handoff latency is high, services like VoIP experience excessive jitter. Bulk of the handoff latency is caused by security mechanisms, such as the 4-way handshake and, in particular, EAP authentication to a remote authentication server. IEEE 802.11r and HandOver KEY (HOKEY) are protocol enhancements that have been introduced to mitigate these challenges and to manage fast and secure handoffs in a seamless manner. 802.11r extends the 802.11 base specification to support fast handoff in the MAC protocol. On the other hand, HOKEY is a suite of protocols standardized by IETF to support fast handoffs. This paper analyzes the applicability of 802.11r and HOKEY solutions to enable fast authentication and fast handoffs. It also presents an overview of the fast handoff solutions proposed in some recent research.

1 Introduction

The last decade in the communications industry has been dominated by mobile services. These services have changed the way we communicate daily. Users are now able to enjoy nonstop mobile services anywhere, anytime and anyplace. And for this reason, user expectations of the mobile services are higher than ever before. Seamless data connectivity, streaming video and Voice over Internet Protocol (VoIP) are now considered as the core applications that will drive the construction of next generation mobile networks [2].

If we now look at the wireless access technologies, we will find that IEEE 802.11 or Wi-Fi has been the prevailing option for wireless access to the internet. Over the years, 802.11 has ruled the wireless world with its ever growing mobility and security amendments. However, as the options are changing, so are the

F.A. Aagesen and S.J. Knapskog (Eds.): EUNICE 2010, LNCS 6164, pp. 118–131, 2010.

demands. Recent research has pointed out that 802.11 suffers from a lack of QoS, specifically in case of the real time applications.

Handoff latency is considered as the main reason behind this technical problem. Current advancements in technology leave no space for latencies and unnecessary variation in delays (jitter) while a Mobile Node (MN) moves around the network and the connection jumps from one Access point (AP) to another. At present, the security protocols in wireless environments are designed in such a way that a MN needs to be authenticated at each AP when it moves around [2]. 802.11 essentially employs the EAP (Extensible Authentication Protocol) as a generic authentication protocol which supports multiple authentication methods [13].

Wireless handoff comprising APs is not a simple process, rather it includes a complex set of operations involving multiple layers of protocol execution. One crucial part of this handoff process is EAP re-authentication, which indeed is a major contributor to the overall handoff latency [13]. IEEE 802.11r [1] is an amendment to the original standard, which thankfully targets re-authentication as one of the key areas to lower the handoff latencies. It performs this by keeping the re-authentication procedure as local as possible, thus reducing the round trips. By contrast, Hand Over KEY (HOKEY) essentially targets the fact that in the existing networks, re-authentication to a new AP requires a full EAP exchange with the initial server which authenticated it previously. This phenomenon introduces extra processing delay and data transit delay in the network, which in turn increases the latency which can be 100 to 300 ms (milliseconds) per round trip [13]. The problem is that real-time applications like VoIP allow a maximum end-to-end delay of 50 ms. This goes as the basic motivation behind the HOKEY protocol. [11].

Both 802.11r and HOKEY target to solve the same problem and that is to achieve fast handoffs by reducing latencies. However, they perform modification at different layers of protocol stack. 802.11r modifies the MAC layer whereas HOKEY modifies the EAP layer. This is because, IEEE has standardized 802.11 MAC layer while IETF has standardized EAP. Interestingly, none of them tries to figure out what is the best way to reduce the handoff latencies.

Handoff latency has also attracted much research which focuses on optimized handoff procedures for 802.11. This paper presents an overview of the standards and proposals that have evolved to optimize 802.11 handoffs. Sec. 2 describes the 802.11 handoff procedure. Sec. 3 includes a brief description of 802.11r and HOKEY in terms of improvements and changes for fast handoffs. Sec. 4 summarizes a performance overview of 802.11r and HOKEY from the existing literature. Sec. 5 presents some research ideas that are not standardized yet [10,4,12,9,5,8,3,6] for 802.11 handoff improvements by reducing the latency. Finally, Sec. 6 concludes the paper.

2   802.11 Handoffs

This section briefly presents the key operations of 802.11 handoffs. During the mobility period, when a mobile node moves to the periphery of the current

Access Point's (AP) coverage, it is necessary for the mobile node to attach itself to another AP with better coverage to continue the communication. This process of communication handover is simply known as handoff. It essentially incorporates a number of message exchanges among the mobile node, the old AP and the new AP which results in physical connection and state transfer from the old AP to the new AP [4]. During this period of handoff from the old AP to the new AP, the mobile station is potentially unable to communicate any type of data traffic. This delay is known as 'latency'. The higher the latency, the worse the QoS is experienced by the end user.

The handoff process is performed in three basic stages: detection, search and execution [4]. As its name implies, 'detection' phase is the phase when the MN detects a need for handoff. It can be determined by different ways and it typically depends on specific network deployment. For example, the MN can decide to execute a handoff if it experiences a lower signal-to-noise ratio (SNR) or if it sees a number of packet losses consecutively. The term 'lower' in the former case may be determined by comparing the SNR with a predefined threshold value. Nevertheless, the handoff can be initiated if a degraded performance is experienced by a MN from the current AP or the neighboring APs are offering better signals.

After the detection phase, the MN starts searching for the new AP among the candidate neighboring APs. In this phase, the MN scans through all 802.11

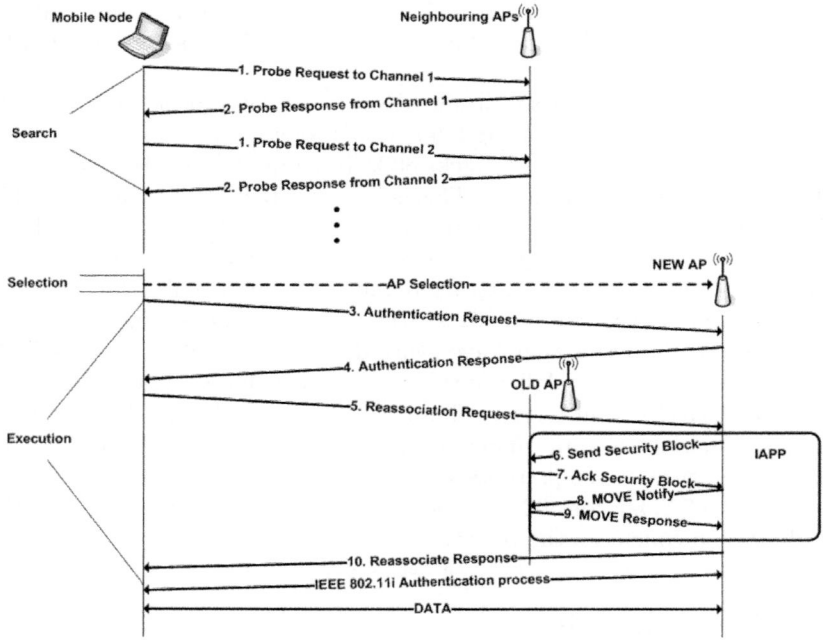

Fig. 1. 802.11 Handoff Example

channels. Fig. 1 shows the search phase in which the MN sends a broadcast probe-request and waits for the probe-response from each of the channel. In the Fig. 1, these messages are numbered as 1 and 2 respectively. Upon receiving the responses, the MN chooses the best candidate AP for the handoff process. The selection is based on the SNR associated with each AP's probe response [4]. It is easy to assume that the SNR of the selected AP should be at least above the SNR of the current AP, to avoid unnecessary handoffs.

The execution phase is performed in two stages. At first, the new AP authenticates the MN using open authentication and then re-associates with the MN. Although, IEEE 802.11i extends the protocol to provide secure authentication. In Fig. 1, messages 3 and 4 show the simplest authentication handshake between the MN and the new AP. This authentication is followed by a re-association between the MN and the new AP. Re-association is performed to finalize the communication transfer. The re-association request includes important practical information such as the MAC address of the MN and the old AP and the Extended Service Set (ESS) identifier.

After the re-association the security state is transferred between the old AP and the new AP using the Internet Access Point Protocol (IAPP). This is a 2 way handshake as shown by the messages 6 and 7 in the Fig. 1. Once the security block is exchanged, the old AP needs to transfer all relevant information and communication states associated with the MN to the new AP. New AP issues a MOVE notify request to old AP asking for this information . The information exchange between old AP and new AP is encrypted and authenticated for security of the context information. Messages 8 and 9 in Fig. 1 show the two way MOVE handshake. Finally, the new AP answers the re-association request of the MN (Message 10) and the handoff process is finalized at this point. The MN can now transfer communication data through new AP.

IEEE 802.11i prescribes the use of EAP/IEEE 802.1x mechanism to overcome the vulnerabilities of snooping and intrusion by 3rd parties. It introduces new methods of key distribution to overcome the weaknesses in the earlier methods. A Pairwise Master Key(PMK) is generated between the MN and Authentication Server (AS) through earlier EAP exchange. The PMK is used to derive Pairwise Transient Key (PTK) via 4-way handshake between the MN and the AP. The 4-way messages are sent as Extensible Authentication Protocol Over LAN (EAPOL) frames. This 4-way handshake ensures the integrity of the MN and the AP and demonstrates that there is no man-in-the-middle. It also synchronizes the PTKs.

One of the major drawbacks of wireless handoffs is the slow transfer of connections from old AP to new AP while moving around. Even when a MN jumps from the coverage of one AP to another in the same mobile domain, a connection needs to be established with the remote AS which may be distant in location. In addition, the number of roundtrips to the remote AS and the number of messaged exchanged between MN and AP also contribute to the delay because of roundtrips and channel acquisition time.

Preauthentication was introduced in 802.1X to mitigate these delays while providing secured authentication at the same time. Preauthentication performs this reduction in latency by caching some of the keying material derived during the authentication step in the neighboring potential new APs to which the MN may roam. In Pair-wise Master Key (PMK) caching, the associated AP and the MN perform the 802.1X authentication and cache the PMK hoping that the MN will associate with this AP in future. In such situations, only a 4-way handshake is needed between MN and AP to create new session keys from the PMK after re-association. The problem of choosing the correct PMK has been solved by using key identifiers in the reassociation request. We will not cover the pre-authentication detail as support for 802.11 preauthentication has been dropped from the 802.11r draft. This is because, although pre-authentication lessens the latency, it is still not sufficient for real time services like voice calls.

3   New Standard Solutions for Fast Handoff

This section presents two of the most powerful protocols that emerged to support fast handoff in mobile WLANs: IEEE 802.11r and HOKEY. As stated earlier, 802.11r is an amendment to IEEE 802.11 and thus, it has been standardized by IEEE. HOKEY is standardized by IETF in the form of RFCs. The following subsections illustrate the basic functions of these two protocol suits in more detail.

3.1   IEEE 802.11r

IEEE 802.11 experienced a major glitch with the specification of Wired Equivalence Protocol (WEP). The level of security provided by WEP was significantly deficient for Wireless applications. IEEE 802.11e is introduced in the process to provide better and enhanced security for real time applications (especially voice). IEEE 802.11i also emerged to mitigate the shortcomings of WEP and to provide enterprise level security. IEEE 802.11i adds stronger encryption and authentication methods for higher data security in Wireless Applications. However, these security amendments achieve stronger security at the expense of large delays which are unacceptable for applications like voice. Mechanisms like re-authentication and QoS re-negotiations during roaming introduce unnecessary and unacceptable large delays resulting in degraded voice quality. As a result, a new IEEE Task group, Task Group r, was formed to address these roaming issues in the 802.11 enhanced security amendments. This group has produced a new Fast Basic Service Set (BSS) Transition standard which is able to minimize unnecessary delays during roaming while still providing the same amount of security as promised by IEEE 802.11i and 802.11e. Let's look at this 802.11r in a bit more detail.

802.11r performs the fast BSS transition in three major ways: integrating the four-way handshake into the 802.11 authentication/association exchange, pre-allocation of QoS resources and most importantly through efficient key distribution. 802.11r also introduces some new fields in the protocol messages such as:

Mobility Domain Information Element (MDIE) and Fast Transition Information Element (FTIE). As the name implies, MDIE is concerned with current mobility domain information and FTIE manages resource reservation and security policy information. FTIE also includes some of the EAPOL-Key messages.

The method in which 802.11r reduces the 802.11i BSS transition time is by piggy-backing the EAPOL-Key messages at the four-way key exchange on top of four existing frames. It is performed by adding security-related information to 802.11 authentication and association request and response. For example, the MDIE and FTIE are contained in beacons, probe responses, association requests and association responses. This piggybacking significantly reduces the overall handoff latency, as PTK derivation step can be overlaid on open authentication and re-association steps. It essentially omits the extra round trips frames for PTK derivation. Details about PTK and 802.11r key hierarchy are discussed in Sec. 3.1.1.

The other optimization technique used by 802.11r is pre-reservation. In this procedure, the MN is allowed to perform the QoS admission control with the new AP before open authentication or re-association. It can be performed in two ways: Over the Distribution Service (OTD) and Over the Air (OTA) [7]. With OTD, the MN communicates with the new target AP via the currently associated AP's distribution services. OTD traffic flow is similar to 802.11i pre-authentication traffic flow and OTD is preferred in 802.11r as it provides pre-reservation capabilities without the interruption of the current traffic flow. QoS provisioning is an additional mechanism used in 802.11r standard to allow fast BSS transitions. In this approach, the pre-reservation is delayed until the association- request/response. It is sometimes appropriate when MN detects that target AP is lightly loaded and the reservation fail [7]. OTA also helps to reduce the latency of full BSS transition if QoS provisioning is included in the delay measurement.

The main aim of 802.11r fast BSS transition is to reduce the security overhead. The most obvious benefit is that bulk of the authentication process is performed before the actual handoff occurs. Once a MN associates itself with an AP residing in a particular subnet, the PMK can be distributed to all the APs that are associated with the subnet or rather we should call it the mobility domain. Hence, when a MN moves across the mobility domain, the PMK is assumed to be present in all the APs and so the time needed to reauthenticate the MN is significantly reduced. This is simply because the latency to communicate with the remote AS is omitted in this case. This prederivation of keys is supported by a new key management system in 802.11r. The following subsection discusses the new key hierarchy in more detail.

3.1.1 IEEE 802.11r Key Hierarchy

Fig. 2 illustrates the 802.11r key management hierarchy. 802.11r key hierarchy consists of two levels of key holders. When a MN performs initial association and full authentication with an AP, the logical entity R0 Key Holder (R0-KH) derives the top level keys PMK-R0, which is in turn derived from the Master Session Key (MSK) and Pre-Shared Key (PSK) from the AS [1]. There is one R0-KH in each

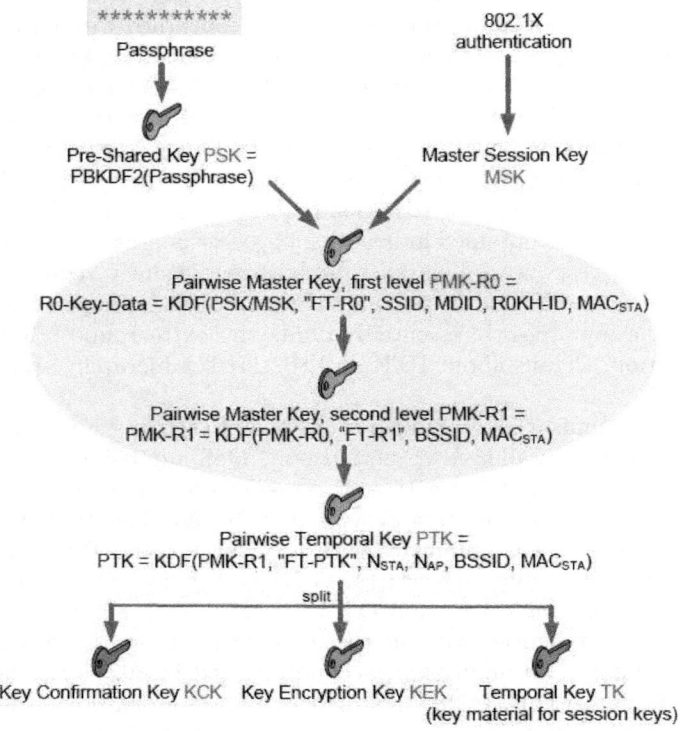

Fig. 2. IEEE 802.11r Key Hierarchy [16]

security mobility domain and one PMK-R0 per MN in each security mobility domain. The next level of keys, denoted as PMK-R1 is derived for associating the MN with the new AP and it will be different for different AP-MN pair. The AP also stores the PTK (Pair wise Temporal Key). The Key holders, which were previously known as wireless switches, may be located within an AP or may be a separate physical device.

802.11r key management system specifies two key domains: Security Domain (SD) and the Security Mobility Domain (SMD)[1]. SD comprises of all the entities shown in the figure namely R0-KH, all associated R1-KH and APs. SMD is a set of SD, in which a R0-KH can derive PMK-R1 for any R1KH. SMD essentially sets the boundary within which a MN is allowed to perform fast BSS transitions.

As shown in the Fig. 2, R0-KHID and R1-KHID are derived initially when a MN is associated with an AP residing in a SMD, for the first time. Then PMK-R1 keys are generated using the R0-KHID and R1-KHID. At this moment, the R0-KH is able to distribute the PMK-R1 to all other R1-KH [1]. Alternatively, it may distribute the keys on demand. Now, if a MN moves to another AP in the SMD, the R1-KH associated with the new AP already possesses PMK-R1 and so no IEEE 802.1X authentication is necessary. One interesting thing to notice here is that 802.11r does not specify any protocol for key distribution amongst the key holders and the APs. However, essentially it is assumed that a secure connection exists between APs and key holders.

3.2   HOKEY

IETF is on the way of standardizing another group of effective handover keying protocols known as HOKEY. The aim of HOKEY is twofold- firstly, support handovers from AP to AP and secondly support roaming between different operators. HOKEY has enhanced the EAP protocol to achieve low latency handoffs and method-independent fast re-authentication. To match the terminology of IETF EAP specifications, this section will denote the MN as 'peer'.

The EAP keying hierarchy requires two keys to be derived by all key generating EAP methods: the Master Session Key (MSK) and the Extended MSK (EMSK). In common scenarios, an EAP peer and an EAP server authenticate each other through an EAP authenticator. Successful authentication results in a derivation of a Transient Session Key (TSK) by the EAP peer using the MSK [15]. To avoid unnecessary delays and roundtrips, it is desirable to avoid full EAP authentication when a peer moves from one authenticator to another. Although some EAP methods utilize state information from initial authentication to optimize the re-authentication, most of the method specific re-authentications cost 2 round trips at minimum, with the original EAP server [13].

EAP Re-authentication Protocol (ERP) is designed to provide method independent re-authentications with lower handover latencies. The main idea behind ERP is to permit a peer and the server to verify the possession of keying material which has been previously obtained from an EAP method [15]. And more importantly, this EAP is a single-round trip exchange between peer and server. EAP exchange is independent of lower layer.

The design of EAP is pretty simple. A full EAP exchange is performed whenever a peer tries to connect to any network for the first time. As a result, MSK is available to the EAP authenticator at this point and the peer and server also derive EMSK. EMSK or DSRK (Domain Specific Root key) is used to derive re-authentication Root Key (rRK) which is available both at the peer and the server[15]. Furthermore, an additional key: re-authentication Integrity Key (rIK) is derived from rRK which is used to prove the possession of keying material during ERP exchange.

Fig. 3 shows a typical ERP exchange. As the authenticator cannot be sure whether the peer is able to perform ERP beforehand, at first it sends an EAP-Initiate/Re-auth-Start message. It also serves the purpose of advertising the capability of the authenticator to support ERP. If the peer is capable of performing ERP exchange, it replies back with an EAP-Initiate/Re-auth message. If the peer does not know anything about the EAP-Initiate/Re-auth-Start message, it will not respond to it. So, after a certain period, of time the authenticator will initiate EAP by sending EAP-Request/Identity message. Similarly, an authenticator can also fall back to EAP if it does not support ERP.

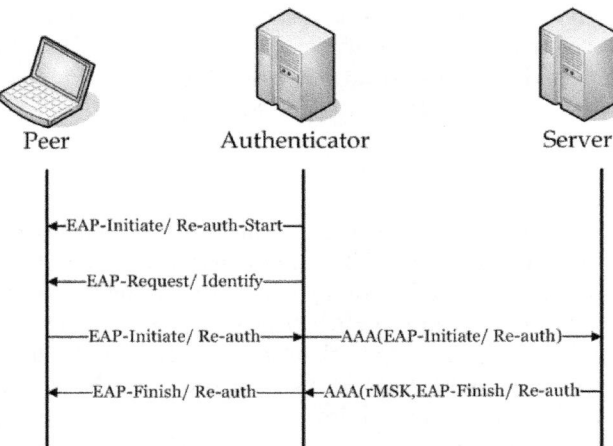

Fig. 3. ERP Exchange

For the sake of discussion, we assume here that the peer supports ERP and responds with an EAP-Initiate/Re-auth message. This message contains two important pieces of information: keyName-NAI (Network Access identifier) and rIK. The authenticator uses the keyName-NAI field to send the message to the appropriate server. Server uses the keyName-NAI to look up the rIK [15]. After successful verification of rIK, server derives rMSK (re-authentication MSK) from the rRK and a sequence number supplied by the peer with the previous response. This rMSK is then transported along with the EAP-Finish/Re-auth message by the server. rIK is used to protect the integrity of all these messages. Upon receiving the response, the peer verifies the integrity of the message using rIK and generates rMSK.

The above discussion also gives a picture of the key hierarchy used in ERP. At each re-authentication an rMSK is established between a peer and the authenticator which serves the same purpose of MSK. To prove the possession of rRK, rIK is used which is derived from rRK. rRK is derived from EMSK or DSRK. While using ERP, HOKEY is also able to support roaming [6]. Roaming can be performed when the new network has a roaming relationship with MN's home network. One important thing to notice here is that, HOKEY makes use of the EMSK rather than reusing any existing key materials of 802.11i keying hierarchy.

4   Performance of 802.11r and HOKEY

802.11r and HOKEY both target more or less the same goal: keeping the authentication during the handoff local to the access network and thus reducing the handoff latency. However, they operate at different layers of the WiFi protocol stack. As depicted in Fig. 4, 802.11r tries to optimize the 802.11 protocols

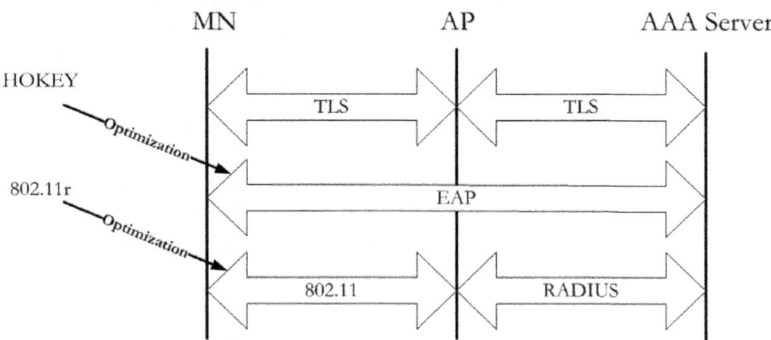

Fig. 4. HOKEY and IEEE 802.11r Optimization

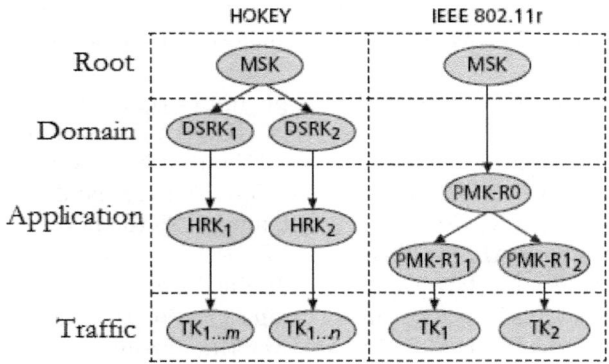

Fig. 5. Key Hierarchies for HOKEY and IEEE 802.11r [6]

at the MAC layer whereas, HOKEY tries to optimize EAP which operates at link layer. HOKEY extends the AAA (Authentication, Authorization and Accounting) architecture to distribute security credentials without performing a full EAP authentication. By contrast, 802.11r focuses on passing security credentials directly between APs as a MN transitions from one AP to other.

Both 802.11r and HOKEY change 802.11i key hierarchy in different ways as depicted in Fig. 5. As a result, the relative complexity of the hierarchies also varies. 802.11r hierarchy does not support domain-level keys but HOKEY supports domain-level keys and as such HOKEY supports inter-operator roaming. Traffic keys are generated via the original four-way handshake in HOKEY and via a modified MAC protocol in 802.11r [6].

Key hierarchies play a crucial role in overall system security. In HOKEY, keys are essentially distributed along the AAA hierarchy. So if a AAA entity is compromised, all keys underneath will be compromised. In 802.11r, keys are handled centrally in R0KH. As discussed earlier, R0KH can be an AP or a separate physical device. In [6] it is argued that, R0KH will be a simpler target for key compromise than an AAA entity. The reason behind this can be that 802.11r centralizes all the keys to the edge of the network. So, 802.11r can be considered as weaker in security model than HOKEY.

[6] also presents a nice comparison of the handover performance of HOKEY and 802.11r. Let,

N = Number of roundtrips required to perform a particular EAP method

Tw = Transmission latency between the MN and AP

Tc = Latency between two close devices in wired LAN such as AP to AP communication and

Ta = Latency between various components and AAA server.

Now, let's look back to initial 802.11i handover: We have already mentioned that most of the EAP method specific authentications cost 2 round trips at minimum, with the original EAP server. So, the time to complete the EAP authentication should be 2N(Tw + Ta). Then, it will take Ta time to distribute the MSK to the AP and finally the four way handshake between MN and AP will take 4Tw time. So, the total 802.11i handover time sums up to 2N(Tw + Ta) + Ta +4Tw.

HOKEY handover time includes the single roundtrip execution of ERP which automatically delivers Handover Root Key (HRK) . Thus HOKEY omits the time to distribute MSK to the AP from 802.11i initial handover. So, for HOKEY the handover time will be 2(Tw + Ta) + 4Tw. 802.11r handover comprises of handover request from MN to AP, key distribution to the new AP and final key handshake at the new AP. It sums up to 2Ta + 2Tc + 2Tw.

The typical values for Tw ,Ta and Tc obtained from several network deployments are: Tw =15μs,Ta=20μs and Tc=5μs [6]. So, HOKEY handover time results in 130μs and IEEE 802.11r handover time results in 70μs. These handover times are far less than a full IEEE 802.11i authentication using an EAP method. From the figures, we can easily compare that 802.11r requires less time for an intra-domain handoff than HOKEY. However, only HOKEY improves the performance of cross-domain handovers.

The analysis results found in [6] seems optimistic. To get a rough idea about the handover performance of 802.11r and HOKEY, we performed a simple experiment. As shown in the Fig. 6 (A), in case of 802.11i, the AP obtains the keying materials from the AS over the internet and the rough time is 10-30 ms depending on the location of the AS. In case of 802.11r , the keying materials are fetched and stored in R0KH which is located within a host in the same domain or LAN(Fig. 6 (B)). In case of HOKEY, the location of the proxy server is not mentioned explicitly but it is some where in the inernet but close to the AP (Fig. 6 (C)). We denoted the servers as AAA Home (AAAH) and AAA Foreign (AAAF) servers in the figure. We collected ping time from a host in Computer Science and Engineering Department, TKK to www.helsinki.fi, to get an idea about the 802.11i handovers. The average round trip time was 9.164 ms. For HOKEY, we pinged www.cs.hut.fi which was close to our department and the average round trip time was 1.625 ms. Lastly, to get an idea about 802.11r

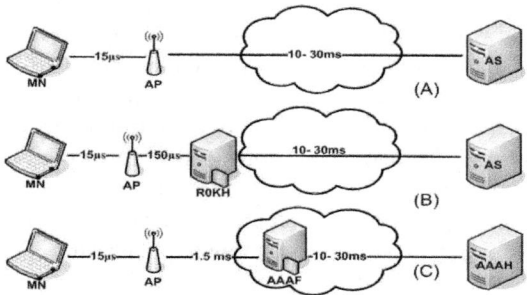

Fig. 6. Experiment of Handover Times in Different Cases

handovers, we pinged a host on the same LAN and it took .157 ms on an average. So, we can get a rough idea about the optimization of 802.11r and HOKEY from this simple experiment.

HOKEY and 802.11r both offer roughly same performance. However, their infrastructure requirements are different and as a result usage scenarios are also different. HOKEY will be more effective for service provider networks whereas IEEE 802.11r suits well for WLAN environments having low latency constraints. Both the standards solve the problem of handover latency but they implement it in different ways and at different protocol layers. Choice of either one will strictly depend on particular network specifications and other constraints.

5   Other Proposed Improvements

Much research efforts have been devoted towards reducing the authentication delay and thus reducing latency during handoffs in wireless network. This section tries to present some of the better ideas from those research which can be a good contribution towards reducing the handoff latency in wireless network.

Tuomas Aura et. al. have proposed a new protocol for re-authentication of a MN to different APs or different wireless networks in [2]. This protocol passes credential information to a MN which has recently been attached in a successful association with an AP. The MN uses these credentials as a proof of its previous good behavior and when the MN tries to attach with a new AP it presents these credentials to it. One interesting property of this protocol is that the computations are mainly based on keyed one-way functions. So, it can result in very low computation overhead. Although, this approach takes the risk of passing credentials of one MN to other APs, it has the potential to result in a good solution with low complexity computations.

In [10] the authors have performed an explorative analysis to find the reasons of unnecessary handoffs. They collected data from both 802.11a and 802.11g networks over a period of five days. Their analysis reveals interesting results. They showed that most of the handoffs are triggered based on packet-loss information and this packet-loss information can be adverse in densely populated networks. Although, we are not always dealing with densely populated networks, but even then, handoff mechanisms should be adaptive to congestion losses.

Minho shin et.al. [11] have focused on reducing the probing time by reducing the number of probed channels. They have used neighbor graphs to improve the AP discovery process. However, this approach scales poorly as it requires each AP to store its neighbor graphs. [12] introduces a new selected scanning algorithm combined with caching mechanism. They have used a dynamic channel mask to reduce the probing to a subset of channels rather than all the channels. Finally, [14] shows that handoff latency can be significantly reduced by using shorter beacon intervals and active scanning. All of the above mentioned ideas can be very useful. However, many of them have been implemented and tested in constrained environment. It will be interesting to see how they perform in real networks.

6   Conclusion

Handoff latency is a critical mobility issue for real-time applications such as VoIP over WLAN. It is very important to maintain an acceptable QoS during the handoff process while also supporting advanced security standards. This paper presents a brief overview of the two effective suites of protocols which have been standardized to reduce the handoff latency during wireless handover: IEEE 802.11r and HOKEY. Although both of the standards have evolved to solve more or less the same problem, they optimize different parts of authentication properties and use cases. As a result, they operate differently. However, as far as effectiveness is concerned, both of the protocols promise to lessen handoff latency while maintaining the same security level as that of IEEE 802.11i. This paper also discusses some other potential research proposals that can achieve better results if those proposals are combined with 802.11r or HOKEY. To conclude, we hope that, HOKEY and 802.11r will be a potential and significant step forward towards efficient wireless networks for real-time applications.

References

1. I. S. 802.11r TM. IEEE Standard for Information Technology - Telecommunications and Information Exchange Between Systems - Local and Metropolitan Area Networks - Specific Requirement. Part 11: Wireless LAN Medium Access Control (MAC) and Physical Layer (PHY) Specification. Amendment 3: Specifications for Operation in Additional Regulatory Domains (July 2008)
2. Aura, T., Roe, M.: Reducing Reauthentication Delay in Wireless Networks. In: SE-CURECOMM 2005: Proceedings of the First International Conference on Security and Privacy for Emerging Areas in Communications Networks, pp. 139–148. IEEE Computer Society, Washington (2005)
3. Bangolae, S., Bell, C., Qi, E.: Performance Study of Fast BSS Transition using IEEE 802.11r. In: IWCMC 2006: Proceedings of the 2006 international conference on Wireless communications and mobile computing, pp. 737–742. ACM, New York (2006)

4. Bargh, M.S., Hulsebosch, R.J., Eertink, E.H., Prasad, A., Wang, H., Schoo, P.: Fast Authentication Methods for Handovers between IEEE 802.11 Wireless Lans. In: WMASH 2004: Proceedings of the 2nd ACM international workshop on Wireless mobile applications and services on WLAN hotspots, pp. 51–60. ACM, New York (2004)
5. Bianchi, G.: Performance Analysis of the IEEE 802.11 Distributed Coordination Function. IEEE Journal on Selected Areas in Communications 18(3), 535–547 (2000)
6. Clancy, T.: Secure Handover in Enterprise WLANs: CAPWAP, HOKEY, and IEEE 802.11r. IEEE Wireless Communications 15(5), 80–85 (2008)
7. Goransson, P., Greenlaw, R.: Secure Roaming in 802.11 Networks. Newnes (2007)
8. Salowey, V.J., Nakhjiri, M., Dondeti, L.: Specification for the Derivation of Root Keys from an Extended Master Session Key(EMSK). RFC 5295, The Internet Engineering Task Force (August 2008), http://ietf.org/rfc/rfc5295.txt
9. Pack, S., Choi, Y.: Fast Handoff Scheme Based on Mobility Prediction in Public Wireless LAN Systems. IEE Proceedings Communications 151(5), 489–495 (2004)
10. Raghavendra, R., Belding, E.M., Papagiannaki, K., Almeroth, K.C.: Understanding Handoffs in Large IEEE 802.11 Wireless Networks. In: IMC 2007: Proceedings of the 7th ACM SIGCOMM conference on Internet measurement, pp. 333–338. ACM, New York (2007)
11. Shin, M., Mishra, A., Arbaugh, W.A.: Improving the Latency of 802.11 Hand-offs using Neighbor Graphs. In: MobiSys 2004: Proceedings of the 2nd international conference on Mobile systems, applications, and services, pp. 70–83. ACM, New York (2004)
12. Shin, S., Forte, A.G., Rawat, A.S., Schulzrinne, H.: Reducing Mac Layer Handoff Latency in IEEE 802.11 Wireless LANs. In: MobiWac 2004: Proceedings of the second international workshop on Mobility management & wireless access protocols, pp. 19–26. ACM, New York (2004)
13. Clancy, T., Nakhjiri, M., Narayanan, V., Dondeti, L.: Handover Key Management and Re-Authentication Problem Statement. RFC 5169, The Internet Engineering Task Force (March 2008), http://ietf.org/rfc/rfc5169.txt
14. Velayos, H., Karlsson, G.: Techniques to Reduce the IEEE 802.11b Handoff Time, vol. 7, pp. 3844–3848 (June 2004)
15. Narayanan, V., Dondeti, L.: EAP Extensions for EAP Re-authentication Protocol (ERP). RFC 5296, The Internet Engineering Task Force (August 2008), http://ietf.org/rfc/rfc5296.txt
16. Aura, T.: Lecture Notes on Network Security: WLAN Security. Helsinki University of Technology (TKK) & University College London (UCL) (2008)

Introducing Perfect Forward Secrecy for AN.ON

Benedikt Westermann[1] and Dogan Kesdogan[1,2]

[1] Q2S*, NTNU, 7491 Trondheim, Norway
[2] Chair for IT Security, FB5, University of Siegen, 57068 Siegen, Germany

Abstract. In this paper we discuss AN.ON's need to provide perfect forward secrecy and show by an estimation of the channel build up time that the straight forward solution is not a practical solution. In the remaining paper we propose an improvement which enables AN.ON to provide perfect forward secrecy with respect to their current attacker model. Finally, we show that the delay, caused by our improvement, does not decrease the performance significantly.

1 Introduction

Anonymity systems are used by various users with manifold motivations. Some users just want to preserve their privacy, while other users need anonymity systems to circumvent censorship systems. While the state of being anonymous is for some users just a positive side effect that however is not needed, it is of major importance for other users to avoid serious consequences. The latter group is for instance represented by a whistler-bowler who wants to inform the public or a law enforcement agency about a serious crime within an organisation. Obviously, the use of his real identity or an easy to trace pseudonym is not a good idea if the whistle-bowler would face serious consequences due to his action.

Therefore, the whistler-bowler has a strong motivation to be anonymous during his action and naturally wants to be anonymous also after he has provided the information to a third party. Especially the part of staying anonymous is important. Thus, it should not be possible after a week or even several years to reveal the identity of the user of an anonymity network. Therefore special care must been taken to protect the identity of the users after their actions have been taken. The simplest solution is not keeping any records of the user's communication. Although, this solution is simple and represents the usual policy of the operators of nodes in an anonymity network, it does not hinder a third party or a malicious node to record the transfered messages.

The probable most popular anonymity systems are Tor[1] and AN.ON[2]. Both systems provide different solutions to anonymize users. This leads naturally to different requirements, preconditions and problems. One problem AN.ON faces is the lack of *perfect forward secrecy*. Thereby perfect forward secrecy is the

---

* "Center for Quantifiable Quality of Service in Communication Systems, Center of Excellence" appointed by The Research Council of Norway, funded by the Research Council, NTNU and UNINETT. http://www.q2s.ntnu.no

F.A. Aagesen and S.J. Knapskog (Eds.): EUNICE 2010, LNCS 6164, pp. 132–142, 2010.

property that if a long-term private key used in a key establishment protocol gets compromised it does not affect the security of session keys that had once generated with help of the compromised private key[3]. Due to the lack of perfect forward secrecy the anonymity of all users can be revoked even weeks after their connections took place due to compromised private keys. Obviously, a period of several weeks provides an attacker with various possibilities to compromise the keys. He can mount various targeted attacks, for example he can blackmail operators or can attack the servers itself to gain access to private keys. Once he has accomplished this and has a record of the previous communications the attacker can deanonymize all users with all their connections during the usage period of the compromised keys. Obviously, if this situation occurs it is without doubt a threating situation for the users. Hence, countermeasures are necessary to protect the users against such attacks.

In this paper we propose how this kind of attacks can be prevented by introducing perfect forward secrecy for AN.ON without decreasing its performance notably.

The paper is structured as followed. In Section 2 we describe Tor and AN.ON in more detail. Based on the description we point out the differences between AN.ON and Tor, the implications of the differences and the challenges for AN.ON regarding perfect forward secrecy. In Section 3 we describe the idea and propose an improvement to counter the described threat which arises due to compromised keys. Additionally, we estimate the additional delay that our improvement would introduce to the channel build up time. The security implications are discussed in Section 4. Section 5 concludes the paper.

2   Two Popular Anonymity Networks

In this section we describe the functionality of AN.ON [2] and Tor [1] and discuss their differences.

2.1   Tor: The Onion Routing

In Tor [1], the most popular low latency anonymization network, a user sends packets over so-called *circuits*. A circuit is a user selected path through the Tor network and it consists out of several nodes. A node in Tor is called *Onion Router (OR)*. By routing the message over several ORs it is achieved, that only the user is aware from whom to whom his message travels. This state is called *relationship anonymity*. Tor as well as AN.ON aim to provide this kind of anonymity for their users against the network operators or a local attacker.

A user who wants to send a message anonymously with Tor has to establish a circuit. During a circuit setup, a user authenticates a selected OR and creates a session key with help of a DH key exchange. Additionally, he can extend the circuit to another OR. Usually, a circuit consists out of three ORs. After the circuit is successfully established, a user can tunnel over the circuit various (data) *streams*. A stream is an end-to-end connection between the user and the final destination, for example a web server. With the first packet in a stream a user informs the last OR in the circuit about the final destination which subsequently establishes a TCP connection to the destination.

The *telescopic* path-building design with its different layers of encryption prevents the first OR within a circuit to see the final destination of a stream[1]. Each OR in the circuit can remove exactly one layer of the encryption. This ensures that only the last OR can read the final destination.

The last OR in a circuit can link the different streams of a user, since the streams are tunneled over a single circuit which is only used by a single user. On the long run the linkability of the streams can threaten the anonymity of users. The Tor client, also called *onion proxy (OP)*, changes therefore periodically the circuits of its user.

To authenticate the ORs a custom made protocol is used. The protocol uses a DH key exchange which is authenticated with help of a known public encryption key. The protocol was proven to be secure under the random oracle model in [4].

2.2   AN.ON: Anonymity Online

Another popular anonymity system is AN.ON[2]. Nowadays it is also called Jon-Donym. On the first glace it seems that AN.ON and Tor are quite similar. Both systems use layered encryption to achieve anonymity for their users. A closer look shows that AN.ON and Tor are quite different. One important difference is, that the sequence of servers in AN.ON is fixed and cannot be chosen by the users. The user only has the option to choose from a set of predefined sequences of nodes. A sequence of nodes is determined by the nodes' operators and it is called a *cascade*. Figure 1 depicts a cascade. In AN.ON a node is named *mix*.

Similar to Tor, AN.ON's users encrypt data in layers and send it along a cascade. Data intended for the same TCP connection is sent over a so-called *channel*. Every time an application requests a new TCP connection, a new channel has to be established. Each mix removes exactly one layer of encryption from a packet. Due to the fix sequence of the mixes it's important to provide unlinkability for channels of a user especially with respect to the last mix. Otherwise an attacker

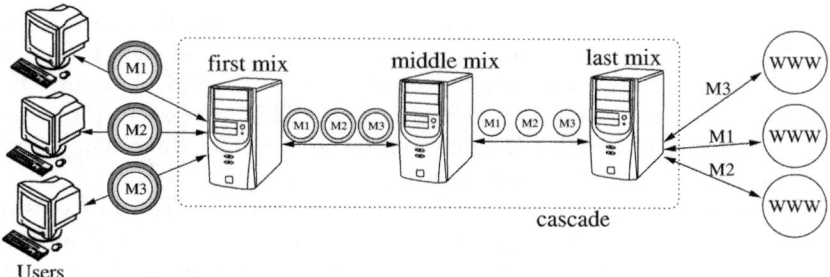

Fig. 1. Sketch of a Cascade in AN.ON

could deanonymize a user and all his prior channels by just attacking a single channel. Thus, for each channel established by a user, it is necessary to generate new keys in a way such that the mixes can neither link the key nor the channel to prior keys and channels respectively.

When a user wants to establish a new channel, for example if he wants to request a website, he generates two new session keys for each mix by uniformly selecting two 128 bit words from all possible 128 bit words. The generated keys are sent along the cascade and they are encrypted with the corresponding public encryption keys of the mixes. Thus, only the first mix in a cascade can link different channels with each other due to the user's IP address. This procedure fulfills therefore the unlinkability requirement of AN.ON.

2.3   The Challenge of Introducing Perfect Forward Secrecy

As mentioned above, Tor uses a DH key exchange to generate session keys between the OP and the ORs. In Tor the forward secrecy is achieved by the properties of the DH key exchange. Additionally, the fact that a DH key exchange results in a different key even though one side uses the same exponent, helps also to protect against replay attacks.

Unfortunately, AN.ON does not perform a DH key exchange. Contrary to Tor, AN.ON uses the cascade principle instead of the free routing principle and requires therefore the unlinkability property for each channel/stream. Thus, if AN.ON used a DH key exchange in way Tor does, it would require a user to perform a DH key exchange for each TCP connection an application requests. This would introduce an additional delay. The delay however greatly influences the user experience regarding the performance. For example various channels are needed to retrieve a single web page. Most often these channels are created in a sequential way. An additional delay of 100 ms can therefore extend a complete website retrieval by several seconds.

In order to roughly estimate the dimension of the additional delay, which would be introduced when every channel requires a DH key exchange, we conducted a small experiment with AN.ON. In this experiment we measured the build up times for a channel from the user to a destination at a cascade of length three. To do so, we connected to a cascade via the AN.ON client. After that our measurement started. The following procedure was repeated 200 times. We began to stop the time and requested a new channel over the cascade back to ourself. As soon as a TCP connection was established by the exit node, we stopped the time measurement. Please not, that this does not mean that we received already a confirmation about successful establishment on the other side of the channel. In total we measured the time it took to establish one TCP connection, to transmit three mix packets with a size of 998 bytes as well as the time the processing of the packets took. In addition we measured in a separate measurement the round trip times between the last mix in the cascade and our measurement node.

For the measurement we chose the payment cascade "Koelsch-Rousseau-SecureInternet"[1]. We collected 200 samples with an average channel buildup time of 229 ms with a 0.95-confidence interval of $[225.7\ ms, 232.56\ ms]$ and a standard deviation of 48.5 ms. For the round trip time we measured 52 ms on average from the last mix to the server. Thus, we can roughly estimate that 78 ms are used to establish the TCP connection from the last mix to the server under the assumption that no retransmissions of packets for the TCP handshake were necessary. The processing and the transmission of three mix packets took therefore 151 ms.

A DH-Key exchange for each new channel is likely to introduce $(n+1) \cdot n - 2$ additional messages as depicted in Figure 2 assuming that the client does not need to establish each time a key exchange with the first mix. In the current protocol n mix packets are exchanged between the mixes to establish a connection to the retriever. If we assume that each additional message introduces an additional delay of 20 ms[2] an additional delay of 200 ms would be introduced for a cascade with three mixes. This is more as twice as much as an establishment in the current version. Additionally, this estimation is quite rough, it does not consider the extra load on the mixes which is introduced due to the asymmetric cryptography. The estimation strongly indicates that the straight forward solution is not a suitable solution. Another solution would be, to perform a DH key exchange only to the middle mix for every channel. However, this would already introduce 4 new messages leading to an estimated additional delay of 80 ms. This delay is significant and therefore this solution is also not practical. Especially considering two facts. Firstly, the anonymity depends on the number of active users. Secondly, there is a linear correlation between the delay and the number of users[5]. In other words, the anonymity, a low latency anonymity system can provide, depends among others on the performance of the system. Thus, the solution of performing a DH key exchange for every new channel is not acceptable.

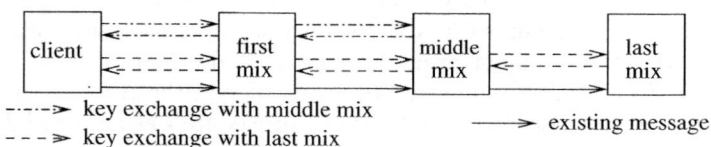

Fig. 2. Additional messages if a DH key exchange would be introduced

The challenge in AN.ON is to introduce a DH key exchange with neither risking the anonymity of the user nor decreasing the performance significantly.

---

[1] http://anonymous-proxy-servers.net/de/status

[2] This is an optimistic value considering that 3 messages took already 150 ms.

3   Introducing Perfect Forward Secrecy

A DH key exchange provides the property of perfect forward secrecy. AN.ON does not use a DH key exchange, instead it uses the private keys directly to decrypt the session keys sent by the clients and therefore it cannot provide perfect forward secrecy. In this section we present an approach which introduces a DH key exchange without influencing significantly the build up phase of a new channel which is an important performance parameter for low latency anonymity networks.

3.1   Observable Information

The idea of our approach is based on the fact, that different mixes have different knowledge about the channels of users. In a cascade we can distinguish between three types of mixes. The first type is the *first mix*. This mix sees the IP addresses of the users and it is the entry point for the cascade. This mix does not see the final destination. *Middle mixes* represent the second type and they are only present in cascades having more than two mixes. They neither see the final destination of a message nor the user's IP address. The last type of mix is the *last mix* that performs the request on behalf of the user. This type of mix sees the final destination of the messages, but it does not see the user's IP address.

As mentioned before, it is important in a cascade that the last mix cannot link single channels of a user. Therefore, a user should generate new session keys for every new channel. However, we claim that the channels of a user have to be unlinkable only with respect to the last mix, but they can be linkable for the first and the middle mixes. In the case of a first mix, the mix can already link user's channels due to the user's IP address and the user's established TCP connection respectively. Hence, a newly generated key cannot establish the unlinkability property with respect to the first mix.

Contrary to the first mix, middle mixes cannot link the channels of their users. Compared to the information that is available to a middle mix, the first mix has at least the same information available. If a middle mix can link different requests, the information that is available to the middle mix is still a strict subset of the information available to the first mix. This implies, that if a middle mix can revoke the relationship anonymity due to the introduced linkability of a user's requests, then a first mix is able to do the same, since it has at least the same information available. Therefore, assuming that an attacker only controls a single mix, we do not weaken the security of AN.ON.

Therefore, we propose to introduce DH key exchanges between a user and middle mixes as well as the first mix. The DH key exchanges are performed once directly after a user has established a connection to a cascade. The generated key serves as a master key in order to derive the different session keys for every new channel. Note, that reusing the same keys for de- and encryption is dangerous as for example demonstrated in the case of the old AN.ON protocol in [6].

Due to the fact, that the master key is once generated when a user connects, it introduces only once an additional delay, namely when the user connects to a cascade.

3.2   Used Protocol: Tor Authentication Protocol

In order to perform an authenticated DH key exchange we reuse Tor's authentication protocol (Tap)[7]. AN.ON fulfills the requirements of the protocol completely, and therefore no modifications of the protocol are necessary. Thus, the risk of introducing vulnerabilities in the design is lower. Additionally, the Tap is proven to be secure under the random oracle model[4].

In this paper we use the protocol as it was formally described and proven in [4]. For the protocol it is assumed that a user knows the public encryption key of the node. Let p and q both primes with $q = \frac{p-1}{2}$ and g a generator of the subgroup $\mathbb{Z}_p^*$ of order q. Let l_x the maximum length of the exponent that is chosen from the interval $[1, \min(q, l_x) - 1]$ [4].

In the first steps of the protocol a user chooses uniformly an exponent x. This exponent is used to calculate g^x which is sent encrypted with the node's public encryption key to the node. The node decrypts the message and checks if the decrypted message m is within the interval $[2, p - 2]$. If so, the node selects uniformly an exponent y from the interval $[1, \min(q, l_x) - 1]$. The node computes g^y. Additionally, it computes the hash of m^y. Both results are sent back to the user. The user can now compute the key K himself and can verify that the node uses the same key by hashing his own key and comparing it with the received hash (b).

3.3   Key Derivation Function

As mentioned above the generated key K is not used directly to encrypt data. Instead, the session keys have to be derived by this generated master key K. If we assume that the packets are received in the same order as they were sent, we can use a standard *key derivation function (kdf)* like KDF1 from the ISO/IEC 18033-2:2006 standard [8].

Due to the fact, that the first mix directly communicates with a user and hence the packets arrive in the same order as they were sent, it is safe to use such a key derivation function. This key derivation function is based on a counter which is hashed together with the master key. If AN.ON uses SHA-1 as a hash function which produces 20 bytes and the required key length is 16 bytes, then the ith key can be generated by taking the first 16 bytes of SHA-1(K, i). Thereby i is represented by a fixed length integer limiting the maximal number of keys that can be derived. In [8] i has a length of 4 bytes and thus at most 2^{32} keys can be generated. Due to the fact,that AN.ON requires two keys, one for the *forward direction* and one for the *backward direction*, two keys have to be generated for each channel. The forward direction describes the processing of packets which are conveyed by the client and the backwards direction represents the other case, namely the processing of packets that are sent by the last mix. For example, keys with an odd counter can be used for the forward direction while keys with an even counter are used for the other direction.

A positive side effect of this procedure is, that replays of an old message do not lead to a successful decryption of the messages. The reason for this is that different keys are used for the en- and the decryption. While the key, used for encrypting the message, is based on an old counter, the mix uses a newer counter to generate the key for the decryption. Thus, both keys are different and the decryption fails. Currently, AN.ON has to protect by other means against replays[2,9].

Usually, AN.ON's mixe do not reorder the packets, since this would result in a decreased performance. However, it is possible to activate this feature for a mix. In the case that a previous mix reorders packets, the above mentioned kdf cannot be used for the middle mixes. Therefore, another kdf has to be used.

In the current protocol, a user sends two session keys to every mix whenever a new channel has to be established. One key for the forward direction and a second key for the backward direction. We name the key used for the forward direction k_f and the key for the backward direction k_b.

To minimize the necessary changes in the software, we use the sent keys k_f and k_b to derive the new session keys. We propose the following function, thereby K represents the master key generated with the Tor authentication protocol. Since the required keys are shorter than the hash, only the first 128 bits are used for the keys:

$$k_{nf} = H\left(K||k_f\right)[0, 127]$$
$$k_{nb} = H\left(K||k_b\right)[0, 127]$$

As no sequence number is involved in this key derivation, the protocol is again vulnerable to replay attacks. However, the replays are limited for the time a middle mix knows the master key. The master key should be deleted immediately after a user disconnects from the cascade.

In order to counter a replay attack, one of the described replay protections can be used.

3.4   Estimation of the Additional Delay

As discussed earlier, our new approach should not significantly lower the performance of AN.ON. Due to the fact, that the DH key agreement is performed once when the user connects to the cascade, the additional delay is of minor importance. Thus, even a delay of several seconds is acceptable. The important question is, if the key derivation as well as the lookup of the master key have a significant influence on the build up time of a channel. To answer this question we performed a simulation of the additional operations necessary to calculate the session keys with help of the key derivation function.

In our simulation we measured the time to lookup a key from a hash table with a size of 2000000 entries. For each round we populated the hash table with 1000000 randomly selected (master) keys, 16 bytes for each key, which where stored under a 4 byte identifier in the hash table. The identifiers for the keys were also randomly chosen. The chosen values represent reasonable values regarding the current protocol. After we have populated the hash table, a random identifier was picked and the corresponding key was retrieved. This key was then hashed together with a 4 byte key representing k_f and k_b respectively. SHA-1 was the used hash function. For each round we picked 200000 different identifiers and generated two session keys each. For every new round we generated and populated a new hash table.

In total we simulated 100 rounds. On average the generation of the two session keys together with the master key lookup took $4.53\mu s \approx 0.004ms$ with an 0.95-confidence interval of $[4.506\mu s : 4.549\mu s]$. The simulation was performed on a Intel Core2Duo P8400 CPU. This simulation indicates that the additional overhead, introduced by our improvement, is negligible considering a total delay of over $100ms = 100000\mu s$.

4   Discussion

In Section 3.1 we discussed the information which is observable by a mix. The last mix can observe the final destination and the content of a message. This is the reason to require the unlinkability property for AN.ON. A possible, but risky solution would be to perform a DH exchange periodically with the last mix. The problem with this approach is, that the last mix could correlate different key exchanges with each other. For example, a user who terminates a "session" is likely to be the initiator of one of the next key exchanges, since he uses the same cascade for his subsequent channels. Hence, we do not propose this solution in the case of AN.ON. Contrary to AN.ON, in Tor the whole circuit is changed and therewith also the exit node. Hence, it is unlikely that one of the next circuits, observed by the exit node, is originated by a user who recently has closed a circuit and therefore the linkability of requests is only limited to a short period in the case of Tor.

One problem arises when mixes collude, namely the last mix with its predecessor. Due to our approach the collaboration of these two mixes can lead to the identification of a single user with all of his connections from his current session. This happens for example, if he sends an identifying information in one of his channels. This could not even link every visit from the current session to him, but also his previous sessions. This is the case if he visits regularly unpopular sides in a single session. This risk is not present in the current approach. However, also in the current version of AN.ON a collaboration of two mixes, namely the first and the last mix, is critical and leads to a deanonymization of a user. In this case, a user must not even send an identifying information over one of his channels. To mount the attack the first and the last mix have to introduce artificial timing or burst patterns to correlate in and outgoing packets. The feasibility of such a traffic conformation attack was shown in [10] in the case of Tor. Thus, neither our nor AN.ON's current version can protect against a collaboration of two mixes.

Due to the fact that no DH key exchange is performed between the user and the last mix of a cascade, the perfect forward secrecy does not hold for the last mix and therewith is also limited for the whole cascade. The interesting question is, in which situations AN.ON can provide anonymity even if the used private encryption keys has been compromised at some time after the connection took place. In the original version of AN.ON, an attacker who recorded every packet on the first mix can deanonymize every user under the compromised private key assumption. The attacker sees the IP address of all users and since he knows every private key, he can decrypt all session keys and subsequently all messages of all users that have once passed the mix during the usage period of the compromised key. Hence, the attacker can see the final destination of the messages as well as the IP address of the sender. Therewith he has revoked the relationship anonymity of all users.

With the introduced approach, the attacker needs to control the first $n - 1$ mixes to mount the attack. Hereby n is the length of the cascade. If an attacker wants to mount the attack, he has to control the first $n - 1$ mixes while the connection takes place. In this case he can easily correlate the user's messages with the outgoing messages at the $n - 1$th mix by decrypting every intermediate message. The reason that the attacker does not need to control the last mix, is due to the assumption that the attacker will eventually compromise the private encryption key of the last mix. Therefore, he can decrypt the messages of the last mix later and therewith retrieve the final destination.

Obviously, our improvement does not provide any additional protection if we consider a cascade with length 2. Here the attacker only needs to control the first mix to mount the attack.

However, an attacker who has the power to control $n - 1$ mixes in a cascade when the connection takes place can usually mount simpler attacks requiring less resources to deanonymize users.

Another problem occurs if the last middle mix gets the private key of the last mix. In this case the middle mix has the possibility to link different sessions to a user if the user transmits identifying information over one of his channels. Thus, the middle mix can mount the same attacks a last mix could mount due to the introduced linkability. However, the middle mix cannot directly identify every user under the compromised key assumption. Thus, the impact of the attack is less.

5   Conclusion

In our paper we argue that the lack of perfect forward secrecy exposes all users of AN.ON to a great risk if an attacker compromises the long term keys of the mixes. By introducing perfect forward secrecy in AN.ON up to the middle mixes, this thread is diminished. To perform a DH key exchange for every channel and thereby keeping the unlinkability property of AN.ON is unfortunately not possible due to practical reasons. The additional delay is simply to high.

By allowing middle mixe to link different channels to the same originator we do a trade-off between different risks and their impacts. While all users are affected by compromised keys in the original version, even though they did everything correct, only a couple of users are affected with our improved version. Therefore we claim that the linkability of channels regarding middle mixes reduces the overall risk.

Additionally, our approach does not introduce a significant performance decrease. Our estimations indicate that the introduced delay is less than 0.01 %. Since a mix can predict the session keys due to the knowledge of the master key, it might be possible to eliminate several asymmetrical cryptographic decryption operations. Thus, it is even possible that our approach decrease the overall build up time of a channel.

With our improvement an attacker cannot deanonymize earlier connections of all users by compromising the long term private keys of the mixes. Thereby, we improve the security of AN.ON.

References

1. Dingledine, R., Mathewson, N., Syverson, P.F.: Tor: The second-generation onion router. In: USENIX Security Symposium, USENIX, pp. 303–320 (2004)
2. Berthold, O., Federrath, H., Köpsell, S.: Web MIXes: A system for anonymous and unobservable Internet access. In: Federrath, H. (ed.) Designing Privacy Enhancing Technologies. LNCS, vol. 2009, pp. 115–129. Springer, Heidelberg (2001)
3. Mao, W.: Modern Cryptography: Theory and Practice. Prentice Hall Professional Technical Reference (2003)
4. Goldberg, I.: On the security of the Tor authentication protocol. In: Danezis, G., Golle, P. (eds.) PET 2006. LNCS, vol. 4258, pp. 316–331. Springer, Heidelberg (2006)
5. Köpsell, S.: Low latency anonymous communication - how long are users willing to wait? In: Müller, G. (ed.) ETRICS 2006. LNCS, vol. 3995, pp. 221–237. Springer, Heidelberg (2006)
6. Westermann, B., Wendolsky, R., Pimenidis, L., Kesdogan, D.: Cryptographic protocol analysis of an.on. In: Proceedings of the 14th International Conference of Financial Cryptography and Data Security, Tenerife, Spain (2010)
7. Dingledine, R., Mathewson, N.: Tor protocol specification (visited Feburary 3, 2010)
8. ISO/IEC 18033-2: 2006: Information technology – Security techniques – Encryption algorithms – Part 2: Asymmetric ciphers. ISO, Geneva, Switzerland (2006)
9. Köpsell, S.: Vergleich der Verfahren zur Verhinderung von Replay-angriffen der Anonymisierungsdienste AN.ON und Tor. In: Dittmann, J. (ed.) Sicherheit. LNI, vol. 77, pp. 183–187. GI (2006)
10. Øverlier, L., Syverson, P.: Locating hidden servers. In: Proceedings of the 2006 IEEE Symposium on Security and Privacy. IEEE CS, Los Alamitos (2006)

Mobility-Aware Drop Precedence Scheme in DiffServ-Enabled Mobile Network Systems

Bongkyo Moon

Dongguk Univ-Seoul, Dept. of Computer Science and Engineering,
Seoul 100-715, South Korea
bkmoon@dongguk.edu

Abstract. In DiffServ-enabled mobile network systems, TCP sender may time out if packet loss occurs at handover event. Thus, TCP window size may be reduced temporarily due to packet loss. Under overloaded traffic situation in local IP-managed network, furthermore, the sending rate of TCP packets that a mobile host generates right after handover may not be enough to keep its contract service rate on SLA (service level agreement). Therefore, giving temporal priority to the packets in a handover flow can compensate for the reduction of packet sending rate after handover. In this paper, we propose a mobility-aware drop precedence scheme in order to alleviate the performance loss from temporal disconnection or the reduction of sending rate.

1 Introduction

In differentiated services (DiffServ) model, the RED (random early detection) mechanism is typically capable of dividing the available bandwidth fairly among TCP data flows that belong to the same AF (assured forwarding) PHB (per-hop behavior) class, as packet loss automatically leads to a reduction of the packet sending rate in an TCP flow. In a RIO(RED with In and Out) queue, the packets of a flow are marked IN if the temporal sending rate at packet arrival is within the contract profile of the flow. Otherwise, the packets are marked OUT. The RIO mechanism actually starts to drop incoming OUT packets randomly with a certain probability in order to inform TCP sources of congestion after the average queue length of the buffer reaches the lower minimum threshold. It also starts to probabilistically drop IN packets when the average queue length exceeds the upper minimum threshold. However, due to the sawtooth variation of the TCP window, a flow has to transmit a certain amount of OUT packets in order to realize its reservation. In general, a connection with a larger reservation has a larger window, and hereby, it is obliged to transmit more OUT packets. Hence, it may not be easy to realize the reservation since OUT packets are very likely to be dropped. Until now, several studies have been done in order to solve this problem [1][2][3][4].

In a DiffServ-enabled mobile network system, meanwhile, the TCP sender may time out if packet loss occurs at handover event. Thus, TCP window size

F.A. Aagesen and S.J. Knapskog (Eds.): EUNICE 2010, LNCS 6164, pp. 143–154, 2010.

may be reduced temporarily due to the packet loss. Furthermore, if the managed IP network in the wireless mobile network architecture is overloaded and congested right after handover event, most of the low priority packets may be dropped in the same manner as the ordinary packets that belong to the same AF PHB class. Hereby, the packet-sending rate of TCP flow may be reduced again. Moreover, the fairness issue of the RED (random early detection) queue may not be temporarily kept, and the flow may fail to realize its reservation. In particular, in the case of streaming audio or video, service disruption might occur. Consequently, the sending rate of TCP packets that a MS (mobile station) generates right after handover may not be enough to keep its contract service rate on SLA (service level agreement)[10]. Therefore, giving temporal priority to the packets of a handover flow can compensate for the reduction of packet sending rate in the managed IP network.

In this paper, we propose a mobility-aware drop precedence scheme in order to alleviate the performance loss from temporal disconnection or the reduction of sending rate. First, the DiffServ-based mobile network model is presented as an example. Second, a packet classification model for the mobility-aware drop precedence scheme is explained, and then AF PHB buffer with Markov chains is modelled. Finally the performance measures and results for the proposed scheme are presented.

2   Mobile Network System Model

2.1   An Example of Mobile Network System

In this section, we will consider the mobile WiMAX network as an example of the mobile network model. There are actually four main components in the mobile WiMAX network architecture: mobile station (MS), access service network - gateway (ASN-GW), connectivity service network (CSN), and core network. The Fig. 1 presents an example of the network architecture for Mobile WiMAX system. In this figure, the MS communicates with ASN-GW using wireless access technology, and also provides MAC processing functions, Mobile IP, authentication, packet retransmission, and handover. The ASN-GW also provides wireless interfaces for the MS, and takes care of wireless resource management, QoS support, and handover control. Hence, MSs can move efficiently between ASN-GWs, provide smooth ASN-GWs transitions with minimal loss of packet data units (PDUs), and also provide fast ASN-GWs transition with guaranteed QoS. It means that when the MS roams from one ASN-GW to another, IP stack working on the top of L2 layer stays unaware on the roaming of MS. The CSN actually plays a key-role in IP-based data services including IP packet routing, security, QoS and handover control. In order to provide mobility for MS, the CSN supports handover between the ASN-GWs while Mobile IP provides handover between the CSNs[5][6].

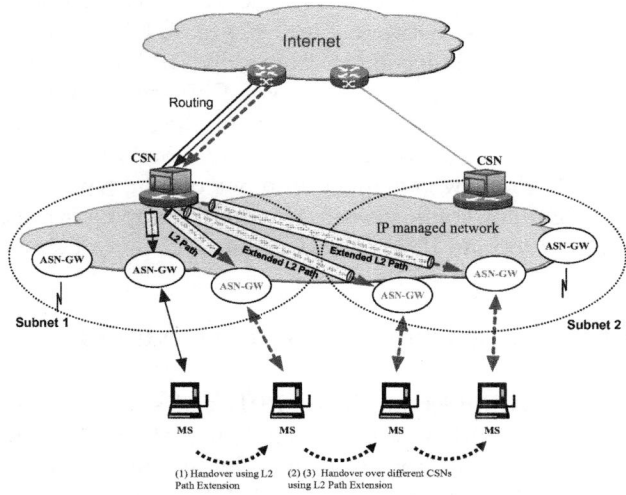

Fig. 1. Mobile WiMAX Network Diagram with Handover Scenario

2.2 DiffServ Approach in Mobile Network System

Fig. 2 shows a logical view of the DiffServ-based end-to-end IP QoS approach
in Mobile WiMAX network architecture. In this figure, IPv6 packet is firstly
encapsulated with 802.16 MAC header between MS and ASN-GW and then the
encapsulation header is replaced with Ethernet header between ASN-GW and
CSN. Meanwhile, an appropriate link path is set up through tunnel creation in
order to meet the mobility requirements as shown in Fig. 1. Furthermore, the

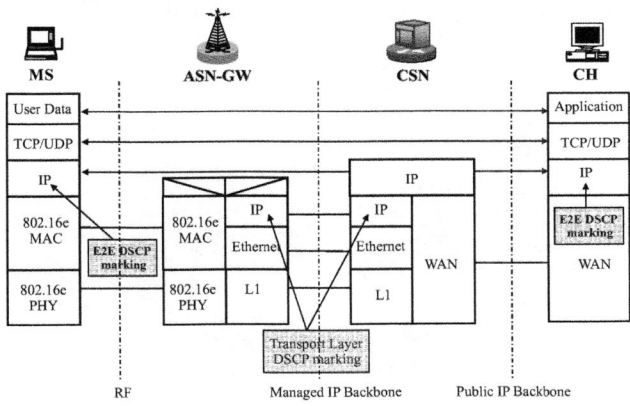

Fig. 2. DiffServ-based IP QoS approach in Mobile WiMAX Network Architecture

mobility support by tunnel creation can make route path controlled over physically heterogeneous access network. Since the underlying routing path control
due to mobility is actually independent from global IP layer, both IPv4 and IPv6
protocols can be used at the same time. For DiffServ in Mobile WiMAX network system, meanwhile, a new application flow which arrives in IP layer will be
firstly parsed and classified according to the definition in DSCP (Diff-Serv Code
Points). Eventually the flow is mapped into one of four types of services (UGS,
rtPS, nrtPS or BE). For traffic classification and mapping strategies, AF rules
are actually defined to map IP layer service into MAC layer services[7][8][9].

3   Mobility-Aware Drop Precedence Model

Fig. 3 shows a simple DiffServ with AF PHB nodes and several traffic source
groups in a mobile network system. In this figure, the CSN becomes a boundary node. It is essential to keep drop precedence policy of DiffServ model in
L2 extension tunnel between CSN and ASN-GW in the managed IP network
of Fig. 1. Here, two phase DiffServ mechanism for end-to-end QoS approach
should be deployed. That is, the first phase should work between the first-hop
router in local cloud, which the ASN-GW is directly connected to, and the
boundary node (CSN). The second phase should work between the CSN and the
CH(corresponding host) over the core network. In the second phase, actually the
CSN becomes the global first-hop router.

We now assume that a MS is able to recognize the handover events and thus
mark the DSCPs of packets it creates during handover. That is, the packets
that a MS generates, during a handover period, can be marked by new mobility-
aware (MA) tags with higher priority than the colours in three-drop precedence.
In addition, when traffic exits from the managed IP network via a CSN, the packets marked with MA tag can be mapped to the normal three-drop precedence.
Meanwhile, when a MS is a receiver, it informs the CSN of its location during handover in a mobile network system. After that, the CSN can re-mark the

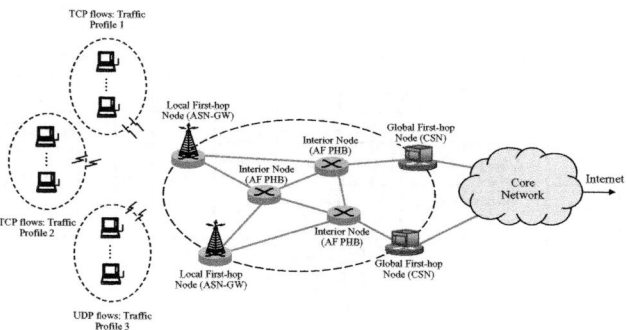

Fig. 3. Simple DiffServ Scenario with AF PHB nodes in Mobile Network System

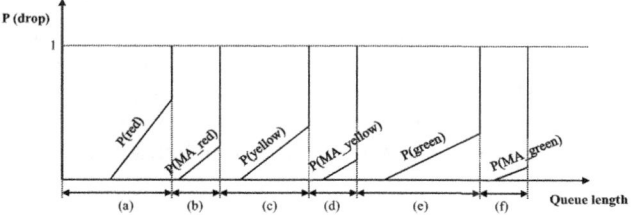

Fig. 4. Drop Probabilities of Mobility-aware Drop Precedence Scheme

incoming packets from the core network with MA tags in order to give priority to the handover flow.

In an MA drop precedence in managed IP network, there may be six possible subscription levels: 1) Only *red* packets are dropped [Fig. 4(a)]; 2) every *red* packet is dropped, and some of MA_*red* [Fig. 4(b)] packets are dropped; 3) every *red* and MA_*red* packet is dropped, and some *yellow* [Fig. 4(c)] packets are dropped; 4) every *red*, MA_*red* and *yellow* packet is dropped, and some of MA_*yellow* [Fig. 4(d)] packets are dropped; 5) every *red* ,MA_*red*, *yellow* and MA_*yellow* packet is dropped, and some *green* [Fig. 4(e)] packets are dropped; and 6) every *red*, MA_*red*, *yellow*, MA_*yellow* and *green* packet is dropped, and some MA_*green* [Fig. 4(f)] packets are dropped.

4   Modelling of MA-Drop Precedence Scheme

For simplicity of analytic model, one AF PHB node in a single link is considered. This model is used to compute steady state throughputs of the flows in the AF PHB node. The results can be used to analyze fairness and priority issues of the MA marking scheme. That is, the fairness issue includes how the AF link bandwidth is fairly shared between the different flows together with the MA marking in a mobile network system. The priority issue includes how the MA marking scheme can compensate mobility-sensitive TCP sources for the sending rate reduction that may be caused by handoff events. The analytic models in this section heavily rely on previous works [13][14].

4.1   Packet Classifier Model

The DSCP field provides the necessary information for classifying the packets at a DiffServ boundary node. In this section, the packet classifier is modelled with a flow conditioner mechanism which splits a flow into several subflows according to the levels of drop precedence. The subflows are marked with different levels of drop precedence depending on the traffic intensity characteristics and the handover rates of the flow.

Let $L = \{1, ..., N_l\}$ denote the set of TCP flow groups. Each flow group consists of N_l identical flows. The sending rate and traffic profile associated with the flow together with the flow conditioner mechanism determine how a flow is split into different drop precedence levels. Let x_l denote the sending rate of the flow associated with group l. Assume that flows in group l are marked with the same AF class, and thus they are buffered in a queue in the AF PHB node. The flow conditioner splits the flow intensity into six drop precedence levels with the characteristic function α_h, where α_h is 1 if handover occurs, otherwise it becomes 0.

$$
\lambda_l^{(1)} = \begin{cases} \alpha_h x_l & , 0 \leq x_l \leq v_l^1 \\ \alpha_h v_l^1 & , v_l^1 \leq x_l \leq v_l^2 \\ \alpha_h v_l^1 & , x_l \leq v_l^2 \end{cases} \tag{1}
$$

$$
\lambda_l^{(2)} = \begin{cases} (1-\alpha_h) x_l & , 0 \leq x_l \leq v_l^1 \\ (1-\alpha_h) v_l^1 & , v_l^1 \leq x_l \leq v_l^2 \\ (1-\alpha_h) v_l^1 & , x_l \leq v_l^2 \end{cases} \tag{2}
$$

$$
\lambda_l^{(3)} = \begin{cases} 0 & , 0 \leq x_l \leq v_l^1 \\ \alpha_h(x_l - v_l^1) & , v_l^1 \leq x_l \leq v_l^2 \\ \alpha_h(v_l^2 - v_l^1) & , x_l \leq v_l^2 \end{cases} \tag{3}
$$

$$
\lambda_l^{(4)} = \begin{cases} 0 & , 0 \leq x_l \leq v_l^1 \\ (1-\alpha_h)(x_l - v_l^1) & , v_l^1 \leq x_l \leq v_l^2 \\ (1-\alpha_h)(v_l^2 - v_l^1) & , x_l \leq v_l^2 \end{cases} \tag{4}
$$

$$
\lambda_l^{(5)} = \begin{cases} 0 & , 0 \leq x_l \leq v_l^1 \\ 0 & , v_l^1 \leq x_l \leq v_l^2 \\ \alpha_h(x_l - v_l^2) & , x_l \leq v_l^2 \end{cases} \tag{5}
$$

$$
\lambda_l^{(6)} = \begin{cases} 0 & , 0 \leq x_l \leq v_l^1 \\ 0 & , v_l^1 \leq x_l \leq v_l^2 \\ (1-\alpha_h)(x_l - v_l^2) & , x_l \leq v_l^2 \end{cases} \tag{6}
$$

The traffic profile of a flow is thus determined by two traffic intensity limits v_l^1 and v_l^2, $0 < v_l^1 < v_l^2$. Let $\lambda_l^{(i)}$ denote the traffic intensity of the subflow that is marked with drop precedence i within flow l. As a result, the characteristic function α_h and the traffic profile of the flow define $\lambda_l^{(i)}$ by equations (1) - (6). The traffic conditioner definitions in equations (1) - (6) split a flow into drop precedence levels such that the perceptual distribution of a flow into drop precedence levels can be expressed by

$$
\delta_l^{(i)} = \frac{\lambda_l^{(i)}}{x_l}, \forall i \in I \tag{7}
$$

4.2   AF PHB Buffer Model

Fig. 5 shows an example of an AF PHB node with buffers in a 802.16e network. In this figure, $w1, w2, w3$ and $w4$ are weight factors for packet scheduling according to AF service classes. Each class buffer (FIFO queue) has its own threshold values according to drop precedence levels. In this model, however, only one class buffer is modelled in order to evaluate the effect of the number of threshold values in the proposed scheme.

For simplicity of modelling, it is assumed that packets arrive according to a Poisson process and the packet service times are exponentially distributed. Even though this assumption is not suitable for exact modelling of packet arrival rate in the Internet, a Markov chain model can be used just for evaluating the performance of the proposed scheme compared to the existing three drop precedence scheme. Consequently, one class buffer can be modelled as a queueing system with a Markov model, and hereby, packet loss probabilities of subflows for each drop precedence aggregate can be computed.

Denote the acceptance thresholds for drop precedence level i by K_i, $K_1 = K$, in which K is the size of the buffer. In the class 1 of the Fig. 5, therefore, K_1 is for MA_green, K_2 is for green, K_3 is for MA_yellow, K_4 is for yellow, K_5 is for MA_red, and K_6 is for red. Let $\lambda(i)$ denote the packet arrival rate into drop precedence class i and μ^{-1} the mean value of the packet service time. Defining the cumulative sum of arrival intensities of drop precedence levels accepted into the buffer as $\lambda_i = \sum_{k=1}^{i} \lambda(k)$, the buffer can be modelled as an $M/M/1/K$ queue on the state space $\{m \mid 0 \leq m \leq K\}$, with state dependent arrival intensities. Thus, the stationary distribution of buffer occupancy can be solved from the balance equations of the system shown in (8). Let π_m denote the equilibrium probability for state m. The balance equations for the one buffer behavior can be written as

$$
\begin{cases}
\lambda_6 \pi_0 = \mu \pi_1 & , m = 0 \\
(\lambda_6 + \mu)\pi_1 = \lambda_6 \pi_0 + \mu \pi_2 & , m = 1 \\
\quad \cdots & \quad \cdots \\
(\lambda_5 + \mu)\pi_{K_6} = \lambda_6 \pi_{K_6-1} + \mu \pi_{K_6+1} & , m = K_6 \\
\quad \cdots & \quad \cdots \\
(\lambda_4 + \mu)\pi_{K_5} = \lambda_5 \pi_{K_5-1} + \mu \pi_{K_5+1} & , m = K_5 \\
\quad \cdots & \quad \cdots \\
(\lambda_3 + \mu)\pi_{K_4} = \lambda_4 \pi_{K_4-1} + \mu \pi_{K_4+1} & , m = K_4 \\
\quad \cdots & \quad \cdots \\
(\lambda_2 + \mu)\pi_{K_3} = \lambda_3 \pi_{K_3-1} + \mu \pi_{K_3+1} & , m = K_3 \\
\quad \cdots & \quad \cdots \\
(\lambda_1 + \mu)\pi_{K_2} = \lambda_2 \pi_{K_2-1} + \mu \pi_{K_2+1} & , m = K_2 \\
\quad \cdots & \quad \cdots \\
(\lambda_1 + \mu)\pi_{K-1} = \lambda_1 \pi_{K-2} + \mu \pi_K & , m = K - 1 \\
\mu \pi_K = \lambda_1 \pi_{K-1} & , m = K
\end{cases}
\tag{8}
$$

By solving the linear system of equations in (8), π_m can be defined for $m = 1, \ldots, K$ as a function of π_0,

$$
\pi_m =
\begin{cases}
\left(\frac{\lambda_6}{\mu}\right)^m \pi_0 & , 1 \leq m < K_6 \\
\left(\frac{\lambda_{i-1}}{\mu}\right)^{m-K_i} \left(\prod_{j=i}^{5} \left(\frac{\lambda_j}{\mu}\right)^{K_j - K_{j+1}}\right) \left(\frac{\lambda_6}{\mu}\right)^m \pi_0 & \\
\quad , K_i \leq m < K_{i-1} \quad (i = 2, 3, 4, 5, 6 \quad and \quad K_1 = K)
\end{cases}
\tag{9}
$$

From the normalization condition $\sum_{m=0}^{K} \pi_m = 1$, the equilibrium probability π_0 for the empty buffer state can be solved. As the buffer state probabilities π_m are known, the packet drop probability $P(i)$ for packets aggregated to drop

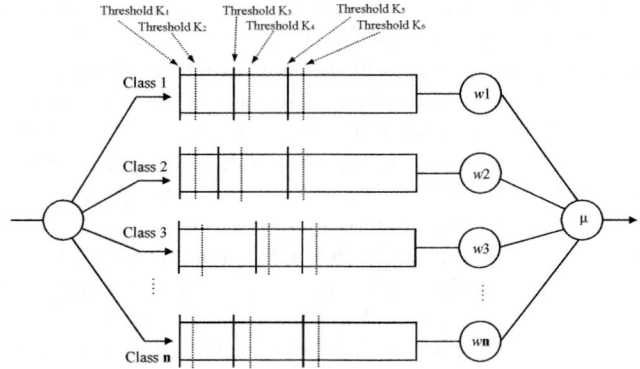

Fig. 5. Example of an AF PHB Node with n Buffers in a Mobile Network System

precedence i can be solved by summing up the probabilities of states $m \geq K_i$,

$$P(i) = \sum_{m=K_i}^{K} \pi_m. \tag{10}$$

5   Performance Measures

From the definition in equation (7), the total traffic intensity (packet arrival rate) of drop precedence level i in a buffer with the same AF class can be obtained as

$$\lambda(i) = \sum_l n_l \delta_l^{(i)} x_l. \tag{11}$$

Based on the packet drop probabilities $P(i)$ for drop precedence level i from equation (10), the packet loss probability of a flow can be computed. The packet loss probability q_l of a flow is defined as

$$q_l = \sum_{i=1}^{6} \delta_l^{(i)} P(i). \tag{12}$$

It is assumed that TCP congestion control follows the differential equations for aggregates of TCP flows as described in [15]. TCP throughput is thus proportional to average sending rate x_l that depends on the round trip time (RTT) and the packet loss probability q_l of a flow where

$$x_l = \frac{1}{RTT} \sqrt{\frac{2(1-q_l)}{q_l}}, \quad l \in L_{TCP}. \tag{13}$$

For each TCP group, equation (12) relates the TCP average sending rate with the packet drop probability via the AF PHB buffer. As the packet drop probability depends on the flow group l, the equation must be formulated for each TCP traffic profile group separately.

6   Results and Discussions

In this section, one AF PHB class buffer model is used in evaluating how packet loss probability and average sending rate between the normal TCP flow and handover flow is achieved with the AF PHB mechanism. The buffer size is set to $K = 39$ and the drop precedence limits for the buffer are $K_3 = 13$, $K_2 = 26$ and $K_1 = 39$. The model was evaluated with the buffer service intensities of both $\mu = 1.0$ and $\mu = 2.0$. The compensation for handover flow packets is implemented using reservation trunk . The buffer reservation rates (R) for MA-tagged packets in the buffer model are fixed as both 30% and 60%.

Fig. 6 shows the packet loss probability of handover flow in the AF PHB node as a function of packet sending rate with the buffer service intensity of $\mu = 1.0$. In this figure, the MA-drop precedence scheme with reservation rate (e.g. 30% and 60%) for handover flow has a smaller packet loss than the three-drop precedence scheme in the case where the load ratio (packet sending rate/buffer service intensity $\mu = 1.0$) is less than 1.1. This is because the three-drop precedence does not provide any priority mechanism for handover flow. However, when the load ratio becomes considerably greater than 1.0, there is little difference in packet loss probability. This means that MA-drop precedence scheme does not get better results when network congestion level is high.

Fig. 7 shows the packet loss probability of handover flow for a MA-drop precedence scheme when the buffer service intensity μ is 1.0 and 2.0. In this figure, when μ is 2.0, the MA-drop precedence scheme has much lower packet loss probability than when μ is 1.0. In the cases when μ is 2.0, and the load ratio approaches 1.0 (that is, when sending rate approaches 2.0), however, packet loss

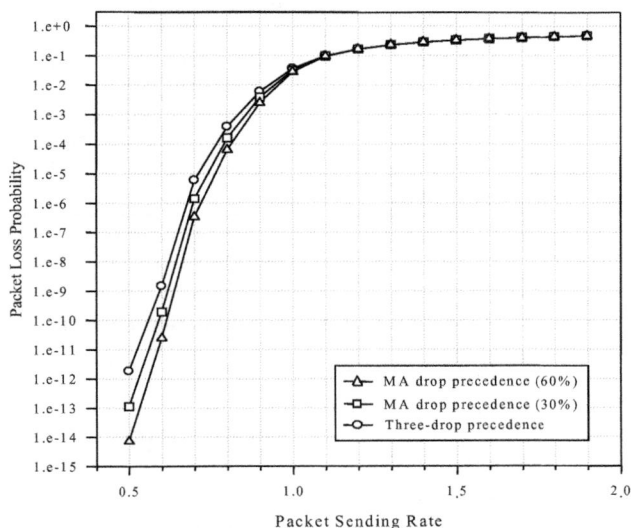

Fig. 6. Packet Loss Probability of Handover Flow in AF PHB Node ($\mu = 1.0$)

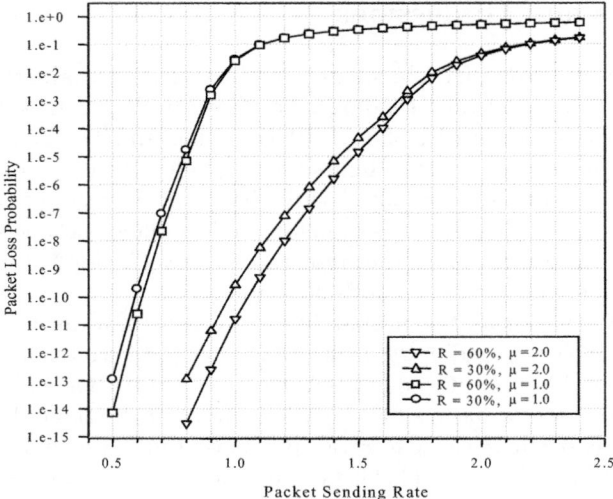

Fig. 7. Packet Loss Probability of Handover Flow in the AF PHB Node

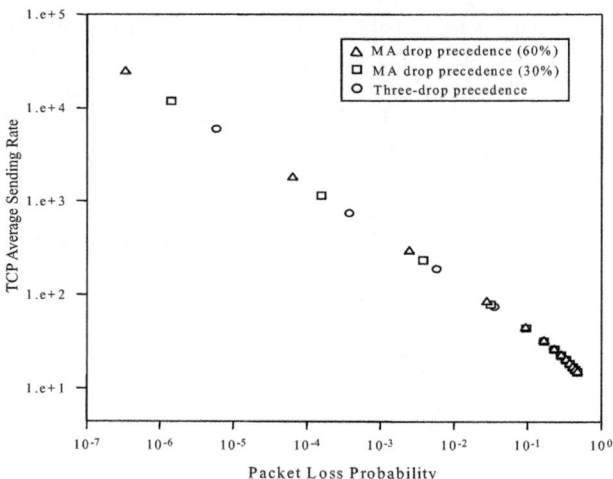

Fig. 8. Average Sending Rate of TCP Flow as a Function of Packet Loss Probability with $\mu = 1.0$ and $RTT = 0.1$

probabilities become nearly equal to each other although reservation rates for handover flow are both 30% and 60%.

Fig. 8 shows the average sending rate (proportional to throughput rate) of a TCP flow as a function of packet loss probability with buffer service intensity $\mu = 1.0$ and the round trip time $RTT = 0.1$. Buffer reservation rates (R) of both 30% and 60% are used for handover flow packets. In this figure, MA-drop

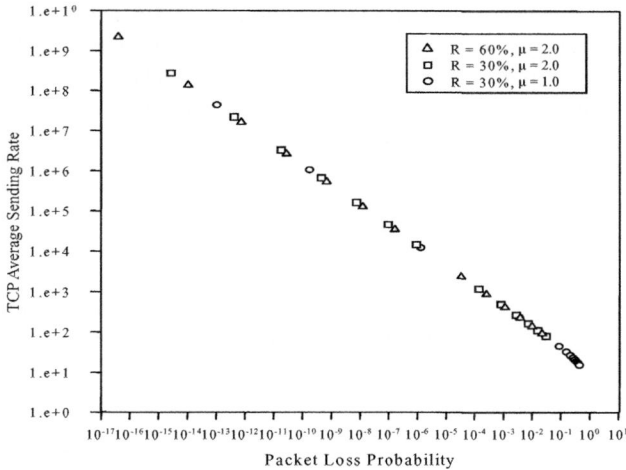

Fig. 9. Average Sending Rate of TCP Flow as a Function of Packet Loss Probability with $\mu = 1.0$ and $\mu = 2.0$, and $RTT = 0.1$

precedence scheme for TCP handover flows can achieve much higher throughput rate than existing three-drop precedence scheme as long as packet loss probability is less than or equal to 0.1. Moreover, the reservation rate of 60% gives higher throughput rates than that of 30%. This result shows that the proposed scheme can compensate for the reduction of TCP window size due to frequent host mobility. However, a high reservation rate can make some buffers unused, and thus wasted, in the case where handover rate is low.

Fig. 9 shows that the reservation rate of 60% can, when μ is 2.0, give much higher throughput rates than that of 30%. On the other hand, a reservation rate of 30% with $\mu = 1.0$ gives lower throughput rates compared to the case with $\mu = 2.0$ and, in addition, most of them fall together in the area less than 50 in sending rate and greater than or equal to 0.1 in packet loss probability.

Consequently, when the network load is low, a TCP flow could achieve much higher throughput rate (proportional to average sending rate) than the contract rate. Hence, more packets in the flow are marked with a high drop precedence. Under more congested network environments, meanwhile, the sending rate of TCP flow tends to reduced less than the contract rate due to more packet loss. Hereby, the more packets in the flow can be marked into the low drop precedence levels. Therefore, when network congestion level is considerably low or high, packet differentiation in drop precedence is difficult to achieve.

7   Conclusions

In this paper, a mobility-aware drop precedence scheme is proposed in order to alleviate the reduction of sending rate in the managed IP network of mobile network systems. The proposed scheme could compensate for the reduction of TCP window size due to frequent mobility under wireless mobile environment. However, a high buffer reservation rate for the packets of handover flow can make some buffers unused and thus wasted in the case where handover rate is low. The proposed scheme is focused on the DiffServ-aware managed IP network of mobile network systems.

Acknowledgements

This work was supported by the Korea Research Foundation Grant funded by the Korean Government (MOEHRD, Basic Research Promotion Fund, KRF-2006-331-D00357).

References

1. Feng, W., Kandlur, D., Saha, D., Shin, K.: Understanding TCP Dynamics in a Differentiated Services Internet. IEEE/ACM Transactions on Networking (April 1998)
2. Yeom, I., Narasimha Reddy, A.L.: Modeling TCP Behavior in a Differentiated Services Network. IEEE/ACM Transactions on Networking 1(9) (Feburary 2001)
3. Clark, D., Fang, W.: Explicit Allocation of Best Effort Packet Delivery Service. IEEE/ACM Transactions on Networking (August 1998)
4. Heinanen, J., Finland, T., Guerin, R.: A Two Rate Three Color Marker. Internet Draft (May 1999)
5. Andrews, J.G., Ghosh, A., Muhamed, R.: Fundamentals of WiMAX: Understanding Broadband Wireless Networking. Prentice Hall, Englewood Cliffs (2007)
6. WiMAX Forum, WiMAX End-to-End Network Systems Architecture (Stage 3), Network WG, January 11 (2008),
   http://www.wimaxforum.org/technology/documents
7. Chen, J., Jiao, W., Guo, Q.: An Integrated QoS Control Architecture for IEEE 802.16 Broadband Wireless Access Systems. Bell Labs Research China (2005)
8. Mai, Y.-T., Yang, C.-C., Lin, Y.-H.: Cross-Layer QoS Framework in the IEEE 802.16 Network. In: ICACT 2007, Feburary 2007, vol. 3, pp. 2090–2095 (2007)
9. Jiao, W., Chen, J., Liu, F.: Provisioning End-to-End QoS Under IMS Over a WiMAX Architecture. Bell Labs Technical Journal 12(1), 115–121 (2007)
10. Blake, S., Black, D., Carlson, M., Davies, E., Wang, Z., Weiss, W.: An Architecture for Differentiated Services. IETF RFC 2475 (December 1998)
11. Madanapalli, S., Patil, B., Nordmark, E., Choi, J., Park, S.: Transmission of IPv6 packets over 802.16's IPv6 Convergence Sublayer. IETF Internet-Draft (June 2006)
12. Jeon, H., Jeong, S., Riegel, M.: Transmission of IPv6 packets over Ethernet CS over IEEE 802.16 network. IETF Internet-Draft (September 2006)
13. Nyberg, E., Aalto, S., Virtamo, J.: Relating flow level requirements to DiffServ packet level mechanisms. COST279 TD (01) 04 (October 2001)
14. Kuumola, E.: Analytical Model of AF PHB Node in DiffServ Network. Mat-2.108, Networking Laboratory, Helsinki University of Technology (October 2001)
15. Kelly, F.: Mathematical Modelling of the Internet. In: Proceedings of Fourth International Congress on Industrial and Applied Mathematics, pp. 105–116 (1999)

Theoretical Analysis of an Ideal Startup Scheme in Multihomed SCTP

Johan Eklund[1], Karl-Johan Grinnemo[2], and Anna Brunstrom[1]

[1] Department of Computer Science, Karlstad University, Sweden
[2] School of Information and Communication Technology, KTH, Sweden

Abstract. SCTP congestion control includes the slow-start mechanism to probe the network for available bandwidth. In case of a path switch in a multihomed association, this mechanism may cause a sudden drop in throughput and increased message delays. By estimating the available bandwidth on the alternate path it is possible to utilize a more efficient startup scheme. In this paper, we analytically compare and quantify the degrading impact of slow start in relation to an ideal startup scheme. We consider three different scenarios where a path switch could occur. Further, we identify relevant traffic for these scenarios. Our results point out that the most prominent performance gain is seen for applications generating high traffic loads, like video conferencing. For this traffic, we have seen reductions in transfer time of more than 75% by an ideal startup scheme. Moreover, the results show an increasing impact of an improved startup mechanism with increasing RTTs.

1   Introduction

New applications with a diversity of requirements regarding timeliness and robustness continuously appear in computer networks. This is challenging for the traditional communication protocols and the Stream Control Transmission Protocol (SCTP) [20], with its multihoming ability has become an attractive alternative to address these challenges. The protocol is not a totally new invention; it has in fact inherited some crucial features from the most commonly used transport protocol in computer networks, TCP [17]. One of these features is the congestion control, particularly the slow-start mechanism, where the initial congestion window is set to a small value, successively increased as acknowledgements of successful transfers are returned. This startup mechanism was introduced to preserve network stability, and to respect fairness between competing connections over the same path. On the other hand, if there is spare bandwidth in the network, slow start increases the time it takes before the traffic may utilize the bandwidth.

The multihoming feature of SCTP implies that a connection, a so-called association, may consist of more than one path, all available from the start of a session. This mechanism was included in the protocol to increase robustness in case of failure, as the transfer may proceed utilizing one of the alternate paths,

F.A. Aagesen and S.J. Knapskog (Eds.): EUNICE 2010, LNCS 6164, pp. 155–166, 2010.

without having to establish a new association. In this case, the path switch is called a failover.

After standardization, new mechanisms have been added to SCTP. One of these additional mechanisms is the so-called Dynamic Address Reconfiguration, (DAR) [21] mechanism, whereby IP addresses may be added or deleted to/from an existing association dynamically. This feature was added to enable maintenance of a network interface without disturbing the association, a so-called hotswap. Furthermore, this feature enables SCTP to be an alternative for seamless mobility. Particularly, the DAR mechanism can be used to support handover for wirelessly connected terminals moving between different networks.

One aspect that may have an impact in all the above mentioned scenarios, failover, hotswap and handover between different paths, is that the transfer has to go through the slow-start phase every time the association is switched to a new path. The small initial congestion window on the new path may result in a sudden reduction of throughput and increased message transmission delays, which may severely impede the current transfer.

Today, several techniques exist to estimate the available bandwidth of a network path. These estimates are based either on data from traffic sent over the network, for example, the active packet-pair technique [16], or on passive network monitoring [8]. In a multihomed association, this estimate could be performed before an expected path switch, since alternate paths are available prior to the path switch.

Some research on improving the startup of the transfer has been conducted. In a radical proposal by Liu et al., they propose removal of the slow-start phase from the congestion control to let the transport layer start the transfer at whatever rate [15]. Another proposal to achieve an improved initial sending rate in controlled networks, called Quick-Start for TCP and IP, is standardized in RFC 4782 [5]. A proposal to increase the initial congestion window of TCP (and SCTP) has recently been proposed by Chu et al. [4]. Some studies have also been conducted to improve startup on the alternate path in multihomed SCTP. In a study by Zheng et al. [14], they estimate the bandwidth on the alternate path by utilizing a packet-pair technique [16]. Based on the estimate, they enlarge the initial congestion window of the alternate path at the time of handover. Further, they regard the latency of the new path in relation to the former path, to reduce the risk of packet reordering. Another study, proposing a handover scheme called ECHO, was conducted by Fitzpatrick et al. [11], where they aim at improving the handover scheme SIGMA [6] for traffic in a wireless environment. ECHO improves the quality of service by evaluating the Mean Opinion Score (MOS) [10] before handing over to the new network. All the above mentioned studies target mobile clients and Voice over IP (VoIP) traffic.

Fracchia et al. present a modification, WiSE [18], to make SCTP more suitable for wireless networks. WiSE aims at always choosing the best path for the transfer by estimating the bandwidth of all existing paths. Further, WiSE possesses a mechanism to utilize the bandwidth estimate to differentiate losses due to congestion from losses due to radio channel errors. It uses this mechanism to properly adjust the congestion window. An extension of WiSE is presented by Casetti et al., called AISLE [3], which aims to optimally distribute traffic across overlapping WLANs. Unlike WiSE, AISLE triggers a potential path switch on both timeouts and fast retransmits. Both these proposals target bulk transfer.

In this paper, we analytically examine the potential performance gains of using a bandwidth-aware startup scheme for multihomed SCTP. We do this by setting the initial congestion window on the alternate path to the available bandwidth. We consider three scenarios, where a path switch may occur. Further, we present relevant traffic patterns for these scenarios. The main contribution of the study is that we quantify the theoretically feasible performance gains of using a bandwidth-aware startup scheme. We do not consider a particular bandwidth estimation technique, but rather make the assumption that the estimated bandwidth is correct and stable. Although, this assumption does not match a real scenario, it allows us to analytically derive and quantify the upper bounds of the performance gains, in a range of relevant traffic scenarios.

Intuitively, applications with real-time requirements may be affected by an extra delay after a path switch. Our results indicate that the benefit of using an improved startup scheme is most considerable for this kind of applications. That is applications generating high traffic loads, especially in cases with long RTTs. For this traffic we have noticed improvements in transfer time by more than 75%. Furthermore, the results show that the traffic pattern affects the results in two ways; packets generated at high frequency may have to wait a long time to be transmitted, while large messages may increase the congestion window fast.

The rest of this paper is structured as follows. In Section 2, different use cases and traffic characteristics are discussed. Section 3 presents analytical results from comparing an ideal startup to slow start for different traffic scenarios. Finally, Section 4 ends the paper with a discussion and some conclusions.

2   Scenarios and Traffic Patterns

To evaluate the benefit of an ideal startup mechanism, we match the three path switch scenarios, failover, hotswap and handover, to relevant traffic types. To be able to perform an appropriate evaluation it is indeed important to extract the characteristics for the different traffic types.

2.1   Scenarios

The three scenarios where a path switch may occur have different characteristics.

Failover. This scenario occurs as a consequence of a path failure. In many cases, the failover is preceded by a failure detection period. During this period no data reaches the destination without retransmissions, which degrades performance and may build up a queue at the sender. In situations where the failure happens to the network interface or to the first or last hop link, the failure is usually detected immediately, and a path switch is not preceded by the aforementioned failure detection period.

Hotswap. This is a scenario where a network path is switched intentionally, due to maintenance or upgrading of a network component. This event is a planned switch, which is why it is not preceded by any failure detection period. In this scenario, the only throughput degradation is due to the slow-start mechanism after startup on the alternate path.

Handover. This scenario occurs in a mobile scenario, where a wirelessly connected terminal moves and switches from one network to another. In this scenario, the path switch is usually conducted as the performance, according to application specific aspects, on the new path, exceeds the performance of the current path. In the handover scenario, throughput degradation could occur before handover as well as during startup on the alternate path.

The failover and hotswap scenarios refer primarily to managed telecom networks carrying signaling traffic. Therefore, not all traffic patterns are applicable for all path-switch scenarios. The handover scenario is applicable for user data, while the other scenarios are more applicable for control traffic. A view of the connection between scenario and applicable traffic is shown in Table 1.

Table 1. Traffic types representative of different scenarios

	Failover	Hotswap	Handover
Traffic Type	signaling	signaling	real-time/bulk

2.2   Characteristics for Different Traffic Types

Signaling Traffic. The traffic generated by signaling applications normally consists of small messages with slightly varying size. The messages are generated at irregular intervals, usually in bursts [2,19]. Each signaling message carries its own piece of information, which is why subsequent messages are generally independent. The requirement on signaling traffic is that the messages should reach the destination within specific times, so both reliability and timing issues are relevant.

Real Time Traffic. Some major applications generating real-time traffic are VoIP and video conference applications. The crucial aspect of this type of traffic is timeliness, i.e., that data reaches the destination within specified time. However, robustness is less critical since a single lost data message only imposes marginal impact on user experience. Applications for VoIP traffic falls into two categories: those with silence detection and those without. If silence detection is used, the data to be transferred is generated only during so-called talk spurts, while applications without this feature generate traffic continuously. In this work, we focus on the latter case, which more or less generate traffic at a constant bit rate (CBR).

The traffic generated during a video conference is also usually transferred at a constant rate. Data from these applications, called frames, carry information blocks from the video. The frame size varies according to the amount of information sent. This traffic is more complex to model as the generated bit rate is variable (VBR).

Bulk Traffic. Bulk traffic is characterized by the availability of all data at session start. The data is sent on the link as soon as possible. Traditionally, timeliness is not an issue for bulk data. However, this has to some extent changed in recent years. For example, timeliness is an issue for interactive web transfers and streaming media. The traffic generated by these applications has characteristics in common with bulk traffic. Thus, timeliness can be important also for bulk-like transfers.

3   Analysis of an Ideal Startup Mechanism

In this section, we compare the impact of the traditional slow-start mechanism with that of an ideal one. For the ideal mechanism, the initial congestion window is set to a size that utilizes all available bandwidth.

3.1   Assumptions

To enable an analytical approach and a clear presentation, we have made a few simplifying assumptions:

- Lossless links are assumed. In fact, losses only occur after failure on the primary path.
- The available bandwidth is constant, and for non-bulk traffic bandwidth is not a limiting factor.
- A symmetric network is assumed, i e., all available paths have the same bandwidth.
- Transmission time for packets is negligible, and does not vary with packet size.
- Overhead for headers are not regarded in the calculations.
- When the slow-start mechanism is used, the initial congestion window is set to 4500 bytes[1]

These assumptions somewhat estrange the situation from a real scenario, but despite these assumptions, we believe the results will be relevant to identify the scenarios where a more efficient startup is most beneficial, and to point out the magnitude of the performance improvement.

3.2   Parameters and Metrics

As mentioned before, different applications generate different types of traffic which results in different message sizes generated at different frequency. Signaling traffic is in this study modeled as of 250 bytes packets, a plausible average size of a signaling message. Signaling traffic in the access network is usually not

---

[1] According to RFC 4960, the initial congestion window is MTU-dependent. To simplify the calculations we have chosen 4500 as the size of the initial congestion window, since 4500 bytes is an even multiple of the assumed path MTU (1500 bytes) and of the packet size used for signaling messages in our study (250 bytes).

very intense. In the core network, on the other hand, traffic from several signaling endpoints may be aggregated into a common association where the traffic intensity may be quite high. In this study, we model signaling traffic inter arrival times between 1 ms and 0.5 s to cover both situations. Since the aim of our study is to compare the slow-start mechanism to an ideal situation, the delay of an arriving message will depend on the queuing delay created by previously arrived messages. Thus, the quantity of sent data, and not the distribution of data between the different messages is the most important. Therefore, we simplify the modeling by approximating VBR traffic to CBR, using the average message size. In this study, we approximate the signaling traffic with CBR traffic with messages of uniform size.

Concerning VoIP traffic, the frequency and the size of the packets vary depending on codec used. One commonly used codec is G.711, which typically packetizes data into messages of 80 bytes, transmitted every 20 ms [9]. We have used this traffic pattern to model VoIP traffic in the study.

As mentioned in Section 2, video traffic is sent as VBR traffic, where the frame sizes vary according to several parameters. A recent codec is H.264 [22], which generates different types of frames with great variability between different frames. Depending on how the codec is configured, different traffic patterns are generated. In this study, we will utilize a high definition trace taken from the "horizon" talk show, provided by researchers at Arizona state university for network evaluation purposes [7]. With the same motivation as for signaling traffic, we approximate the video traffic to CBR. From the trace, we have calculated the mean frame size to be about 6000 bytes and on average 30 frames are generated per second. This will be the input representing video traffic in this study. All the above mentioned combinations have been calculated for a range of RTTs varying from 5 to 250 ms. A complete view of the parameters used in this study is found in Table 2.

Different applications do not only generate different traffic patterns, they value different traffic properties. Thus different metrics should be used for different applications. For bulk traffic, the total transfer time is important, while for signaling traffic, the Message Transfer Time (MTT) of single messages is crucial. Real-time applications, like VoIP or video conference also expect smooth and timely delivery. For this reason, the results for the different scenarios will be presented in different forms related to the specific application requirements.

Table 2. Parameters

Traffic	Message size (Bytes)	Message interval (ms)	RTT (ms)
Signaling	250	1-500	5-250
VoIP	80	20	5-250
Video	6000	33	5-250
Bulk	Bulk	Bulk	5-250

The path MTU is in all scenarios set to 1500 bytes. Further, for signaling traffic and real-time traffic, Nagle's algorithm [12] and delayed acknowledgement [1,20], may be an issue. In this study, these mechanisms are assumed to be disabled.

3.3   Impact on Signaling Traffic

For signaling messages, the MTT for a single message is of importance. Further, as discussed in Section 2, two scenarios are relevant for signaling messages; the failover scenario and the hotswap scenario. In the failover scenario, several packets may, depending on the traffic pattern, be queued at the sender during the failure detection period. Thus, at the startup, after failover, the traffic may consist of several packets waiting to be transferred. These messages will be bundled before transfer and several packets may be transferred in connection to each other. Therefore, the startup behavior in a failover scenario may be comparable to the startup for bulk traffic, described in Subsection 3.5. In the hotswap scenario, it is possible to model the MTT for a packet. When utilizing the ideal startup behavior, as there is no queuing delay, the MTT will be equal to the propagation delay, RTT/2, for all messages. In case the slow-start mechanism is used, it is possible to model the MTT under the conditions that all messages have uniform size, and that the congestion window is not decreased due to long idle periods. The MTT for a message with number n^2, m_n, is shown in Eq.(1).

$$MTT(m_n) = \frac{RTT}{2} + max(0, D_n) \qquad (1)$$

It is seen in Eq.(1) that the MTT is at least RTT/2, which is the delay for transfer over the link. Further, there may be an extra delay, D_n, which occurs when the message is queued at message arrival as a result of the congestion window being full. In these cases, the message will have to wait at the sender before transmission.

The above mentioned delay D_n depends on three components, seen in Eq.(2).

$$D_n = p_n RTT - \alpha_n + \Delta_n \qquad (2)$$

These components are:

- The number of RTT's before the message may be sent, p_n, which is calculated according to Eq.(3).

$$p_n = \left\lfloor log_2 \left(\frac{ns - 1 + w_i}{w_i} \right) \right\rfloor \qquad (3)$$

  p_n is dependant on the message number n, the message size s (bytes) and on the size of the initial congestion window w_i (bytes).
- The arrival time of the message, α_n, (since all data is not available initially).

---

[2] The first message in the transmission on the alternate path is given the ordinal number one.

– An offset, Δ_n, which is given in Eq.(4) and may occur if the message does not fit in the initial congestion window.

$$\Delta_n = \begin{cases} 0, & n \leq w_i \\ \alpha_{\Phi_n}, & n > w_i \end{cases} \qquad (4)$$

Δ_n depends on the arrival times of messages sent in the initial congestion window, which makes it possible to calculate Δ_n by transposing the current message n to the matching message Φ_n in the first window, according to Eq.(5).

$$\Phi_n = \left\lfloor \frac{n - w_i \sum_{j=0}^{p-1} 2^j}{2^p} \right\rfloor + 1 \qquad (5)$$

Thus it is possible to extract the arrival time of message Φ_n, α_{Φ_n}.

An illustration of the formulas is seen in Fig. 1. In the illustrated scenario, we assume that four messages fit into the initial congestion window, w_i, while messages 5-12 will have to wait until the first RTT is completed to be transferred. Taking message 7 as an example, it has to wait until one RTT is completed, thus $p_7=1$. Moreover, message 7 is sent when the acknowledgement for message 2 arrives and Φ_7 is equal to 2. Thus, Δ_7 shown in the figure depends on the arrival time of message 2.

An initial congestion window of 4500 bytes restricts the number of outstanding packets containing signaling messages of size 250 bytes to 18 when slow start is used. In case the RTT is greater than 17 times the message interval, messages are delayed at the sender until the first message is acknowledged according to Eq.(2). Then, the first queued message is transferred together with the next waiting message, since the congestion window is increased by the size of one message per acknowledgement.

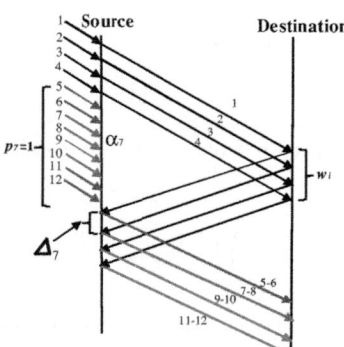
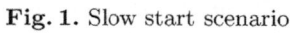
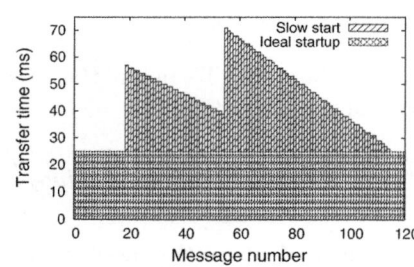

Fig. 1. Slow start scenario

Fig. 2. MTT versus message number in a hotswap scenario. Message interval 1 ms, RTT 50 ms.

Figure 2, shows the MTT as a function of message number in a hotswap scenario, when utilizing slow start as well as for an ideal startup mechanism. The figure represents messages sent at high intensity, with an average message interval of 1 ms and an RTT of 50 ms, plausible for an intra-continental association. In the figure, it is seen that all messages representing an ideal startup mechanism will have an MTT of exactly RTT/2. Further, in case slow start is used, the messages that fit in the initial congestion window will have an MTT of RTT/2. Moreover, it is seen that the first message outside this window has an extreme increase in MTT, from waiting for the first acknowledgement, while for the following messages, the MTT will decrease due to later arrival, until the congestion window is filled again. At this moment, a new peak in MTT will appear. After three round trips, the congestion window has opened up enough to transfer all messages immediately.

For signaling messages, it is important to reach the destination within a certain time bound. From this aspect, the Maximum Message Transfer Time (MMTT), for a sequence of messages is important. In the upper part of Table 3, some MMTTs for signaling traffic are seen in relation to message interval and round-trip times. From the table, it is evident that only when RTTs are quite long, and message intervals are small, like in the core network, the slow-start mechanism has a significant impact on the MMTT for signaling traffic.

3.4   Real-Time Traffic

When considering media traffic with real-time requirements, timeliness demands have to be reached, and a smooth flow of data between the endpoints is desirable. Also for real-time traffic, the formulas in Eq. 1 to 5 are valid to model MTT. The situation for VoIP is essentially the same as the situation for signaling traffic. The major difference is that the VoIP messages are usually smaller than signaling messages. An initial congestion window of 4500 bytes enables 56 VoIP messages to be outstanding. Thus, queuing of messages due to the restricted

Table 3. Maximum message transfer times (ms)

Traffic	Message Interval (ms)	RTT (ms)		
		20	50	100
Signaling, slow start	1	12	71	224
Signaling, slow start	3	10	25	96
Signaling, slow start	5	10	25	60
Signaling, slow start	10-500	10	25	50
Signaling, ideal startup	1-500	10	25	50
VoIP, slow start	20	10	25	50
VoIP, ideal startup	20	10	25	50
Video, slow start	33	30	65	190
Video, ideal startup	33	10	25	50

initial congestion window will occur only if more than 56 messages are generated before the first acknowledgement arrives. Thus, slow start does not imply a real problem for this traffic, since delays only occur in cases where RTTs exceed 1100 ms., which is unacceptable for a VoIP call anyway. The lack of impact is also seen in Table 3.

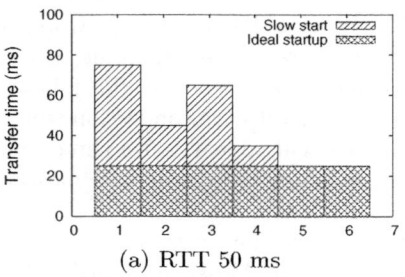
(a) RTT 50 ms

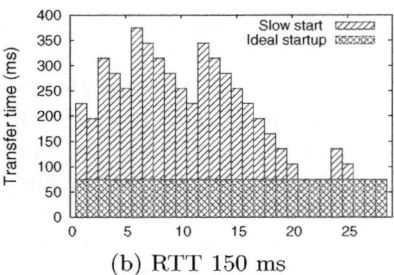
(b) RTT 150 ms

Fig. 3. MTT versus message number, traffic representing video conference

For the video trace we have used [7], the average message size is about 6 Kbytes, which means that with a congestion window size of 4500 bytes the initial message will, in the slow-start case, not fit in the initial congestion window. Further, the second part of the message will not be transmitted until the first part is acknowledged. Intuitively, this means a dramatic impact on the transfer delay, especially if the RTT is long.

Figure 3, displays the transfer times for the first messages transferred after handover for two different RTTs, representing plausible intra- and intercontinental transfers. As in the previous results, it is seen that when using an ideal startup mechanism, the delay stays constant at RTT/2. Utilizing the slow-start mechanism, on the other hand, brings significant extra delay to the transfer until the congestion window has opened up. In the figure, it is seen that the MTTs as well as the time it takes until data is transferred smoothly over the new path, is related to the RTT. In Figure 3(a), it takes about four messages before data is transferred without extra delay. If the RTT is longer, like in Figure 3(b), it takes about 20 messages to reach stable transfer times. Further, the situation in Figure 3(b) shows a reduction of the maximum MTT of more than 75% from using an ideal startup mechanism compared to slow start.

3.5   Bulk Traffic

For bulk traffic, the latency for transfer of an entire file is of interest. To calculate the latency, we have used formulas for TCP presented in [13]. These formulas are applicable also for SCTP, since the protocols utilize a similar congestion control mechanism. The formulas in [13], include latency for the connection phase, which is not applicable for a failover scenario, since all available paths are established before path switch in SCTP. Thus, we have removed the latency for the connection phase in our calculations.

Figure 4 presents some results related to bulk transfer. The different subfigures represent different bandwidths and RTTs, and in both subfigures the number of sent bytes as a function of rounds is displayed for both slow start and for an ideal startup mechanism. It is seen that the difference between the different startup schemes increases in the initial rounds, and becomes constant after some rounds. The constant difference implies that the congestion window has reached the bandwidth of the link in the slow-start case.

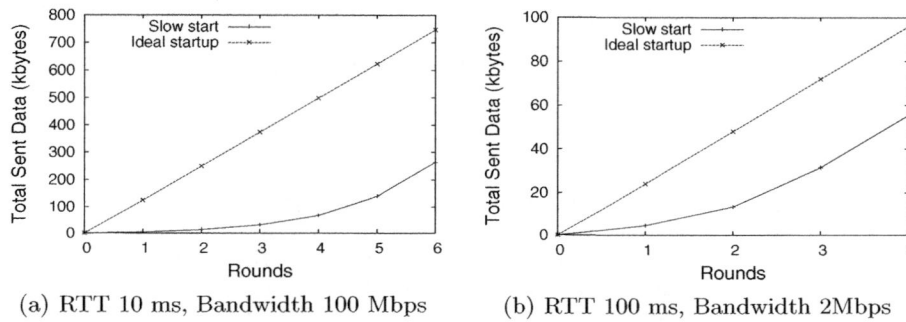

(a) RTT 10 ms, Bandwidth 100 Mbps   (b) RTT 100 ms, Bandwidth 2Mbps

Fig. 4. Transferred data as a function of Rounds

As mentioned in Section 2.2, timing aspects are generally not crucial for bulk traffic. Still, recall that the impact on bulk traffic could be representative also for bulk-like traffic as well as for signaling traffic after a failover, where timing aspects may be of importance.

4   Conclusions

In a multihomed association, awareness of the available spare bandwidth on the alternate path could serve as basis for a more efficient startup mechanism than slow start. In this paper, we have analytically studied the impact of the slow-start mechanism in comparison to an ideal startup mechanism. We have identified typical scenarios where the negative impact of slow start may be crucial for the user. The results from our calculations show that the positive effect from an improved startup mechanism is most prominent for CBR or VBR traffic with large payload and small intervals between messages. For this traffic, the impact of a small initial congestion window grows drastically as RTT increases. This scenario may be representative for a mobile user taking part in a video conference moving between networks.

The analysis in this paper is made assuming ideal conditions, to point out and quantify the impact of not utilizing full network capacity after a path switch. For future work, we intend to complement our analytic results with experimental data.

References

1. Allman, M., Paxson, V., Blanton, E.: RFC 5681: TCP Congestion Control (September 2009)
2. Andersen, A.T.: Modelling of Packet Traffic with Matrix Analytic Methods. PhD thesis, Technical University of Denmark, DTU (1995)
3. Casetti, C., Chiasserini, C.-F., Fracchia, R., Meo, M.: AISLE: Autonomic interface selection for wireless users. In: Proceedings of WOWMOM 2006, Washington, DC, USA, pp. 42–48. IEEE Computer Society Press, Los Alamitos (2006)
4. Chu, Y., Dukkipati, N., Cheng, Y.: Increasing TCP's Initial Window. Internet draft, Internet Engineering Task Force, draft-hkchu-tcpm-initcwnd- 00.txt (February 2010)
5. Floyd, S., Allman, M., Jain, A., Sarolahti, P.: RFC 3758: Quick-Start for TCP and IP (January 2007)
6. Shaojian, F., Ma, L., Attiquzzaman, M., Yong-Jin, L.: Architecture and performance of SIGMA: a seamless mobility architecture for data networks. In: Proc. of ICC 2005, Seoul, Korea (May 2005)
7. Trace file and Statistics: H.264/AVC, Horizon Talk show,
   http://trace.eas.asu.edu/h264/horizon
8. Yihua, H., Brassil, J.: NATHALIE An Adaptive Network-Aware Traffic Equalizer. In: Proc. of ICC 2007, Glasgow, UK (June 2007)
9. ITU-T. Recommendation G.711 Pulse Code Modulation(PCM) of voice frequencies. ITU-T (1972)
10. ITU-T. Methods for Subjective Determination of Transmission Quality, P.800. ITU-T (1996)
11. Fitzpatrick, J., Murphy, S., Atiquzzaman, M., Murphy, J.: ECHO A Quality of Service Based Endpoint Centric Handover Scheme for VoIP. In: Proc. of the Wireless Communications and Networking Conference, Las Vegas, USA (April 2008)
12. Nagle, J.: RFC 896 Congestion Control in IP/TCP Internetworks (January 1984)
13. Kurose, J., Ross, K.: Computer Networking -A Top Down Approach Featuring the Internet, 3rd edn. Addison Wesley, Reading (2002)
14. Zheng, K., Liu, M., Li, Z.-C., Xu, G.: SHOP: An Integrated Scheme for SCTP Handover Optimization in Multihomed Environments. In: Proc. of the Global Telecommunication Conference 2008, New Orleans, LA, USA (December 2008)
15. Liu, D., Allman, M., Jin, S., Wang, L.: Congestion Control without a Startup Phase. In: Proc. of PFLDnet Workshop, Los angeles, CA, USA (February 2007)
16. Hu, N., Steenkiste, P.: Estimating Available Bandwidth Using Packet Pair Probing. Technical report, Carnegie Mellon University (2002)
17. Postel, J.: RFC 793: Transmission Control Protocol (September 1981)
18. Fracchia, R., Casetti, C., Chiasserini, C.-F., Meo, M.: A WiSE extension of SCTP for wireless networks. In: Proc. of ICC 2005, Seoul, South Korea (May 2005)
19. Scholtz, F.J.: Statistical Analysis of Common Channel Signaling System No. 7 Traffic. In: 15th Internet Traffic Engineering and Traffic Management (ITC) Specialist Seminar, Wurzburg, Germany (July 2002)
20. Stewart, R.: RFC 4960: Stream Control Transmission Protocol (September 2007)
21. Stewart, R., Xie, Q., Tuexen, M., Maruyama, S., Kozuka, M.: RFC 5061: Stream Control Transmission Protocol Dynamic Address Reconfiguration (Septermber 2007)
22. Weigand, T., Sullivan, G.J., Bjontegaard, G., Luthra, A.: Overview of the H.264/AVS video coding standard. IEEE Transactions on Circuits and Systems for Video Technology, 560–576 (July 2003)

The Network Data Handling War: MySQL vs. NfDump

Rick Hofstede, Anna Sperotto, Tiago Fioreze, and Aiko Pras

University of Twente
Centre for Telematics and Information Technology
Faculty of Electrical Engineering, Mathematics and Computer Science
Design and Analysis of Communications Systems (DACS)
Enschede, The Netherlands
r.j.hofstede@student.utwente.nl, {a.sperotto,t.fioreze,a.pras}@utwente.nl

Abstract. Network monitoring plays a crucial role in any network management environment. Especially nowadays, with network speed and load constantly increasing, more and more data needs to be collected and efficiently processed. In highly interactive network monitoring systems, a quick response time from information sources turns out to be a crucial requirement. However, for data sets in the order of several GBs, this goal becomes difficult to achieve. In this paper, we present our operational experience in dealing with large amounts of network data. In particular, we focus on MySQL and NfDump, testing their capabilities under different usage scenarios and increasing data set sizes.

1 Introduction

Computer networks are growing in size and complexity, resulting in a network load that is constantly increasing [1, 2]. In such a scenario, network monitoring can provide vital information about the health of the network's communication infrastructure. Such information can then be used to achieve a satisfactory network operability. Therefore, network monitoring is a crucial activity in many network management solutions.

Several techniques for monitoring network traffic are available today, each one having a particular purpose and highlighting different aspects of network traffic information. Some of them focus on collecting information about individual packets, such as Tcpdump [3], while others focus on information about flows (*i.e.*, metadata information about sets of packets), such as NetFlow [4].

Independent of the involved level of abstraction, most of these techniques rely on storage points in order to collect and analyze network data. On this subject, diverse solutions are available, such as databases [5, 6], single-file data storage [3, 7] or multiple-file data storage [8]. In this paper, we report about our operational experience with MySQL [5] and NfDump [8] while dealing with data sets of several GBs. MySQL has the advantage to be a well-known Database Management System (DBMS) and to offer the full potentiality of the SQL language. It is extensively used by Web applications [9], but it has also been employed as

F.A. Aagesen and S.J. Knapskog (Eds.): EUNICE 2010, LNCS 6164, pp. 167–176, 2010.
© IFIP International Federation for Information Processing 2010

an information source for network traffic monitoring [10]. On the other hand, NfDump specifically targets the problem of network data storage and processing. It indeed stores network data into binary files without the context of a DBMS. NfDump is well known in the network community and commonly employed to collect network information [11] [12].

Both information sources (MySQL and NfDump) deal relatively well with small amounts of data, but few is known when they have to handle larger amounts. We therefore want to give an answer to the following research question: *What are the differences in performance between MySQL and NfDump when handling large data sets?* In order to give an answer, we measured the response time of the two systems on increasingly large data sets. For this comparison, we used 24 hours of network traffic from the University of Twente (UT) [13] network, which was converted to both information sources' storage formats.

The remainder of this paper is structured as follows. In Section 2 we review the current state of the art on works considering different network information sources. Section 3 describes the data set used for our information source analysis. In Section 4 we describe the used methodology by presenting our measurements. Section 5 presents the results of our comparison. Finally, we close this paper in Section 6, where we draw our conclusions.

2   Related Work

In literature, several studies on the performance of information sources for network monitoring have been proposed. Siekkinen *et al.* [14] presents a DBMS-based solution, called InTraBase, which provides the infrastructure for analysis and management of data and metadata obtained from network measurements. The authors compare the processing times of the InTraBase approach to a file-based approach, namely Tcptrace [7]. They conclude that for relatively small source files, Tcptrace is the best choice. However, for improved manageability and scalability, a DBMS approach based on PostgreSQL [6] is advocated. Similarly to them, our investigation is also based on a DBMS approach, MySQL. However, we concentrate on a multi-file approach, NfDump. It is also worth mentioning that our considered network data set is three times larger (around 30 GB) than the data set considered in Siekkinen's work (10 GB).

Similarly to us, Kobayashi *et al.* [15] presents a comparison between a file-based approach for storing network traffic information and a DBMS approach. NfDump was used as a representative of the file-based approach, while MySQL and PostgreSQL were the tested implementations of the DBMS approach. The comparison was divided into two categories: 1) storing time, and 2) search and display time. Regarding storing time, the authors concluded that a file-based approach was in general faster than the DBMS approach. For the search and display times, the file-based approach was relatively slow compared to the DBMS. While the research in [15] is concentrated on short term data storage, we are interested in analyzing performance of information sources when handling historical data. In addition, we will focus our measurements only on search time.

Besides handling large amounts of network data by using an ordinary DBMS or a file-based approach, also decentralized setups can be used to speed up the query process. One approach is described in [16], where decentralized nodes perform data aggregation, which leads to shorter query execution times. However, in our work, we do not take distributed solutions into consideration.

3   Data Collection Process

This section describes how we captured, processed and stored the network traffic used in our experiments. After that, we provide statistics about this data set.

The University of Twente (UT) main router exports network information about flows[1] in the form of NetFlow v9 records. The NetFlow records were collected using a *nfcapd* process, part of the NfDump suite [8]. It creates binary files of 5 minutes of network data, in the order in which it is captured. After that, the data was post-processed and imported into a MySQL database, also ordered by time. NfDump offers a routine for reading and displaying *nfcapd* binary files.

The MySQL schema used for storing network data follows the description of a flow. Each flow is stored as one flow record, ordered by its start time and having the following attributes: *flowid*, *start_time*, *end_time*, *duration*, *ipv4_src*, *port_src*, *ipv4_dst*, *port_dst*, *protocol*, *tos*, *packets* and *octets*. Besides these attributes, a MySQL table has a standard index on a table's primary key, which is *flowid* in our case. Moreover, our table also contains some other indexes to improve the speed of operations on this table. The available indexes are as follows: *start_time*, *tuple* (*ipv4_src*, *port_src*, *ipv4_dst*, *port_dst*, *protocol*), *packets*, *octets* and *duration*.

The data set used in our measurements consists of exactly 24 hours of network data, collected on September 18th, 2008, between 0:00 and 23:55. It consists of roughly 445.000.000 flow records. The disk space needed for MySQL data is 31 GB, while its indexes need around 43GB of disk space. On the other hand, the NfDump data needs about 22 GB of disk space.

4   Methodology

In this section we present the methodology used to measure query response times on our data set. We define query response time as the time between invoking a query on a data set and having retrieved the complete result set. However, we do not print the result set to the screen. All measurements have been performed on a machine with an Intel Dual Core Xeon 3070 CPU, 4GB DDR2 RAM and 2 SATA disks in RAID0 (striping). The first subsection will describe various approaches used in our data analysis, while the second subsection defines several case studies considered in our comparison.

---

[1] We consider a flow as "a set of packets passing by an observation point in a network during a certain time interval and having a set of common properties" [17].

4.1   Approaches to Data Analysis

MySQL and NfDump were compared by measuring their query response times when handling large amounts of network data. While from NfDump's side no optimization can be set up, MySQL provides the use of indexes, which results in more efficient access of data. In some cases indexes can considerably increase a database table's performance. The indexes are generally transparent to the end user, but it is possible to force or forbid the use of any of these. If an index does not exist or is not used, MySQL will perform a sequential scan over the whole table. However, if the index exists, MySQL can quickly determine the position to seek for in the middle of the data file without having to look at all the data.

Since indexes can affect the performance of a query, we tested MySQL both with and without indexes. This leads us to a total of four considered approaches:

1. *MySQL*: MySQL decides whether or not to use an available index.
2. *MySQL with indexes*: MySQL is forced to use a specified index.
3. *MySQL without indexes*: MySQL is forced to ignore any index.
4. *NfDump*: NfDump behaves as it is designed for.

4.2   Case Studies

The four approaches aforementioned were observed while triggering different types of operations on the considered network data. These operations, from now on addressed as case studies, are as follows:

1. *Listing*: This case study causes the information sources to do a sequential scan over the data set, without doing any calculation or filtering. Actually, there is a filtering action on the *start_time* attribute, but since we assume this as our basic situation, we consider it as part of the listing.
2. *Listing + Filtering*: The result of this case study only contains the flow records of the first case study that have destination port 22. We chose to filter on this port because this traffic is fairly distributed over the day and thus also over the entire data set. The amount of traffic with destination port 80 for instance, would depend too much on the time of the day.
3. *Grouping + Filtering 1*: This case study groups all flow records of the "Listing + Filtering" case study by their source IP addresses and calculates the sum of octets for each group.
4. *Grouping + Filtering 2*: Instead of grouping by just one attribute, this query groups by five attributes and calculates the sum of octets for each group. We want to verify here whether grouping by multiple attributes will require significantly more time than the previous case study.

In order to find out how MySQL and NfDump behave depending on the size of our data set, we tested their performance on time-incremental basis (with data slices of one hour of network data). In this case, incremental means that the first test is executed on a data set from 12 AM to 1 AM, the second on a data set from 12 AM to 2 AM, and so on. The data set is thus becoming larger with

Table 1. Queries representing our case studies

Case study	MySQL query	NfDump query
Listing	SELECT * FROM table WHERE start_time BETWEEN x AND y	nfdump -M data_dir -T -R nfcapd.date1:nfcapd.date2 -o long
Listing + Filtering	SELECT * FROM table WHERE start_time BETWEEN x AND y AND port_dst = 22	nfdump -M data_dir -T -R nfcapd.date1:nfcapd.date2 -o long "dst port 22"
Grouping + Filtering 1	SELECT ipv4_src, SUM(octets) FROM table WHERE start_time BETWEEN x AND y AND port_dst = 22 GROUP BY ipv4_src	nfdump -M data_dir -T -R nfcapd.date1:nfcapd.date2 -o long -a -A srcip "dst port 22"
Grouping + Filtering 2	SELECT ipv4_src, port_src, ipv4_dst, port_dst, protocol, SUM(octets) FROM table WHERE start_time BETWEEN x AND y AND port_dst = 22 GROUP BY ipv4_src, port_src, ipv4_dst, port_dst, protocol	nfdump -M data_dir -T -R nfcapd.date1:nfcapd.date2 -o long -a -A srcip,srcport, dstip,dstport,proto "dst port 22"

every query execution, until the full data set is used. This results in a total of 24 tests per approach, for each case study.

Table 1 shows the syntax of the queries representing the case studies. All SQL queries have a *start_time BETWEEN x and y* statement in their *WHERE* clause. This statement is needed to select the current time-incremental data set to be tested. The value of x is constant in all our tests. The same can be done with NfDump using the *nfcapd.date1:nfcapd.date2* statement, where the input files of the data set have to be specified. It is important to notice here that it is not possible to filter on a flow's *start_time* or *end_time* using NfDump's query syntax. The only way to do this, is by selecting another input file set.

In the particular case of MySQL, we need to know which index MySQL is going to use for a specific query. MySQL's "EXPLAIN" command shows which indexes *could* be used and which indexes are *actually going to* be used. We used this information to force or forbid indexes in our case studies. Finally, note that all query response times were measured without considering screen printing.

Table 1 only reports the query for the *MySQL* approach and the used command for the *NfDump* approach. For the remaining approaches (*MySQL with indexes* and *MySQL without indexes*) the query is the same plus the use of the "FORCE INDEX" and "IGNORE INDEX" keywords.

5   Results

This section presents the results of our measurements. They show the response times of each approach for each case study. The plots describe the relation between the size of the data set (on the horizontal axis) and the query response time in minutes (on the vertical axis). All tests were executed three times, to reduce the effect of disturbances (*e.g.*, unforeseen processes running on our test machine). The plots are based on the averages of each pair of three query executions. Moreover, error bars were added representing the standard error value:

$$e = \frac{\sqrt{\frac{\sum (X_i - \bar{X})^2}{(n-1)}}}{\sqrt{n}}$$

5.1   Listing and Listing + Filtering

Figure 1 shows the results of the "Listing" case study. A first observation, related to the *MySQL without indexes* approach, is that the result is not constant. Since MySQL would have to do a sequential scan over the whole data set, which would take the same amount of time for each query iteration, this is contradictory to what we expected. Instead, there seems to be an almost linear relationship between the size of the data set (hours of network data) and the query response time, namely between 7 and 21 hours of data.

We can thus conclude that there is an overhead in writing the result set, even though we took care to avoid screen output to reduce the impact of such operations on the response times. Before and after the period of linearity, the data density will be less. Otherwise, the whole trace would be linear. This can be explained due to the fact that during night (*i.e.*, from 9 PM to 7 AM) there is much less traffic transiting within the UT network.

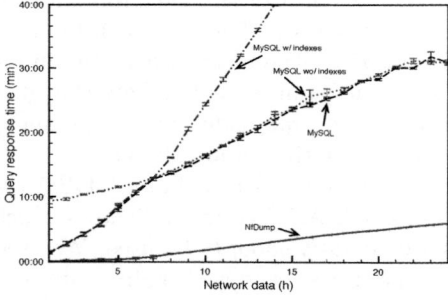
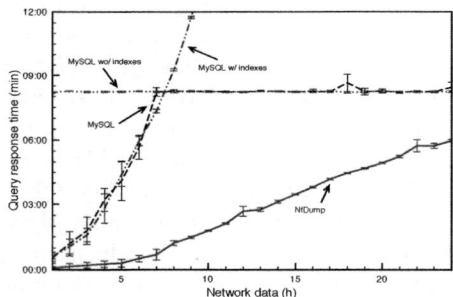

Fig. 1. Listing queries   **Fig. 2.** Listing + Filtering queries

A second observation is that after using a data set of approximately 7 hours of network data, *MySQL with indexes* needs more execution time than *MySQL without indexes*. The reason for that comes from the fact that MySQL has to read too much data from the disk (*i.e.*, from the index and the data), which takes more time than doing a full sequential scan over the data set. Moreover, *MySQL* makes the right choice about when to use the index: exactly from the point where the query response time of *MySQL without indexes* is less than with a forced use of the index, *MySQL* chooses to discard the index.

The queries that were executed using NfDump had the shortest query response times during all queries of the "Listing" case study. The reason for this is that NfDump uses small input files, which are concatenated as specified in the queries. In contrast to this, MySQL has to process the whole data set during every query execution.

On its turn, Figure 2 shows the results of the "Listing + Filtering" case study. Using *MySQL without indexes*, the result is now indeed a (constant) horizontal line. Since the query response times are much less than in the "Listing" case study, we can conclude that MySQL is first filtering on *dst_port = 22*, before a selection based on the start times of the flows is made. That result set is much smaller (231.098 flow records) than the average result set of the "Listing" case study (18.445.740 flow records).

After about 7 hours of network data, *MySQL* decides not to use indexes anymore. This is, according to the query response times, the correct decision. We can also notice that until that point, the standard error value is much larger than after it. As Schwartz *et al.* describe in [18], the MySQL query cache is completely in-memory, which is managed by MySQL itself. This means that the response time of the second and third test until 7 hours of data is shorter because the results are already in memory. After 7 hours of data, the response times stabilized because MySQL cannot take advantage of its cache anymore.

Once more, NfDump has the shortest query response times. Its behavior is closely related to MySQL's behavior in the "Listing" case study: from 7 AM until 9 PM there is a linear relationship between the size of the used data set and the query response time. Before and after that period, the data density is relatively less. Therefore, NfDump's query response times are not linear to the size of the used data set over the whole trace.

5.2 Grouping + Filtering 1 and Grouping + Filtering 2

Figures 3 and 4 show the results of the two "Grouping + Filtering" case studies.

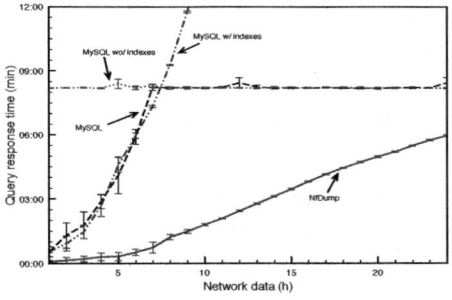 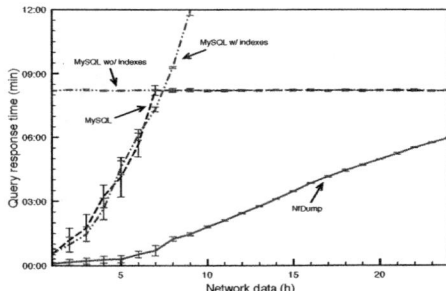

Fig. 3. Grouping + Filtering 1 queries   **Fig. 4.** Grouping + Filtering 2 queries

As previously mentioned, we expected the "Grouping + Filtering 2" case study in general to take longer to complete than "Grouping + Filtering 1".

However, our results show that this is not the case: both case studies take exactly the same execution time. Moreover, these times are also identical to the response times of the "Listing + Filtering" case study.

The "Grouping + Filtering" case studies require the data set to be sorted by all attributes specified in the *GROUP BY* clause. Our results suggest that once the data is retrieved, grouping by five attributes is not more costly than grouping by one. Since the response times for both case studies are identical, we can conclude that data retrieval and the disk operations related to it are the bottleneck of the data set manipulations. Likewise previous case studies, NfDump outperforms MySQL once more.

5.3   Double-Sized Data Set

All figures presented before suggest that there might be an intersection between MySQL's and NfDump's query response times, somewhere between 24 and 48 hours of network data. To get more insight into this speculation, we tested all case studies on a 48 hour data set. The results of the "Listing + Filtering" case study can be found in Figure 5. We do only discuss the "Listing + Filtering" case study here, since the results of the others show the same behavior compared to their 24 hour counterparts. Additionally, Figure 6 was created using the response times on 24 and 48 hours of network data of both *MySQL wo/ indexes* and *NfDump*. The two lines were created using linear regression. As shown in Figure 6, MySQL's and NfDump's query response times will never intersect, since the response times are diverging from each other.

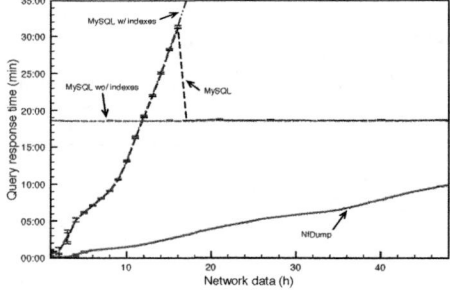
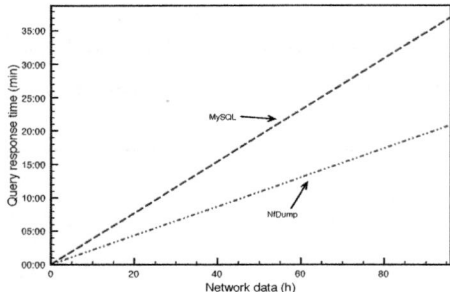

Fig. 5. Listing + Filtering [48h data set] queries   **Fig. 6.** Relationship between MySQL's and NfDump's response times

By comparing the result of the 24 hour data set with its 48 hour counterpart, we can make some interesting observations. First, the behavior of *MySQL wo/ indexes* is the same, but its query response times are doubled. On the contrary, *NfDump*'s query response times are not doubled (*i.e.*, 6 minutes compared to 10 minutes, on a 24 hour and a 48 hour data set, respectively).

A second observation is that after 12 hours of network data, *MySQL* is not able to correctly decide to ignore the available index after that point. Note that on a 24 hour network data set, *MySQL* was able to make the correct decision. Remarkable is that the point where *MySQL* decides to discard the index (around 16 hours of network data), is around twice that point on a 24 hour data set.

More importantly, it can be observed that there is still no intersection between MySQL's and NfDump's query response times. For a complete data set, the response time of *MySQL wo/ indexes* is the shortest for our case studies. If we compare the 48 hour data set to the original 24 hour one, we observe that:

- *MySQL wo/ indexes*' response time grows by a constant factor each time the data set is enlarged. In particular, the response time is doubled here.
- *NfDump*'s query response time grows linearly according to the amount of data processed, but it is not influenced by the size of the complete data set. Moreover, the response time on a 48 hour data set is less than doubled, compared to the response time on a 24 hour data set.

Considering the implementations of MySQL and NfDump, we expect the relationship between the two approaches to be maintained also for larger data sets. Since the query response times are diverging from each other, we assume that MySQL's and NfDump's query response times will most likely never intersect.

6   Conclusions

This paper presented a comparison between the performance of MySQL and NfDump when handling data sets consisting of several GBs of network information. We measured the response time of both systems on a data set of 24 hours of flow data in an incremental manner and keeping into account several usage scenarios. Moreover, since indexes can improve MySQL's performance, we also considered them while defining the usage cases.

Siekkinen *et al.* [14] advocated the use of a DBMS as the best solution for network information for the sake of data management. Differently, our measurement results indicate NfDump as being the best solution to query large network data sets when observing response time. In all our comparisons NfDump outperformed MySQL. Therefore, we advise, when the response time is a striking issue, to make use of NfDump or similarly designed multiple-file based approaches.

Despite NfDump outperformed MySQL in all our comparisons, we observed that MySQL's response time was constant after considering 7 hours of network data, whereas NfDump's response time increased. In order to verify if NfDump's response times would intersect MySQL's on a larger data set, we enlarged our data set to 48 hours of data. Even then, NfDump outperformed MySQL in all tests. However, while MySQL's response times were doubled, NfDump's were slightly close to double. Taking this behavior as a pattern to even larger data sets, we can assume that NfDump's response times will most likely be shorter than MySQL's.

Acknowledgements

This research work has been supported by the EC IST-EMANICS Network of Excellence (#26854). Special thanks to Djoerd Hiemstra and Ramin Sadre for their valuable contribution to the research process.

References

1. Steinder, M., Sethi, A.S.: A survey of fault localization techniques in computer networks. Science of Computer Programming 53(2), 165–194 (2004)
2. Casey, E.: Network traffic as a source of evidence: tool strengths, weaknesses, and future needs. Digital Investigation 1(1), 28–43 (2004)
3. Tcpdump/libpcap (April 2010), http://www.tcpdump.org/
4. Claise, B.: Cisco Systems NetFlow Services Export Version 9. RFC 3954, Informational (2004)
5. MySQL (April 2010), http://www.mysql.com/
6. PostgreSQL (April 2010), http://www.postgresql.org/
7. Tcptrace (April 2010), http://www.tcptrace.org/
8. NfDump (April 2010), http://nfdump.sourceforge.net/
9. Liu, X., Heo, J., Sha, L.: Modeling 3-Tiered Web Applications. In: Proc. of the 13th IEEE Int. Symp. on Modeling, Analysis, and Simulation of Computer and Telecommunication Systems, pp. 307–310 (2005)
10. Hofstede, R., Fioreze, T.: SURFmap: A Network Monitoring Tool Based on the Google Maps API. In: Application session proc. of the 11th IFIP/IEEE Int. Symp. on Integrated Network Management, pp. 676–690. IEEE Computer Society Press, Los Alamitos (2009)
11. Li, Y., Slagell, A., Luo, K., Yurcik, W.: CANINE: A combined conversion and anonymization tool for processing NetFlows for security. In: Proc. of 10th Int. Conf. on Telecommunication Systems, Modeling and Analysis (2005)
12. Minarik, P., Dymacek, T.: NetFlow Data Visualization Based on Graphs. In: Goodall, J.R., Conti, G., Ma, K.-L. (eds.) VizSec 2008. LNCS, vol. 5210, pp. 144–151. Springer, Heidelberg (2008)
13. University of Twente (April 2010), http://www.utwente.nl
14. Siekkinen, M., Biersack, E.W., Urvoy-Keller, G., Goebel, V., Plagemann, T.: In-TraBase: Integrated traffic analysis based on a database management system. In: Proc. of the End-to-End Monitoring Techniques and Services, Washington, DC, USA, pp. 32–46. IEEE Computer Society, Los Alamitos (2005)
15. Kobayashi, A., Matsubara, D., Kimura, S., Saitou, M., Hirokawa, Y., Sakamoto, H., Ishibashi, K., Yamamoto, K.: A Proposal of Large-Scale Traffic Monitoring System Using Flow Concentrators. In: Kim, Y.-T., Takano, M. (eds.) APNOMS 2006. LNCS, vol. 4238, pp. 53–62. Springer, Heidelberg (2006)
16. Lim, K.S., Stadler, R.: Real-time views of network traffic using decentralized management. In: Proc. of the 9th IFIP/IEEE Int. Symp. on Integrated Network Management, Nice, France, pp. 119–132 (2005)
17. Quittek, J., Zseby, T., Claise, B., Zander, S.: Requirements for IP Flow Information Export (IPFIX). RFC 3917, Informational (2004)
18. Schwartz, B., Zaitsev, P., Tkachenko, V., Zawodny, J., Lentz, A., Balling, D.J.: High performance MySQL, 2nd edn. O'Reilly, Sebastopol (2008)

Processing of Flow Accounting Data in Java: Framework Design and Performance Evaluation

Jochen Kögel and Sebastian Scholz

Institute of Communication Networks and Computer Engineering (IKR)
University of Stuttgart
Pfaffenwaldring 47
70569 Stuttgart
{jochen.koegel,sscholz}@ikr.uni-stuttgart.de

Abstract. Flow Accounting is a passive monitoring mechanism implemented in routers that gives insight into traffic behavior and network characteristics. However, processing of Flow Accounting data is a challenging task, especially in large networks where the rate of Flow Records received at the collector can be very high. We developed a framework for processing of Flow Accounting data in Java. It provides processing blocks for aggregation, sorting, statistics, correlation, and other tasks. Besides reading data from files for offline analysis, it can also directly process data received from the network. In terms of multithreading and data handling, the framework is highly configurable, which allows performance tuning depending on the given task. For setting these parameters there are several trade-offs concerning memory consumption and processing overhead. In this paper, we present the framework design, study these trade-offs based on a reference scenario and examine characteristics caused by garbage collection.

1 Introduction

Monitoring network characteristics and traffic is vital for every network operator. This monitoring information serves as input for adjusting configurations, upgrade planning as well as for detecting and analyzing problems.

Besides active measurements and passive capturing of packet traces, Flow Accounting is attractive because it is a passive monitoring approach, where information on flows is created in routers and exported as Flow Records using a protocol like Cisco NetFlow [1] or IPFIX [2]. Due to monitoring on the flow level, Flow Accounting provides a good trade-off between the information monitored and the amount of data to store and process. Flow Accounting is mainly used for reporting and accounting tasks, but can also provide input for anomaly detection or extraction of network characteristics [3].

Processing of Flow Accounting data is challenging, since in large networks routers export several hundred million Flow Records per hour. Three common approaches can be distinguished to handle and process this data. First, Flow Records can be stored in a central or distributed database to create reports and

F.A. Aagesen and S.J. Knapskog (Eds.): EUNICE 2010, LNCS 6164, pp. 177–187, 2010.
© IFIP International Federation for Information Processing 2010

offline analysis at a later point in time. Here, the attributes of Flow Records are often reduced in order to save memory and processing effort. Second, Flow Records can be dumped directly into files for offline analysis without a database. Third, Flow Records can be analyzed online directly in memory without storing them.

We developed a flow processing framework in Java that can read Flow Records from files as well as from the network. Thus, it is suitable for offline and online analysis. The framework processes Flow Records in a streaming fashion, i.e. data flows through a chain of processing blocks. Each block can keep data as long as required for its task (typically using a sliding window) before it forwards the processing results. For tasks like joining and sorting Flow Records, the window size parameter for achieving good processing results depends on characteristics of the Flow Accounting data, which in turn depends on router configuration and traffic characteristics. Hence, these factors influence how long data is kept in memory and thus the overall memory consumption. However, the latter cannot be evaluated as independent metric in Java: more memory consumption results in more garbage collector overhead and thus eventually in reduced throughput.

This paper studies these dependencies. Besides, the processing blocks can be assigned to threads in different schemes, allowing to exploit modern multicore computers. The investigation of these threading schemes is also part of this work.

This paper is structured as follows: Section 2 introduces Flow Accounting, Section 3 presents the design of the flow processing framework, Section 4 shows the result of the performance evaluation, and Section 5 concludes the document.

2   Flow Accounting

2.1   Mechanism and Protocols

Flow Accounting is a mechanism present in most professional routers that keeps counters on per-flow basis. The router exports this information as Flow Records and sends them a collector, where the information is processed further. Flows are identified by a key, which is typically the five tuple consisting of source and destination address, source and destination port, as well as transport protocol number. Routers keep a table (*flow cache*) where data on each flow is stored. The router updates the flow information (e.g. byte and packet count) either for each packet (unsampled) or for a fraction of packets (sampled). Among other information, Flow Records contain the five tuple, the counters as well as start and end time in milliseconds.

Several data formats for sending multiple Flow Records in a packet to the collector exist. Mostly Cisco system's NetFlow format is used. The current NetFlow version 9 [1] and its successor IPFIX [2] are flexible formats based on templates, while the fixed format of version 5 dominates current deployments. Since the data is sent via UDP, packets transporting Flow Records might be lost.

For determining the export time t_x of a Flow Record, there are several criteria in Cisco routers. *Inactive timeout*: if for a flow there is no more packet seen for Δt_{inact}, the flow is exported. *Active timeout*: if a flow has been active for a time

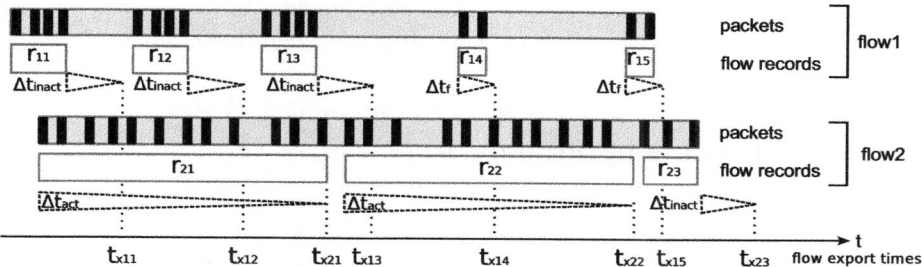

Fig. 1. Flow export: timers of different strategies and resulting record order

period greater than Δt_{act}, the flow is exported. *Fast timeout*: if after Δt_f a flow contains less than n_{fast} packets, the flow is exported. *Cache clearing*: if the cache runs full, the router exports flows earlier than defined by the timeouts.

Some routers determine flow ends also by tracking TCP connection state. The timeouts are illustrated in Fig. 1 for two flows. We can see that information on flows can be distributed across different records with long breaks in between and that records arrive at the collector neither sorted by start nor by end time. These properties have to be considered for processing algorithms that work on a limited window of Flow Accounting data.

2.2   Existing Processing Tools

In large networks several hundred million Flow Records arrive at the collector per hour. Processing or storing this amount of data is challenging. There are several commercial tools that collect and analyze Flow Accounting data, such as IsarFlow, Arbor Peakflow, Lancope StealthWatch or Cisco Multi NetFlow Collector. These tools follow the common approach of using a database (DB) for offline analysis. In order to cope with the data rate, load balancers distribute the traffic across several collectors that form a distributed DB. These tools often aggregate data as soon as possible (e.g. on time intervals) to reduce storage requirements. Thus, evaluations that need fine grained timing information are not possible. There are also free tools for collection and basic processing of Flow Accounting data, such as *flowtools* and *nfdump*. Free reporting and query tools are *nfsen* and *SiLK*.

Evaluation algorithms that rely on fine grained information are typically processing and memory intensive, thus stream-based approaches have advantages over DB based tools. Processing network monitoring data in a streaming fashion is close to the domain of data stream management systems (DSMS) or complex event processing. Related tools are Gigascope [4] for packet trace processing, the TelgraphCQ DSMS [5] or the network monitoring specific CoMo project [6]. Other tools focus on the dispatching of received NetFlow data [7] or on Flow Query Languages [8]. The presented tools are often limited to a certain domain and not suitable for extracting and correlating network characteristics or data from different data sources. A data processing pipeline in Java is presented in [9]. Its focus is on processing large objects (e.g. images), thus it is unsuitable for Flow Accounting data.

3   Data Processing Framework

3.1   Architecture

We developed a framework for processing of Flow Accounting data, especially for fine grained analysis of Flow Records e.g. for extraction and correlation of network metrics. This includes tasks like matching records from different sources, calculation of derived metrics, correlation and statistics. In order to handle the huge amount of data, we use a stream-based approach that capitalizes on modern multicore architectures. We decided to take Java as programming language due to the garbage collection feature of the Java Virtual Machine (JVM), which simplifies modular design, and its built-in utility classes for concurrent programming.

Our framework is based on interconnected processing blocks that form a data processing chain. Processing blocks exchange messages containing references to objects (Fig. 2). Each block has at least to implement a data source or a data sink interface, while combinations with several interfaces are also possible (e.g. several inputs/outputs). A processing block is derived from a generic block according to its role. Fig. 2 shows an exemplary chain with two *SourceBlocks* reading NetFlow data from files or the network and forwarding the data to *SinkSourceBlocks* that perform data aggregation to *JoinedFlows*. The two pipelines are merged in a *CorrelationBlock*, which creates correlation result objects that are statistically evaluated in two *SinkBlocks*.

Processing chains are constrained to a directed acyclic graph. While cycles could make sense e.g. for feedback loops to compensate time offsets in the data, this is currently not supported. We designed the framework in a way that threads can be assigned to one or several processing blocks. If two processing blocks run as different threads, they are connected via blocking FIFO queues, as shown in Fig. 2 for the configuration with the maximum number of threads.

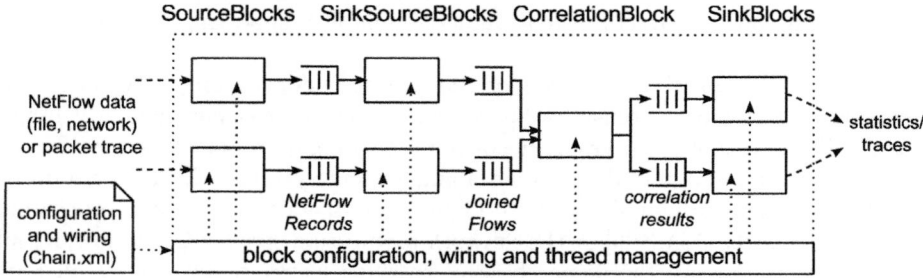

Fig. 2. Exemplary processing chain showing basic types of processing blocks

At each source interface, an arbitrary number of processing blocks can connect, such that all of them will get references to the objects passed on and can process them independently of each other. We avoid race conditions possibly caused due to concurrent access by not modifying objects after they left the block where they have been created. Due to garbage collection provided by the JVM, no mechanisms to manage the references to objects and freeing memory in case of dropped objects is necessary. This enables a clean design of independent processing blocks.

At startup, the chain is set up by a central component that also performs thread management. Configuration is based on an XML file that describes the chain structure and processing block parameters. For this, we build on the dependency injection mechanisms provided by the Spring Framework [10]. After all objects are created and wired as defined, threads are started and the *Source-Blocks* starts delivering data to the chain. Shutdown is initiated by a shutdown message in downstream direction, e.g. if readers run out of data. If the chain contains parallel paths that are merged in a correlator, this mechanism is not sufficient for proper shutdown of upstream blocks that still have data. In such cases, upstream shutdown notification is performed by deregistering connections from upstream blocks, which will then shut down.

3.2   Processing Blocks

In terms of processing tasks there are two basic classes of processing blocks: *window-based blocks* that keep data over a sliding window, and *window-less* blocks that perform processing on data objects immediately.

Examples for window-less blocks

Reader: read data from disk or network, create objects and send them on. Two versions of the file reader exist: A parallel one that internally uses up to nine threads to create message objects and a single-threaded one.
Statistic: calculates mean values or distribution statistics for time intervals.
Dumper: writes object attributes to disk, e.g. as CSV file.

Examples for window-based blocks

Sorter: sorts data according to start or end time. The window moves according to the timestamps of received data. Stored data with timestamps smaller than the lower window edge is forwarded. Data received with timestamps smaller than the lower window edge is dropped.
Joiner: combines records of the same flow that have been exported separately due to timeouts. A window specifies the *maxDuration*, i.e. how long the block should wait for another record before expiring and forwarding the *JoinedFlow*. *maxWaitingTime* specifies the maximum length of the created JoinedFlows. Without the second parameter JoinedFlows of flows lasting for a very long time would be forwarded to downstream blocks very late.

Correlator: with more than one input these blocks correlate different data streams, e.g. based on timestamps. Typically, timestamps of the data compared are not exactly equal or have an offset resulting from measurement. Thus windows are necessary.

While processing times of records in window-less blocks are rather fix, this time is highly variable for window-based blocks. In a window-based block, a data object can either be dropped or kept (added to internal data structures). Additionally, it can lead to window movement and thus to the expiration of several objects. This leads to a high jitter in processing time and makes buffering between window-based blocks and other threads necessary. Without any or with only small queues, window-based blocks are likely to stall the chain.

3.3   Thread and Message Configuration Parameters

Our framework allows us to configure whether a processing block runs as an independent thread or not. A block that does not run as a single thread belongs to the thread of the upstream block and gets the control flow when it receives data. Obviously, this results in the constraint that readers must always run as a thread since they are the data sources. In correlator blocks it depends on the data from which input the next object is read. Thus they always run as a thread, since using the control flow from upstream block would drastically increase complexity. Using more threads helps exploiting modern multicore architectures, but also comes at the cost of higher memory consumption due to objects present in queues that are used to connect blocks running on different threads. The concept of thread pools is not applicable, since its purpose is to reuse a limited number of existing threads instead of creating them for each arriving task.

We realized that the high number of objects flowing through the blocks leads to a high context switch rate and high CPU time in the operating system (OS). Since Java maps threads directly to kernel threads, the OS is involved in locking and context switch operations. Thus, each object added or removed to queues possibly involves a switch from user mode to kernel mode and back. To mitigate these effects, we introduced burst messages, where several data objects are sent in one message. Due to performance reasons, burst messages are only applied for message exchange between threads. The number of included objects is called burst size. Queue operations happen less often and it is more likely that threads can run for a longer time without being blocked. However, this also comes at the cost of additional memory consumption. The size of burst messages is configurable. We study its impact in the next section.

4   Performance Evaluation

4.1   Measurement Scenario

For performance evaluation we selected a reference processing chain (Fig. 3). The chain reads Flow Records from two routers from files, aggregates Flow Records of

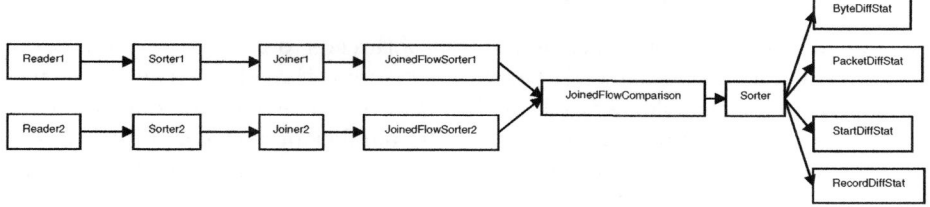

Fig. 3. Joined Flow comparison chain

the same flow in a joiner and correlates these *JoinedFlows*. Statistics evaluate the time, byte and packet difference of *JoinedFlows*. This allows to extract packet loss rates, byte count inaccuracies and network delay ([3]). Where processing blocks require sorted data, sorters are employed. Due to the high number of window-based processing blocks, this chain requires a considerable amount of memory and is thus suitable to study memory effects.

Our performance evaluation centers on the throughput. We try to identify influencing factors by studying system time, user time and the time the garbage collector needs.

For measurement the depicted chain processed two uncompressed files with 65 million Flow Records in total. Each file contained records from only one exporter. The queues between threads had a size of 100 messages. According to the characteristics of our data we set the windows of the blocks as follows: Sorter 1,2: 20 s; Joiner 1,2 *maxWaitingTime*: 5 min; JoinedFlowSorter 1,2: 5 s; last Sorter: 3 s. The measurements were performed on an Intel Xeon X3360 quad core 2.83 GHz processor with a total amount of 8 GB memory. The software configuration included Sun JVM version 1.6 update 15 on Ubuntu 9.10 64 bit with kernel version 2.6.31. The standard garbage collector was used. Measurements with different garbage collector settings showed similar results. The only configuration done with the JVM, was to set the initial and maximum heap size to the same value. Each measurement was done five times. In charts we show the mean values of the results with error bars showing the minimum and maximum absolute value. We measured the real time r, the system time s and the user time u with the /usr/bin/time program. The time the garbage collector needs was obtained with the GarbageCollectorMXBean class provided by the JVM. This time is included in u.

Several parameters likely influence the throughput behavior. The throughput can be increased by using faster CPUs with a larger main memory. We also did measurements on a machine with two quad core Opteron CPUs and 48 GB main memory, where we observed a speedup of up to factor two. We also examined other processing chains and we found proved that other processing chains show a similar behavior. In this work we will focus on the thread assignment and the size of burst messages.

4.2 Benefit of Multithreading

In the following we show which assignment of threads to processing blocks makes sense and how the framework benefits from multithreading. The decision to combine processing blocks needs to be based on their characteristics like the needed processing time, IO intensity or the number of exchanged messages. In general, it is a good solution to combine several processing blocks if their tasks are simple.

We study the seven assignment patterns listed in Tab. 1. All processing blocks that are not listed run in the same thread as their predecessor. We used configurations with the minimum number of threads (assignment A_1) up to the maximum number (assignment F). The patterns B to E try to split the long chains into shorter sub-chains. The parallel reader is the reader with several internal threads, the single reader is the single-threaded version.

Table 1. Thread Assignment Patterns

thread assignment	independent threads
A_1	Reader1 (single), Reader2 (single), JoinedFlowComparison
A_2	Reader1 (parallel), Reader2 (parallel), JoinedFlowComparison
B	Reader1 (parallel), Reader2 (parallel), JoinedFlowComparison, Sorter
C	Reader1 (parallel), Joiner1, Reader2 (parallel), Joiner2, JoinedFlowComparison, Sorter
D	Reader1 (parallel), Sorter1, JFSorter1, Reader2 (parallel), Sorter2, JFSorter2, JoinedFlowComparison
E	Reader1 (parallel), Sorter1, JFSorter1, Reader2 (parallel), Sorter2, JFSorter2, JoinedFlowComparison, Sorter
F	each block is a thread (parallel readers), 14 threads

Fig. 4 shows the throughput in Flow Records per second and the CPU utilization $\rho = \frac{u+s}{r}$ related to different thread assignment patterns. As we can see, throughput as well as utilization are nearly independent of the thread assignment.

The difference between A_1 and A_2 appears mainly in the utilization. But optimizing the utilization is not our goal, since we want to achieve a high throughput. On the basis of the results we could see, that u as well as r is in both cases almost the same. Only s is in A_1 greater, because of blocking since the single-threaded reader cannot create message objects fast enough so that the following threads can work continuously.

In pattern C a lot of data must be processed in the first thread containing the reader and the first sorter. From the result we can see, that blocks processing a high data rate should run as an independent thread.

The performance decreases in F, because the memory consumption increases, especially with higher burst sizes. The reason is that more data must be kept in the queues between the threads. Also the garbage collector must run more often to free memory.

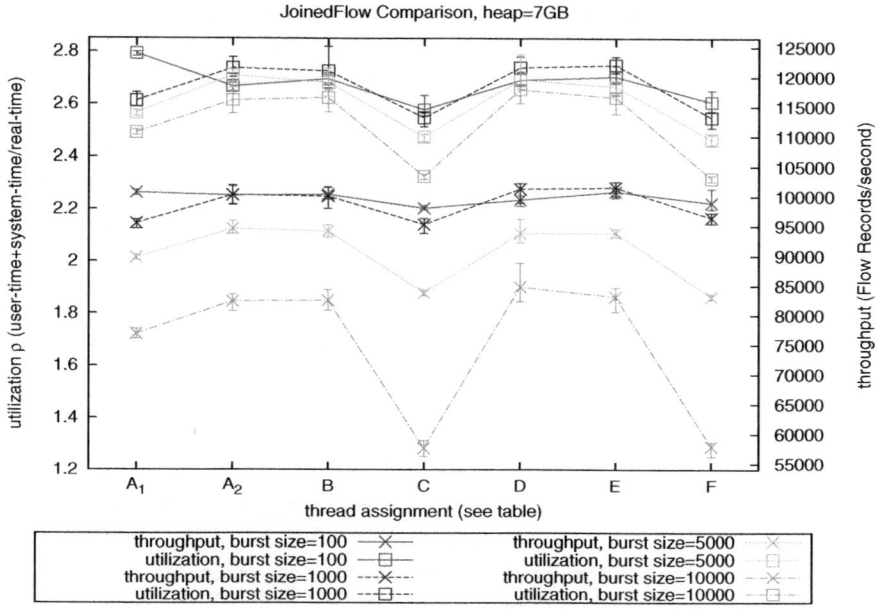

Fig. 4. Benefit of multithreading with different thread assignments

One reason for the small overall utilization is the garbage collector, which stops the execution of the program to perform major collections. Because the standard collector is used, these collections are only done by one CPU core. The collection can be observed when running with low heap size: after several seconds of high utilization (about 3.8), during collections the utilization drops down to 1 for a few seconds. From these measurements we conclude that the burst size has a high influence on the throughput. Therefore, we study the impact of the burst size more detailed in the next section.

4.3 Impact of Burst Messages

Adding and removing messages to and from queues is expensive since locking operations and calls to the OS are required. This leads to high context switch rates and high s if the message rate is high. Atomic operations for locking and switches to the OS and back as well as switches between threads can be reduced by employing burst messages.

The more messages are aggregated into one burst message, the lower s becomes and so the context switch rate. On the other hand, using burst messages with bigger sizes results in a higher memory consumption. Thus a trade-off between the context switch rate and the memory consumption must be found.

Fig. 5 shows the throughput related to the burst size for thread assignment F. As expected, the throughput increases with increasing burst size. Interesting

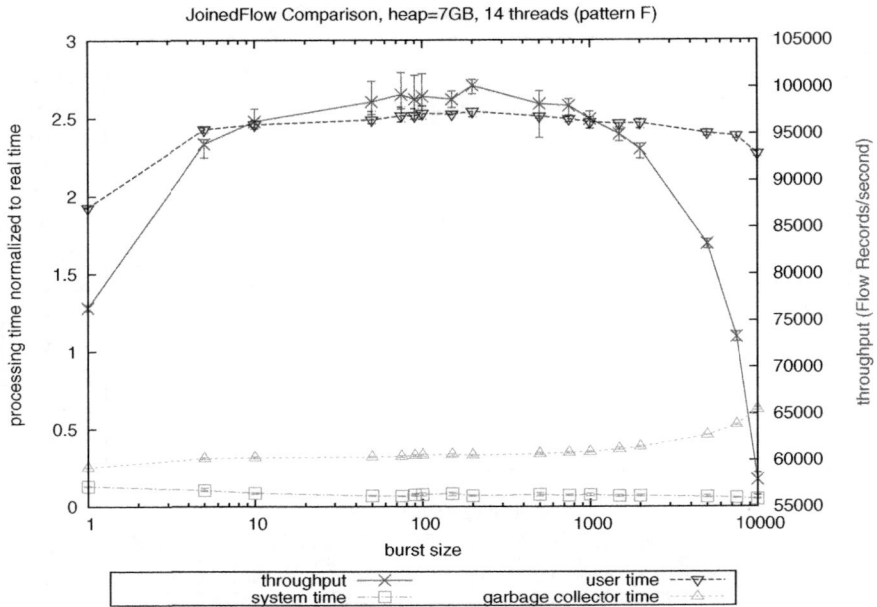

Fig. 5. Impact of burst messages

is the fact, that the normalized user time is more or less constant regardless of the configured burst size while the throughput increases. The reason is that the time axis is normalized to the real time. Thus it shows a kind of utilization. As we see from the results, the burst size should be greater than 100, otherwise the throughput is decreased by up to 25%. Much larger bursts do not lead to better performance, but eventually to performance degradation by up to 50% due to the high memory consumption of larger burst messages. As another effect the increasing garbage collection time can be seen. The reason is the above mentioned higher memory consumption of greater burst messages, so that the garbage collector has to free more memory.

With increasing burst size there is a decrease of s from 1.3% to under 0.5% of the real time. Considering the amount of lock operations required on the FIFO queues, we expected a direct relation of s to the burst size. This cannot be observed. One reason might be the Linux mechanism of fast user space mutual exclusion (futex) [11], which we have found out by using `strace`. So not every access to a FIFO queue results in a context switch, because the mechanism tries to do synchronization in most cases in user space without switching into kernel mode. While this mechanism is cheaper than kernel space semaphores, it is still costly due to atomic operations. Thus, the high locking frequency with low burst sizes has still impact on throughput and contributes to the performance degradation shown in Fig. 5.

5   Conclusion

We presented a framework for processing of Flow Accounting data in Java. The performance evaluation showed that on multicore architectures, using more threads than cores is beneficial despite the additional memory required. Additionally, the usage of burst messages further speeds up processing since less operating system interactions are required. We investigated the impact of the burst size and showed that from a certain size the memory consumption and therefore the overhead introduced by garbage collection reduces throughput. The framework seems to be suitable for online processing of Flow Accounting data received directly from the network.

References

1. Claise, B.: Cisco Systems NetFlow Services Export Version 9. RFC 3954 (Informational) (October 2004)
2. Claise, B.: Specification of the IP Flow Information Export (IPFIX) Protocol for the Exchange of IP Traffic Flow Information. RFC 5101 (Proposed Standard) (January 2008)
3. Kögel, J.: Extracting performance metrics from netflow in enterprise networks. Second Workshop on the Usage of NetFlow/IPFIX in Network Management (October 2009)
4. Cranor, C., Johnson, T., Spataschek, O., Shkapenyuk, V.: Gigascope: a stream database for network applications. In: SIGMOD 2003: Proceedings of the 2003 ACM SIGMOD international conference on Management of data, New York (2003)
5. Chandrasekaran, S., et al.: Telegraphcq: Continuous dataflow processing for an uncertain world. In: CIDR (2003)
6. The CoMo project, http://como.sourceforge.net/
7. Dübendorfer, T., Wagner, A., Plattner, B.: A framework for real-time worm attack detection and backbone monitoring. In: IWCIP 2005: Proceedings of the First IEEE International Workshop on Critical Infrastructure Protection, Washington (2005)
8. Marinov, V., Schönwälder, J.: Design of a stream-based ip flow record query language. In: DSOM 2009: Proceedings of the 20th IFIP/IEEE International Workshop on Distributed Systems: Operations and Management, Venice (2009)
9. Ciccarese, P., Larizza, C.: A framework for temporal data processing and abstractions (2006)
10. Spring Framework, http://www.springsource.org/
11. Franke, H., Russell, R., Kirkwood, M.: Fuss, futexes and furwocks: Fast userlevel locking in Linux. In: The Ottawa Linux Symposium (2002)

Fighting Spam on the Sender Side:
A Lightweight Approach

Wouter Willem de Vries, Giovane Cesar Moreira Moura, and Aiko Pras

University of Twente
Centre for Telematics and Information Technology
Faculty of Electrical Engineering, Mathematics and Computer Science
Design and Analysis of Communications Systems (DACS)
Enschede, The Netherlands
`w.w.devries@student.utwente.nl`, {`g.c.m.moura,a.pras`}`@utwente.nl`

Abstract. Spam comprises approximately 90 to 95 percent of all e-mail traffic on the Internet nowadays, and it is a problem far from being solved. The losses caused by spam are estimated to reach up to \$87 billion yearly. When fighting spam, most of the proposals focus on the *receiver-side* and are usually *resource-intensive* in terms of processing requirements. In this paper we present an approach to address these shortcomings: we propose to *i*) filter outgoing e-mail on the sender side and (*ii*) use lightweight techniques to check whether a message is spam or not. Moreover, we have evaluated the accuracy of these techniques in detecting spam on the sender side with two different data sets, obtained at a Dutch hosting provider. The results obtained with this approach suggest we can significantly reduce the amount of spam on the Internet by performing simple checks at the sender-side.

1 Introduction

Spam comprises approximately 90 to 95 percent of all e-mail traffic on the Internet nowadays [1,2]. To deal with all this unsolicited e-mail, companies have to spend computer and network resources, and human labour hours, which causes economic losses. It is estimated that worldwide spam causes losses from \$10 billion to \$87 billion [3] yearly.

Among the reasons why there is so much spam on the Internet is the cost of sending it: it costs virtually nothing. It is estimated that the transmission of a spam message is 10,000 times cheaper than receiving it [4]. Moreover, e-mail protocols do not impose any security restrictions on *how e-mail is sent*, which turns any machine on the Internet into a potential spam server.

To fight this massive number of spam, recipients rely on *receiver-side* techniques, such as automatic filtering employed by e-mail clients or mail server filters (such as SpamAssassin). Despite being to a certain degree efficient, these techniques present a major drawback: they are very resource-intensive [5]. Due to that, very often Internet Service Providers (ISP's) are forced to be more aggressive in rejecting e-mail messages upfront if they do not comply to very

F.A. Aagesen and S.J. Knapskog (Eds.): EUNICE 2010, LNCS 6164, pp. 188–197, 2010.

rigorous requirements [6], to prevent the filtering mail servers from overloading [6]. In turn, this method may lead to false positives, causing legitimate e-mail to be rejected. As a consequence, such aggressive techniques may compromise the reliability of e-mail communication while still not being able to filter all spam.

In face of these problems associated with sending and filtering spam, in this paper we present a new approach to fight spam, that focuses on *filtering spam on the sender-side* and *is not as resource-intensive* as traditional solutions. If it is so cheap to send spam messages, and so resource intensive to detect on the receiver side, why not tackle the problem by avoiding spam from being sent instead, at the sender side? If we could significantly restrain the transmission of spam, the amount of incoming e-mails would be much lower.

In order to detect spam on the sender-side, we propose the employment of four non-resource intensive techniques in terms of CPU cycles. These techniques focus on detecting spam based on e-mail message's core attributes, such as advertised URLs. In order to evaluate the effectiveness of these techniques, we have conducted a series of tests on a production network of a large Dutch hosting provider.

The rest of this paper is organized as follows: in section 2 we introduce the background and current solutions employed by ISPs to fight spam, and the short-comings associated with each solution. Next, in Section 3 we present our proposal for fighting spam on the sender-side using low resource-intensive techniques. After that, in Section 4 we present an evaluation of the effectiveness of these techniques. To do that, we have used data from real-life networks from a Dutch hosting provider. Finally, in Section 5 we present the conclusions and future work[1].

2   Background and Current Solutions to Fight Spam

In this section we present a study on the techniques used by spammers to conduct their campaigns. Moreover, we provide a review of related work and existing solutions employed to fight spam.

2.1   Background

Currently, three main approaches are employed by spammers to easily send vast amounts of spam at low cost while maintaining most or all anonymity: direct spamming, open mail relay, and botnets [8,9]. Direct spammers are those that use their own mail servers to send larges amounts of spam. Key characteristic of direct spammers is that they do not try to hide their activities and identity, they send large amounts of spam from a limited amount of IP addresses. Some of those people are publicly listed in the Register of Known Spam Operations (ROKSO) [10], so ISPs can recognize and reject them, should they request connectivity. Occasionally direct spammers even resort to hijacking entire IP prefixes [11].

---

[1] It should be noted that an initial version of this paper has been presented at the twelfth Twente Student Conference on Information Technology [7]. However, that was an internal event of the University of Twente, of which the proceedings have not officially been published by any publisher.

On the other hand, open mail relay is when mail servers are configured in a way that allows any user, authorized or not, to send e-mail through it. This allows spammers to send their spam almost anonymously, only the operator of the misconfigured e-mail server could find out about the identity of the spammer. For example, Sendmail version 5 was an open relay by default [8]. Many inexperienced administrators make this configuration mistake, so spammers actively scan the internet for open relays. E-mail server software is fortunately increasingly configured to be secure by default, so this issue is becoming less severe.

Finally, the last and more effective approach to send spam is using botnets. A botnet consists of many compromised hosts (named bots) [12], connected to a central control server maintained by a spammer. Those bots are typically Windows computers infected by a virus and owned by unaware end-users [13,14]. From the control server, the spammer commands the bots to transmit his spam campaigns. This method is rapidly advancing: it provides for a cheap, distributed and completely anonymous way to transmit spam.

In the literature, spamming host are classified according to the number of spam they generate. Pathak *et al.* [15] conducted a research on spammers' behavior by setting up an open relay and generating statistics on the spam they collected in a period of three months. They have observed the prevalence of two sets of spamming hosts: high-volume spammers (HVS) and low-volume spammers (LVS). The LVS is a set containing a high number of hosts, each sending a low volume of spam (like bots in a botnet). In contrast, The HVS is a set containing a low number of hosts, each sending a high volume of spam (like direct spammers).

2.2   Solutions Employed by ISPs to Fight Spam

ISPs are increasingly interested in preventing spam transmissions from their network, mainly due to technical and legal reasons. For instance, bad managed networks eventually results in a bad network reputation. By this ISPs are risking entire IP ranges being blacklisted, making it impossible for them to transmit the legitimate e-mail of their clients. In addition, IP blacklisting as a defensive method is expected to be much less efficient (much higher chance on false positives) in the next few years for ISPs on the receiver's side due to the rise of IPv6 [16]. In regards to legal reasons, several governments implemented legislation that requires the possibility to detect, restrain and penalize spammers. Examples include the American CAN-SPAM act of 2003 [17] and the Australian SPAM ACT 2003 [18].

In practice, the following solutions are used by ISPs to fight spam:

- Blocking SMTP port for end-users;
- Rate limiting (e.g., a maximum number of e-mail messages per hour);
- Filtering out e-mail using tools like SpamAssassin.

The simplest and most radical approach to fight spam is to block the SMTP port for end-users, and allowing connections only to the mail servers of the ISP.

Since SMTP does not impose security restrictions when sending e-mail messages, any computer on the Internet can be used as a mail server [19]. Blocking SMTP port 25 for connections outside the ISP's network block spams to other mail servers, but do not prevent spam from reaching the ISP's own mail severs. The drawback of this approach is that it severely limits the end-user in using different mail servers for different purposes.

Another approach consists in limiting the amount of allowed e-mails per hour that a client can send and the maximum number of recipients. It is a very basic measure and not effective against LVS, since each LVS only sends small amounts of e-mail over a long period of time [20]. In addition, legitimate e-mail activities (such as a newsletter) would also trigger this kind of measure.

The most used approach by ISPs is to filter incoming e-mail using mail filters like SpamAssassin. SpamAssassin performs a series of tests on the header and content of the messages to classify whether they are not spam [21]. It employs Bayesian filtering, DNS blacklist checks, among other techniques. The main disadvantage of this approach is that it is resource-intensive in terms of CPU cycles [22]. In addition, it is usually used on the receiver side, so it is not employed to filter out outgoing messages.

In summary, the main disadvantages of existing solutions are (*i*) being obtrusive to the end-user, (*ii*) being ineffective to modern spam methods, and (*iii*) being expensive. To fill this gap, in this work we propose a lightweight approach to filter out outgoing e-mail.

3   Proposed Solution

After identifying the shortcomings of existing solutions, we now present our proposal, that is based on two pillars: (*i*) on filtering outgoing e-mail on the sender side and (*ii*) using lightweight techniques to check whether a message is spam or not. The main advantage is that we can detect and block spam before it reaches the final destination at a low cost.

Figure 1 presents the architecture of our solution. In this figure, an ISP has a certain number of mail servers that are responsible for handling the e-mail traffic. When they receive the e-mail from their clients, instead of sending the e-mail directly to the other mail servers, they forward it to another host in the same network (the smarthost in the figure). This smarthost is responsible for classifying the outgoing message as spam or not, and filter out the spam ones.

To perform this e-mail message classification, we propose the use of lightweight techniques. The idea is to have very quick checks on each message that still allow good filtering results. To do that, we propose to filter e-mail based on the message's core attributes, since its requirements in terms of CPU cycles are low. Core attributes are described as elements of an e-mail which can not be easily modified or varied, such as advertised URL's. These elements can be used to develop decisive criteria, while ensuring the criteria are light on server resources [22]. For example, checking if an advertised URL is used in spam campaigns can be performed with a DNS lookup, which requires virtually no server processing power.

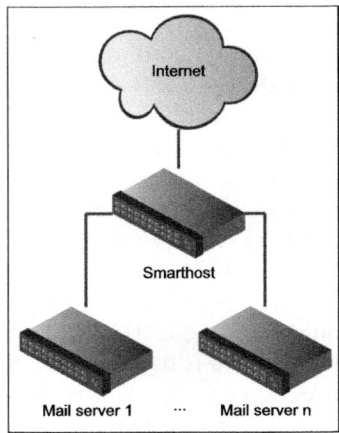

Fig. 1. Overview of new architecture

In this paper we propose the employment of the following techniques to classify a e-mail message:

- **Sender Policy Framework check (SPF):** An ISP's mail server should only transmit e-mail from domains which are actually hosted by the ISP. If an e-mail message is supposed to be sent through ISP mail servers but it does not come from a domain hosted by the ISP, it should be automatically discarded. To proceed with this verification, the filter must only check a DNS text file that contains all domains hosted by the ISP.
- **URL blacklist check:** The smarthost machine should check the URLs in the e-mail messages against several real-time blacklists. These blacklists contains URLs that are used by spammers in their campaigns. Example of these blacklists include: URIBL, SURBL, Day Old Bread, Spamhaus, Outblaze, and AbuseButler. The main advantage of this is that with one DNS lookup the filter can determine if an advertised URL is used for spam or not.
- **Google Safe Browsing (GSB)** [23]: In addition to the URL lists provided by the blacklists, the filter should check the URLs against Google Safe Browsing service. This service allows us to check the advertised URLs in real time with Google's own list of malicious hosts.
- **Full from-address check:** For each message, the filter should check the from-address field, to verify whether it exists or not. This can be quickly performed by connecting to the mail server listed in the MX record of the domain name and issuing a SMTP's RCPT TO command. To this command, the mail server will reply with a SMTP 250 or 550 code [19], which reveals whether the address exist. However, since this so-called sender callout can be quite abusive [24], we advise to only perform this check if the previous two checks suggest the e-mail is ham. This way the check will only be performed on servers owned by the user of this filter. Messages containing non-valid addresses should be discarded.

These four techniques employed by our smarthost are cheap in terms of CPU requirements, especially in comparison with the bayesian techniques. The idea is to use them all in combination to obtain a better classification, thus allowing us to have an scalable mail filter. In the next section we present an evaluation of the effectiveness of the presented techniques.

4   Evaluation the Accuracy of the Filtering Techniques

To gain insight on the effectiveness of the proposed filter, we performed analysis within the infrastructure of a large Dutch hosting provider. Due to privacy concerns, we were unable to test a proof-of-concept implementation because that would mean to inspect user's outgoing e-mail (which requires pre-authorization). To overcome this, we have conducted two different tests on real-life e-mail servers that could also give us the effectiveness of the proposed techniques.

The tests were conducted on two different data sets:

- **Dataset A – SpamAssassin log files:** These logs were obtained by analyzing the incoming e-mail to the mail servers. We have aggregated one week of log files.
- **Dataset B - E-mail Feedback Reports received from America Online:** America Online offers a service to ISP's which enables ISP's to gain insight in the spam which is being sent from their network [25]. For each e-mail which is reported as spam by an America Online user, America Online sends an E-mail Feedback Report to the originating ISP; these reports contain the headers and content of the reported spam e-mail. We aggregated these reports for one month, which means we have a set of outgoing spam e-mail to analyze. The set contains 324 EFR reports; however, it should be noted that this number of reports constitutes a far larger number of actual spam e-mails.

In the next subsections we present the results obtained for each technique.

4.1   On the Accuracy of SPF to Detect Spam

The Sender Policy Framework (SPF) check has been performed on the AOL EFR data set. For each received report, we took the domain part of the `from-address` of the spam e-mail, and we checked if this domain contained a SPF record pointing to a mail server of the hosting provider. If the domain had no SPF record or the SPF record allowed any mail server to send e-mail, we used the proposed fallback mechanism; we resolved the A record of *www.domain.tld*, and checked if this IP address was hosted in the network of the hosting provider.

Table 1 present the results for the accuracy of the Sender Policy Framework approach. As one can see, for the reporting period we received 324 EFR reports for AOL. Out of these, 125 could have been avoided if SPF was employed by the e-mail servers. This allowed us to have an accuracy of 38.6%. However, we

Table 1. Results extracted from E-mail Feedback Reports received in December 2009

AOL EFR Reports	SPF hit	from-address hit	GSB hit
324	125 (38.6%)	324 (100%)	165 (50.9%)

should stress that it does not mean that the approach is not good. In this case it means that 38.6% of spam was sent from non-authorized domains.

To achieve even better results, this technique should be combined with all four presented techniques in this paper.

4.2   On the Accuracy of Blacklists to Detect Spam

To evaluate the accuracy of the URL blacklist check, we have used the Dataset A (SpamAssassin log files for incoming e-mail). With this log, we could obtain the accuracy of employing blacklist for detecting spam simply by comparing the total number of incoming spam (obtained used several techniques by SpamAssassin) and the number of e-mails that were in fact classified as spam only because of the real-time blacklist. Even though we have conduced this test on *incoming e-mail*, we can expect similar results for *outgoing e-mail*, since this method is independent from source/destination.

The results obtained for the accuracy of blacklists for spam classification can be observed in Table 2. For example, on January 5th SpamAssassin has classified 7016 e-mails messages as spam. Out of this total, 51664 were correctly classified as spam using only one technique: the real-time blacklists. This gives an accuracy of 73.6% when detecting spam using only this technique.

Analyzing the results over the 8 consecutive days we have collected the data, we could observe that 36053 e-mail messages were classified as spam only by employing real-time blacklists. With these results, the overall accuracy was 74.4%. Even though it is not as high as we expected, this number can be improved when combining the next techniques together.

Table 2. Results extracted from SpamAssassin logs of 8 days in January 2010

Day	Spam e-mails	URLBL hits	Accuracy
2010/1/2	5,149	3,703	71.9%
2010/1/3	5,493	4,291	78.1%
2010/1/4	5,132	3,991	77.8%
2010/1/5	6,601	4,934	74.7%
2010/1/6	6,600	4,953	75.0%
2010/1/7	7,125	5,110	71.7%
2010/1/8	7,016	5,164	73.6%
2010/1/9	5,331	3,907	73.3%
Total	48,447	36,053	74.4%

4.3 On the Accuracy of Checking Full `from-address` to Detect Spam

The full `from-address` check was performed on the dataset B, the AOL EFR. For each received report, we connected to the mail server listed in the MX record of the `from-address` domain name and issued a SMTP's RCPT TO command. To this command, the mail server either replied with a SMTP 250 or 550 code, which reveals whether the address existed or not. A hit of this check constitutes a non-existing `from-address`.

Table 1 present the results for the accuracy of `from-address` approach. As one can observe, all the e-mail messages (324) reported by AOL contained fake e-mail addresses. This result was better than expected. We could guess that these addresses could have been randomly generated by bots while conducting their spam campaigns. These results suggest that maybe this technique might be enough to process EFR reports from AOL, for example.

4.4 On the Accuracy of the GSB Blacklist to Detect Spam

In order to access the accuracy of the GSB to detecting spam we employed the dataset B. For each received report, we checked each advertised URL against GSB. Due to the manual processing of the reports, and the time between the actual spamming activities and arrival of EFRs, there was a delay of approximately 24-48 hours between the transmission of the spam and our manual processing of the EFRs. Even though, we observed that the accuracy of using GSB we could classify 50.9% of the messages as spam, as can be observed in Table 1.

4.5 On the Overall Accuracy of Combined Techniques

In order to obtain better spam classification results, all the four mentioned techniques should be somehow combined. In this research we could not perform this, since our datasets were generated based on different sources (dataset A was obtained by analyzing outgoing e-mail while dataset B was obtained by analyzing incoming e-mail messages).

However, while deploying our techniques, all the four techniques should be employed to analyze the outgoing traffic so they can be used in combination. However, the most suitable way to combine them should be tested experimentally, which is left as future work. Despite this, we expected much better results when combining these techniques.

5   Conclusions and Future Work

In this paper we have proposed a new solution enabling ISP's and hosting providers to significantly restrain spam transmissions from their network at low cost. Our proposal is based on (*i*) filtering outgoing e-mail on the sender side and (*ii*) using four lightweight techniques to check whether a message is spam or not. The main advantage is that we can detect and block spam before it reaches the final destination at a low cost. Moreover, this solution can be easily deployed at ISPs and hosting providers, requiring only a few modifications.

By analysis within the infrastructure of a large Dutch hosting provider, we evaluated the accuracy of the four different techniques we presented for filtering spam based on core attributes. The two best performing individual checks achieved an effectiveness of 74.4% and 100% respectively. Despite the fact that these results were obtained on relatively small data sets, they are very encouraging: for example, just by checking blacklist, we can potentially reduce more than 70% of spam generated on the sender-side. If ISPs and hosting providers start to use such approach (for several reasons, including law enforcement, good reputation, etc.), we could virtually block at the source most of spam generated nowadays. By doing that, we could reduce the spend on people, hardware and network resources to deal with spam, since most of the messages would not be forwarded to other networks.

As future work, we intend to implement and deploy a prototype to investigate real-time performance and effectiveness of the presented approach. Moreover, we intend to conduct this analysis on a much larger data set, and determine the most suitable way to combine the four detection techniques to detect spam. Next, we compare the detection rate of a setup combining the four detection techniques with the performance of a more traditional approach such as using SpamAssassin solely. We intend to achieve this by letting both setups evaluate the same set of labeled data containing spam and ham messages. Finally, we will monitor and analyze the used system resources.

References

1. Sophos. Only one in 28 emails legitimate (June 2008),
   http://www.sophos.com/pressoffice/news/articles/2008/07/
   dirtydozjul08.html
2. Spamhaus. Effective spam filtering (January 2010),
   http://www.spamhaus.org/effective_filtering.html
3. Soma, J., Singer, P., Hurd, J.: SPAM Still Pays: The Failure of the CAN-SPAM Act of 2003 and Proposed Legal Solutions. Harv. J. on Legis. 45, 165–619 (2008)
4. Lieb, R.: Make spammers pay before you do (July 2002),
   http://www.clickz.com/1432751
5. McGregor, C.: Controlling spam with SpamAssassin. Linux J. 153, 9 (2007)
6. Mori, T., Esquivel, H., Akella, A., Mao, Z.M., Xie, Y., Yu, F.: On the effectiveness of pre-acceptance spam filtering. University of Wisconsin Madison, Tech. Report TR1650 (2009)
7. de Vries, W.W.: Restraining transmission of unsolicited bulk e-mail. In: Proceedings of the twelth Twente Student Conference on Information Technology (2010)
8. Stern, H.: A survey of modern spam tools. In: Proc. of the fifth conf. on email and anti-spam (2008)
9. Sperotto, A., Vliek, G., Sadre, R., Pras, A.: Detecting Spam at the Network Level. In: Oliver, M., Sallent, S. (eds.) EUNICE 2009. LNCS, vol. 5733, pp. 208–216. Springer, Heidelberg (2009)
10. Spamhaus. The register of known spam operations (January 2010),
    http://www.spamhaus.org/rokso/

11. Ballani, H., Francis, P., Zhang, X.: A study of prefix hijacking and interception in the Internet. ACM SIGCOMM Computer Communication Review 37(4), 276 (2007)
12. Fabian, M.A.R.J.Z., Terzis, M.A.: My botnet is bigger than yours (maybe, better than yours): Why size estimates remain challenging. In: Proceedings of the 1st USENIX Workshop on Hot Topics in Understanding Botnets, Cambridge, USA (2007)
13. Chiang, K., Lloyd, L.: A case study of the rustock rootkit and spam bot. In: The First Workshop in Understanding Botnets (2007)
14. Mendyk-Krajewska, T., Mazur, Z.: Software Flaws as the Problem of Network Security. In: Internet-Technical Development and Applications, p. 233 (2009)
15. Pathak, A., Hu, Y.C., Mao, Z.M.: Peeking into spammer behavior from a unique vantage point. In: LEET 2008: Proceedings of the 1st Usenix Workshop on Large-Scale Exploits and Emergent Threats, pp. 1–9. USENIX Association, Berkeley (2008)
16. RIPE Labs. Spam over ipv6 (March 2010),
    http://labs.ripe.net/content/spam-over-ipv6
17. United States of America. Can-spam act of 2003 (2003),
    http://uscode.house.gov/download/pls/15C103.txt
18. Australasian Legal Information Institute. Australian spam act 2003 (2003),
    http://www.austlii.edu.au/au/legis/cth/consol_act/sa200366/
19. Klensin, J.: Simple mail transfer protocol (April 2001),
    http://www.ietf.org/rfc/rfc2821.txt
20. Pathak, A., Qian, F., Hu, Y.C., Mao, Z.M., Ranjan, S.: Botnet spam campaigns can be long lasting: evidence, implications, and analysis. In: Proceedings of the eleventh international joint conference on Measurement and modeling of computer systems, pp. 13–24. ACM, New York (2009)
21. OBrien, C., Vogel, C.: Comparing SpamAssassin with CBDF email filtering. In: Proceedings of the 7th Annual CLUK Research Colloquium (2004)
22. Pras, A., Wanrooij, W.: Filtering Spam from Bad Neighborhoods (under review). International Journal of Network Management (2009)
23. Google. Google safe browsing (January 2010),
    http://code.google.com/apis/safebrowsing/
24. UCEProtect. Sender callouts - why it is abusive (January 2010),
    http://www.backscatterer.org/?target=sendercallouts
25. AOL. E-mail feedback reports for isp's (January 2010),
    http://postmaster.aol.com/cgi-bin/fbl.pl

Degradation Model for Erbium-Doped Fiber Amplifiers to Reduce Network Downtime

Christian Merkle

Lehrstuhl für Kommunikationsnetze, Fakultät für Elektrotechnik und
Informationstechnik,
Technische Universität München,
Arcisstr. 21, 80333 München, Germany

Abstract. The outage of optical components like optical amplifiers and
cross connects reduces the availability of a network and increases the
operational expenditures of network operators. Responsible for this is
the repair process of optical components which takes very long, because
of the large distances in the field. The detection of performance degradation of erbium-doped amplifiers (EDFAs) can be used to reduce the
repair time by changing an EDFA before it fails. With the knowledge of
the degradation state the remaining life time of an EDFA can be calculated, enabling the operator to plan the replacement of EDFAs to avoid
longer outage times. In this paper, an 1,000 km link and a German backbone network are used to simulate the aging of EDFAs with aging curves.
Additionally the operational expenditures are considered to compare the
cost of the repair process of EDFAs.

Keywords: EDFA, aging curve, OpEx.

1 Introduction

In today's optical networks EDFAs are widely used because they can amplify
many signals on different wavelength at the same time. The most commonly used
amplifier in transmission systems is the erbium-doped fiber amplifier (EDFA) [1].
Such amplifiers may fail during their operation, for example due to aging effects
of the pump laser, and lead to transmission failure on a link. Due to the high
bandwidth which are used on backbone network links today an outage of an
amplifier leads to a high loss of data packets transmitted over the failed link.
This also has an impact on the revenues of the provider, who has to pay a penalty
to the customers affected by the link outage. Hence, network operators monitor
their networks to get information of the current network status. If a failure occurs
in a backbone network, an alarm is sent from the network management system
to the network operator center (NOC). Based on the failure alarms, the operator
tries to localize the failure first and solves it afterwards. Today optical amplifiers
are not monitored by the management system. The outage of an EDFA takes
very long because the technician has to travel to the amplifier to repair it.

In this paper the monitoring of the aging of an EDFA is addressed which
enables the operator to plan the repair process and to replace the EDFA before

F.A. Aagesen and S.J. Knapskog (Eds.): EUNICE 2010, LNCS 6164, pp. 198–208, 2010.

it fails. Thus, the repair process can be reduced which increases the network availability. In section 3 the aging of the EDFA performance is discussed and in section 4 two different models, a 1,000 km link and a Germany reference network, are discussed to compare different change strategies for EDFAs. In section 5 the operational expenditures (OpEx) are analyzed for the repair process.

2   Related Work

A detailed description of the functionality of an EDFA can be found in [1]. An EDFA amplifies the signal by using a pump signal from a laser, typically at a wavelength of 980nm or 1480nm. But there are several imperfections that a system designer need to worry about when using an amplifier in a system. One disadvantage is the degradation of the performance due to aging of the components. An algorithm how the aging effects of an EDFA can be detected is described in [3]. The basic concept of the algorithm is to calculate the pump diode current that would be needed to create the measured gain at begin of life (BOL) conditions and to compare the results with the actual pump current. A pump laser of an optical amplifier is used to get a constant amplifier gain. With the monitoring of the actual pump current and the data from the data sheet of an amplifier, a network provider has a rough indication of the remaining life time of an amplifier.

In the literature much research is done in the field of fault identification and localization in optical networks as it is presented in [4], [5], and [6]. The first two papers describe different algorithms to detect the location of soft and hard failures. Because the outage time of a network depends also on the time to find a failure, this is a very challenging task, which can reduce the downtime of network links. In [6] the optimal placement of monitoring equipment is described to reduce the number of generated alarms in a network. The goal of all these methods is to identify the components whose failure has caused the received alarms. In our paper we describe also a monitoring concept, but in our concept the degradation of an EDFA is sent to a management system to replace it before it fails. Hence, the repair process of an EDFA can be planned.

The planning of the repair process is important for a network operator to reduce the OpEx of the network. In [7] a methodology to calculate CapEx and OpEx of a telekom operator is described. Different input parameters like staff cost, energy cost, and the description of the repair process and the provisioning process are identified to calculate the OpEx of a German reference network. The OpEx calculation for the German reference network indicates that the repair processes are the second largest cost factor of the OpEx of a network. Hence, with the optimization of the repair process, the OpEx of a telecom operator can be reduced.

In [8] a total cost of the network ownership model is presented. The paper describes the different factors which contribute to the capital expenditures and operational expenditures. The OpEx are defined as the cost related with the operation of a network. Several types of cost are described which can be distinguished for the reparation sub-process of a network failure. These factors are used in section 5 when the cost of the EDFA repair process are described.

3   EDFA Degradation Model

In this section the degradation of the pump current of an EDFA and the dimensioning of the pump current threshold are described. As mentioned previously, every amplifier requires a pump laser to bring the erbium ions into a higher energy band to amplify an incoming WDM signal [1]. Due to aging effects of the amplifier, the pump power of the amplifier has to be increased to obtain a constant gain. In this paper the pump current is used for the description of the pump power. The aging of an EDFA can be described by aging curves as shown in Fig. 1. The curves show the necessary pump current of an EDFA over a certain time period to achieve the same amplifier gain. There are different aging curves because not every EDFA has the same aging characteristic. It is possible that some EDFAs degrade faster than others. The second characteristic of the aging curves is that they have no constant gradient over the life time of an EDFA. At the begin of life the pump current gradient is higher than at the end of life. The begin of life pump current shown in Fig. 1 was taken from a data sheet of an EDFA. The used pump current of the EDFA at the begin of life is 550 mA. Beside the BOL pump current an EDFA has also a maximum pump current up to which it can be operated without it fails. This parameter is also given by the data sheet of a manufacturer of an EDFA. Here, the maximum pump current which can be used is 670 mA. With the algorithm described in [3] the actual needed pump current can be calculated. With this knowledge the pump current threshold which indicates that the amplifier has to be replaced can be calculated as the difference between the BOL current and the actual pump current. The

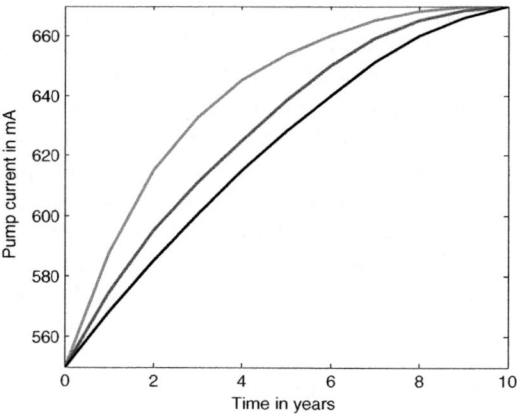

Fig. 1. Different aging curves of EDFAs

network operator has to define the pump current threshold for the EDFAs. The value of the threshold depends on the remaining life time an operator wants to have for the EDFAs. When an EDFA reaches the threshold an alarm message is send to the network operation center (NOC) to inform the operator that it has to be replaced.

The advantage of having a remaining life time for an EDFA is that the network operator does not need to change the EDFA immediately and can plan the repairing process. If an EDFA has a remaining life time of one year, for example, the operator can wait if further EDFAs reach also their degradation threshold in a certain time period after the first EDFA signals its degradation state. Hence, the repair team in the field has the possibility to repair more than one EDFA in one shift. Also the network availability is increased because the traveling time of the repair team does not account to the outage time of an EDFA anymore. Considering only the link breaks is not sufficient to analyze the availability of a backbone network. Also the repair time of the EDFAs and, hence, the downtime of the link are important factors for the network availability. An analysis of the repair time and link breaks is presented in section 4.

Another goal for the operator is to reduce the OpEx of backbone networks. One part of the OpEx is the repair process of failed network components. If the repair time of an EDFA can be planned by an operator, the traveling time can be optimized which reduces the traveling overhead and also the cost of the repair process. Further cost savings can be achieved due to the reduction of spare parts and stock size. The knowledge about the remaining life time of an EDFA can be used to order a new unit not until an old unit reaches the degradation threshold. Hence, only a minimum of spare parts is needed to repair an unforeseen outage of an EDFA for example. The OpEx of the repair process are discussed in section 5.

The signaling of the degradation threshold of an EDFA can be done with a network management protocol like the Simple Network Management Protocol (SNMP). All EDFAs in a network periodically send their actual pump current to a NOC. If an EDFA reaches its degradation threshold, it sends an additional message to the NOC to inform the operator that the degradation threshold was reached. The time interval of the EDFA aging messages are not constant over the EDFA life time. At the begin of life the management messages are sent in a larger interval and when the actual pump current is closer to the degradation threshold the management messages are sent more frequently.

4   Simulation Setup and Results

In this section two different simulation scenarios are analyzed. First, a link with a length of 1000km is considered to simulate two different strategies to replace degraded EDFAs. In the second simulation scenario, a German 50 node reference network is used to analyze the repair process in a larger network. For both scenarios an EDFA spacing of 80 km is used. To simulate the aging of the pump current of an EDFA the curves in Fig. 1 are used. In the beginning of the simulation, one of the aging curves is randomly assigned to every EDFA. The replacement of an EDFA corresponds to a random allocation of a new aging curve. The simulations and the analysis of the results are done with Matlab. For every EDFA model 10 independent iterations have been conducted to avoid statistical aberrations. The main simulation parameters are shown in table 1.

Table 1. Simulation parameters

Simulation parameters	1000km link	German 50 node network
FIT rate of EDFAs	2850	2850
Degradation curves of EDFAs	5	5
Used EDFAs	12	241
Simulated time period in years	30	100
Simulation runs	10	10

4.1 Consideration of a 1000km Network Link

For the analysis of the 1,000 km link 12 EDFAs are used. These 12 EDFAs do not have the same aging process and, hence, degrade in different rates. The fiber at the link also has a failure rate which is given in failure in time (FIT). For the simulations $380, 22 \frac{FIT}{km}$ are used for the fiber. One FIT means one failure per 10^9 hours. It is assumed that two repair teams are at one NOC which is located in the middle of the link at 520 km. Hence, every repair team has to maintain one half of the 1,000 km. The spare parts are also located at the NOC, hence, there is no additional traveling to a stock required. In the following two different replace strategies are analyzed for the 1,000 km link.

In the first simulation scenario, only the EDFAs which reached the degradation threshold are changed either at the end of every year or if a fiber fails. If the link is down due to a fiber failure, it is expedient to also replace all ED-FAs which received the degradation threshold because further downtimes due to EDFA changes are avoided. For the simulations a time period of 30 year is considered to determine the average time period between two replaced EDFAs over a certain time period. In Fig. 2 the total number of changed EDFAs and the number of link breaks in 30 years are shown. Three different pump current thresholds were taken for the simulations. In average 67 EDFAs have to be changed in 30 years when a threshold of 620 mA for the pump current is used. The replacement of these 67 EDFAs lead to 25 link breaks in 30 years which means that there was a link break almost yearly. If a threshold of 660 mA is used for the pump current 4 link breaks occur in 30 years and lead to 13 replaced EDFAs. As expected the higher threshold leads to less replaced EDFAs and, hence, to less link breaks. The disadvantage of a higher threshold is that the rest life time of the EDFAs is shorter compared to threshold of 620 mA. Because the operator does not know how the aging curve of the EDFA exactly looks like, he can only determine a time interval when the EDFA will fail with a certain probability.

A further change strategy of the EDFAs was simulated to compare it with the results of the first scenario. In the second scenario all EDFAs at the link

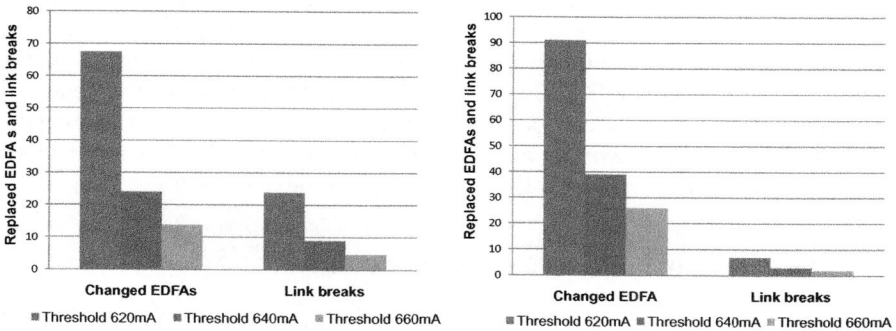

Fig. 2. Number of EDFA outages and link breaks for scenario 1

Fig. 3. Number of EDFA outages and link breaks for scenario 2

are changed if one EDFA reaches the degradation threshold or the fiber fails. In Fig. 3 again the average number of replaced EDFAs and the number of link breaks are shown. For a threshold of 620 mA, the average number of changed EDFAs is 84. Compared with the values of the first scenario it can be seen that, as expected, more EDFAs were changed in 30 years. The advantage of the second change strategy appears if the number of link breaks is considered. The 84 replaced EDFAs lead to 7 link breaks. Hence, in the second scenario the number of link breaks is reduced about 72% in comparison to the first scenario. The number of link breaks are further reduced if the threshold is increased. For a threshold of 660 mA the number of link breaks is 2.

These two cases illustrate that the replacement of more than one EDFA on one link at the same time can reduce the number of link breaks. Hence, if one EDFA has reached its degradation threshold it can be an advantage also to replace other EDFAs on the same link. Therefore, the current value of the threshold of the other EDFAs on the same link can be an indication which of the EDFAs the operator should replace additionally.

To further evaluate the different replace strategies the repair time to replace degraded EDFAs is calculated. For both scenarios, a degradation threshold of 620 mA is assumed. In the first scenario only the EDFAs are replaced which have reached their degradation threshold. The repair time t_{Repair} for both scenario is given by the maximum of the repair times of both teams and the travel time $t_{EDFA_X-EDFA_Y}$ between the EDFAs.

$$t_{Repair} = max(t_{RepairTeam1}, t_{RepairTeam2}) + t_{EDFA_X-EDFA_Y} \qquad (1)$$

The repair time of each EDFA is 2h, as described previously. The traveling time to the first EDFAs can be neglected for both scenarios because it is assumed that the link is disabled not until the repair team has arrived the first EDFA. The travel time $t_{EDFA_X-EDFA_Y}$ depends on the distances between the EDFAs that have to be replaced. For the first scenario, an average distance of 200km was obtained. Hence, the repair time t_{Repair} is 171.42 h in 30 years. When all EDFAs are changed if one or more EDFA have reached their degradation threshold, the total repair time is 126 h. In this model the travel distance is constant and has a length of 420km. The result shows that the repair time and also the downtime of the link is reduced by 26.5% in comparison to the first scenario. Hence, the availabilty of the network can be increased if always all EDFAs are changed on one link. Besides the downtime also the OpEx of network has to be considered. The OpEx of the two scenarios is discussed in the next section.

4.2 Germany 50 Node Reference Network

In this section the Germany 50 node reference network [11] is used to compare the EDFA degradation model with a FIT rate model. The reference network consists of 50 nodes and 87 links and has an average link length of 131 km. The total link length of the network is 8,874.13 km and the number of used EDFAs is 241. There are two NOCs in the network, each responsible for half of the 241 EDFAs. For the simulations a time period of 100 years was used to analyze the total number of replaced EDFAs and the number of EDFAs which can be repaired at the same shift. The difference between the two models is the time when an EDFA is replaced. By the FIT rate model the EDFAs are replaced after they failed as it is done in networks today. We take into account a time period of 100 years to avoid transient effects of the simulations of the degradation model. In this model all EDFAs are considered as new installed equipment in the network which is not true in a real network. Hence, the analysis of the results is done after 10 years, when the first EDFAs are changed. The following assumptions are used for both models: the number of required technician to repair an EDFA is 1 and it takes 2 hours to repair and test an EDFA [7].

Simulation with FIT rate. In this model the number of EDFA outages for a EDFA FIT rate of 2,850 [2] are simulated. The simulation results of this model serve as reference values for the EDFA degradation model which is analyzed in the next section. The failure probability of an EDFA in the time interval t is given by:

$$P(T_{out} \leq t) = 1 - \exp(-\lambda_{FIT} \times t) \qquad (2)$$

Hence, the probability that exact one EDFA fails within one year is 2.47% and the probability that at least one EDFA fails within one year is 99.76%. This failure probabilty leads to 5.77 failed EDFAs per year in average. The average repair time is calculated with equation 3

$$\bar{t}_{Repair} = \frac{\sum_{k=1}^{N}(t_{NOC-EDFA_k} + t_{repair})}{N} \qquad (3)$$

and is 6.69 h for each EDFA. N is the number of failed EDFAs, $t_{NOC-EDFA_k}$ is the traveling time from the NOC to the $EDFA_k$, and t_{repair} is the repair and test time of an EDFA. With this values a whole repair time of 3,857.97 h in 100 years is obtained.

Simulation with degradation threshold. By using the pump current threshold described in section 3, an EDFA can be changed before it fails. This allows the planning of the repair process and the reparation of more than one EDFA in one working shift. This means that a repair team can plan a route between the NOC and the EDFAs they have to repair. The goal of planning the repair process is to repair as much EDFAs as possible without traveling back to the NOC. An example for a repair process with two degraded EDFAs is illustrated in Fig. 4 and in Fig. 5. In Fig. 4 the two EDFAs are repaired independently from

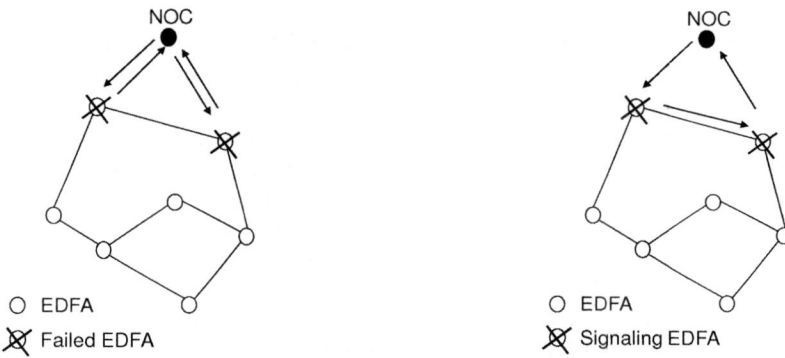

Fig. 4. EDFAs repaired independently   **Fig. 5.** Two EDFAs repaired in one shift

each other. The repair team must always travel from the NOC to the EDFA and back. This is the case in the FIT rate model described above. In Fig. 5 the repair process was planned in advance and the two EDFAs can be repaired by one repair team in one cycle. In order that it is worth to repair the two EDFAs in on cycle the travel time t_{travel} must be shorter than the sum of the single traveling times to the single EDFAs as shown in Fig. 4. The repair time for an EDFA is the same for both cases and can be neglected. The traveling time for the repair process in Fig. 5 is given by following equation

$$t_{travel} = t_{NOC-EDFA_x} + x \times t_{EDFA_x-EDFA_y} + t_{EDFA_y-NOC} \qquad (4)$$

The whole travel time t_{travel} of the technician is the sum of the traveling time from the NOC to the first EDFA $t_{NOC-EDFA_x}$, the traveling time between the EDFAs $t_{EDFA_x-EDFA_y}$, and the traveling time from the last EDFA back to the NOC t_{EDFA_y-NOC}. When the degradation curves in Fig. 1 are used, the average total number of replaced EDFAs is 7.2 per year and in 42% of the 100 years it was possible to repair two EDFAs in one shift. In the other years, the distance between the replaced EDFAs is to large to repair it in the same shift. The results are shown in table 2.

For the degradation model, the number of replaced EDFAs is higher then for the FIT rate model. This is caused because the EDFAs are replaced before they

Table 2. Major results for the German reference network

Parameters	FIT rate model	Degradation model	
Outage probability	2850	aging curves	-
Replaced EDFAs per year	5.77	7.2	-
Total travel time	2703.97h/100years	3036.43h/100years	+10.95%
Total repair time	3857.97h/100years	1546h/100years	-60.03%

fail. A further result is that it was only possible to repair two EDFAs in one shift because of the large distances between a NOC and the EDFA. If more than two NOCs will be used for the simulation it should be also possible to repair three EDFAs in one shift. The small number of NOCs is also a reason for the small number of years in which two EDFAs can be repaired in one shift. The advantage of the degradation model can be seen by considering the repair times of the EDFAs. Because the EDFAs are changed before they fail the traveling time does not account to the total repair time. Hence, the total repair time for the EDFAs is only the repair time of the replaced EDFAs and is 1544 h per 100years. This means that the repair time and hence, the downtime of the network is reduced by 60% in comparison to the FIT rate model. But in the degradation model, the traveling time is higher because more EDFAs had to be replaced than for the FIT rate model. As described above it was possible to repair two EDFAs in one shift in 42% of the 100 years and this lead to a reduced travel distance of 26,520 km. With an average speed of $70\frac{km}{h}$ the saved traveling time is 378.86 h. Hence, the whole travel time is 3,036.43 h per 100 years for the degradation model. Thus, the travel time is increased by 10.95%.

5   OpEx Analysis of the Repair Process

In this section the operational expenditures (OpEx) of the repair process of the EDFAs are discussed. For the OpEx calculations, only the Germany reference network is considered. Important for the cost calculation are the basic salary of a technician, the repair time for an EDFA, and the traveling time in the field. For the calculation, it is assumed that the stock is located at the NOC and the repair team must not drive there to collect the spare parts. In the following, the FIT rate model and the degradation models are analyzed.

5.1   OpEx Analysis for EDFA Outages Given by the FIT Rate

To calculate the cost of the repair process following equations are used:

$$c_{Repair} = (c_{Salary} \times t_{repair}) + c_{EDFA} + c_{stock} \tag{5}$$

$$C_{Stock} = 0.05 \times P_{EDFA} \times N_{EDFA} \tag{6}$$

The cost of the repair process is the sum of the cost for the technician, the cost of the EDFA, and the cost of the stock. Further cost like gas are neglected for OpEx calculations. The basic salary of a technician is assumed to be US $36.19 per hour. This value is based on a network technician salary per year of US $37,000 [10]. The cost of an EDFA is US $11,850[7] and the cost of maintaining the stock is assumed to be 5% of the equipment cost and is given in eqn. 6. How much the operator has to pay to maintain the stock depends on the number of EDFAs in the stock and the price of an EDFA. The analysis of the FIT model has shown that the repair time is 3,857.97 h and that 557 EDFAs had to be replaced per 100 years. Because in average 6 EDFAs are replaced per year it is assumed that 6 backup EDFAs are in the stock. With these values and the equations above the total cost of the repair process are US $70,955.69 per year.

5.2 OpEx Analysis for the EDFA Degradation Concept

For the degradation model the same equations are used to calculate the cost of the repair process. It is assumed that only one EDFA is in the stock because the operator has enough time to order a new EDFA if one needs to be replaced. With the values obtained by the analysis of repair time, the cost of the repair process are US $87,570.15 per year. As it can be seen from the results, the OpEx of the FIT rate model are minor then those of the degradation model. But it has to be mentioned that the penalties that a operator has to pay if he cannot fulfill the service level agreements are not included in the OpEx calculations. Furthermore, the selected pump current threshold for the simulations was 620 mA which means that the EDFAs have in average a rest life time of 6 years. If a higher threshold is used the OpEx for the degradation model should be minor.

6   Conclusion

A degradation concept for EDFAs has been presented that allows the change of EDFAs before they fail. Every EDFA has a specific aging curve of its pump current which is used to determine the remaining life time of an EDFA. The remaining life time depends on the pump current threshold which is set by the network operator. The evaluation of the link model shows that the replacement of all EDFAs on one link reduces the downtime of the link by 26.5%. Furthermore, the degradation model was compared with a FIT rate model of EDFAs where the EDFAs are only replaced when they fail. This was done by using a Germany reference network. The results show that for the degradation model the downtime of network links can be reduced by 60% because the traveling time to the EDFAs can be neclected. But the former change of the EDFAs leads to a higher number of changed EDFAs in comparison to the FIT rate model. Finally, the OpEx for the repair process of the Germany reference network are analyzed. The results show that the repair cost of the degradation model is higher, but service level agreements penalties were not considered in the calculation which increase the OpEx of the FIT model.

Acknowledgement. The author is grateful to Lutz Rapp (Siemens AG) for insightful discussions about the aging of an EDFA.

References

1. Ramaswami, R., Sivarajan, K.N.: Optical Networks - A Practical Perspective. Morgan Kaufmann Publishers, San Francisco (2002)
2. Mello, D.A.A., Schupke, D.A., Scheffel, M., Waldman, H.: Availability Maps for Connections in WDM Optical Networks. Design of Reliable Communication Networks (2005)
3. Rapp, L.: Quality Surveillance Algorithm for Erbium-Doped Fiber Amplifiers. Design of Reliable Communication Networks (2005)
4. Mas, C., Thiran, P.: An efficient algorithm for locating soft and hard failures in WDM networks. IEEE Journal on selected Areas in Communications 18(10), 1900–1911 (2000)
5. Mas, C., Nguyen, H.X., Thiran, P.: Failure location in WDM networks. Optical WDM Networks: Past Lessons and Path Ahead (2004)
6. Stanic, S., Subramaniam, S., Choi, H., Sahin, G., Hyeong-Ah, C.: Efficient alarm management in optical networks. In: DARPA Information Survivability Conference and Exposition (2003)
7. Verbrugge, S., Colle, D., Pickavet, M., Demeester, P., Pasqualini, S., Iselt, A., Kirstädter, A., Hülsermann, R., Westphal, F.J., Jäger, M.: Methodology and input availability parameters for calculating OpEx and CapEx costs for realistic network scenarios. Journal of Optical Networking 5, 509–520 (2006)
8. Machuca, C.M.: Expenditures Study for Network Operators. Transparent Optical Networks 1, 18–22 (2006)
9. Zhang, K.J., Li, S.Y., Liou, L.W., Corless, J.D., Dominguez, M.: Bandwidth-expandable erbium-doped fiber amplets. IEEE Photon. Technol. Lett. 13(4), 281–283 (2001)
10. Payscale: Salary Survey for Job: Computer/Network Support Technician, http://www.payscale.com/research/US/Job=Computer_%2f_Network_Support_Technician/Salary (last seen 2010)
11. SNDlib: Germany 50 node reference network, http://sndlib.zib.de/home.action (last seen 2010)

A Token Based Approach Detecting Downtime in Distributed Application Servers or Network Elements

Sune Jakobsson

Telenor ASA Global Development, Otto Nielsens vei 12,
7004 Trondheim, Norway
sune.jakobsson@telenor.com

Abstract. This paper defines and outlines and proposes a heuristic method to detect unavailability on a practical service created on the Internet. It uses a token that is passed between all the nodes. The algorithm is under trial in a set of servers spanning over multiple administrative domains. This paper confirms that using a simple token passed through the contributing applications servers, is able to detect unavailability without intruding or changing the service that one wants to monitor.

Keywords: Application availability, Web Services.

1 Introduction

Application availability is becoming an increasing concern for users. They use or create services composed from servers origination on different administrative domains, and have hence lost the possibility to observe or monitor the real availability of a given service. Today services built on the Internet make use of data, processing nodes, and services spread on a number of nodes, that belong to different administrative domains, and even spread on multiple continents and countries. The Internet is now of such scale and size, that it is no longer possible to have administrative control over all parts of a service. Virtualization of hardware is common with provides like Google, Sun, Microsoft and others, where users can acquire resources on a pay-per-use basis. Typical resources are application servers and storage, but also access to more specialized services like maps. With the introduction of virtualization all low level faults are masked by the virtualization, but that does not mean that all the challenges with availability are met. Communication, network, routing and node name look-up are still hampering even simple services if their usage is infrequent. Additionally Web Services are becoming the de facto model of interaction on the Internet, where software components are loosely coupled together using standard communication protocols. The lack of firm bindings between the components hence complicates the capacity and resource monitoring. User's perception of services with input and response, is that if there is no response in approximately 8 seconds say, the users perceive the service to be non-operational.

An example of such a service would be an application, showing the location of individuals on a map, as shown in Fig 1. In this example server labeled A, runs the

F.A. Aagesen and S.J. Knapskog (Eds.): EUNICE 2010, LNCS 6164, pp. 209–216, 2010.
© IFIP International Federation for Information Processing 2010

application, and use data or services from the other servers. Server A has an application server that ties the entire service together and the result can be browsed or dispatched to the mobile terminal as a multi-media message. The Server labeled B provides maps, the server labeled C provides the registration and group functions, and relies on the server labeled E, that is a database where the users are registered. The assumption is that all the servers are all managed independently of each other and possibly in different administrative domains. If the usage of such a service is infrequent, and at random times, the challenge is to know when parts of the nodes are unavailable, due to maintenance or failure, or due to network changes or outages. When this knowledge is available, a user could be given the option to use an alternative node configuration, either manually or automated in some way.

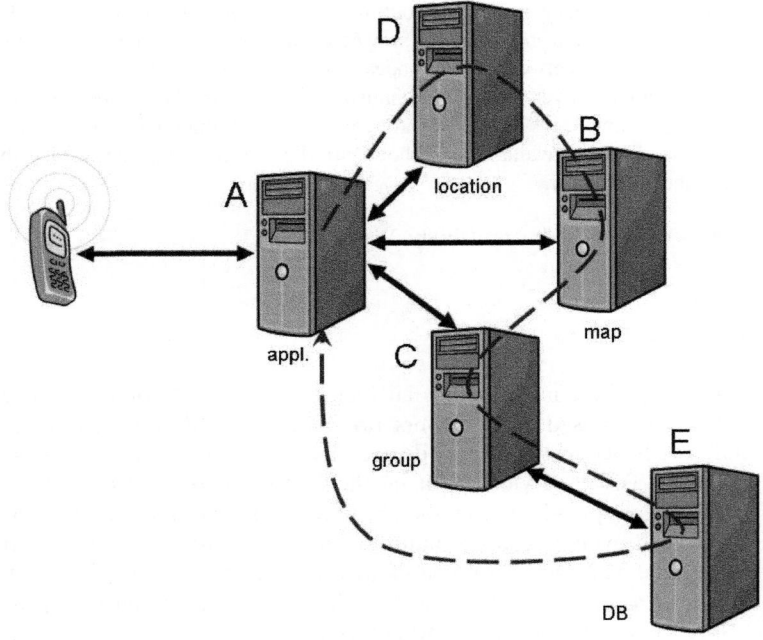

Fig. 1. One example service on the Internet consisting of five application servers

The obvious way would of course to try to use the service at some predetermined interval, but that might involve manual procedures, and could also come at a cost, if there is a fee for the usage of the nodes or their services. It is reasonable to assume that the nodes use some higher level communication protocol, like HTTP and that the nodes expose themselves as Web Services.

The paper starts with a high level background of application availability, and then describes the approach in detail. Traditional approaches to this subject have often focused on solving this on a local domain, but this paper suggest a alternative approach to solve the problem both in a local domain and in multi administrative domains. This is done by circulating a small data structure, referred to as a token.

Then the algorithm is explained, backed by some practical results, showing occasions where a given service would suffer form either application or network outages.

2 Related Work

Application availability and SW availability has and is studied in detail by many researchers and organizations, and there are many techniques to accomplish this. However most of them are focused on redundancy, watchdogs, probing or monitoring on a local domain or cluster of nodes [3]. In virtualization schemes HW resources are distributed and clustered across many users. At the lowest levels like physical storage, disk drives provide magnetic storage, but in order to obtain high availability they are often configured in a redundant configuration like for example RAID, where the data elements always exist in multiple copies in case of failure. At higher levels like operating systems, or application serves, the resource usage is distributed with load balancing techniques, and monitored for their responses, in order to detect failures, and necessities for switching over to other resources. Porter [1] gives a good overview of challenges to pinpoint failures in huge server parks on the Internet.

3 The Approach

Given the Internet picture and typical usage, like the example outlined in the introduction, there is a need for a mechanism or service that is a subset of the real service, yet using or exercising the basic properties of the connectivity and functionality used in the service. Assuming that the nodes are able to communicate and capable of for example HTTP communication patterns, passing a small data object around like shown with the dotted arrow in Fig. 1 would reveal if the contributing servers and the network communicate. Here a node, in this example node A, that also has the information of the participating servers or nodes, creates a list of http addresses, and some administrative data, and passes this to the next server or node on the list. The receiving server or node does the same but uses the next entry on the list. The final entry on the list can be the originating server or node, but this is only necessary in particular cases, where trip time measurements are desirable.

3.1 Assumptions

The "TCPDUMP" tool [6], which is part of most operating system, or can be installed, would be able to capture any or selected traffic between two IP endpoints, and pass the log files to a central node for further processing and analysis. However the capture process can be resource intensive, and would not cover the aspects of the application level availability. In order to be able to monitor a given node at application level, some software must be installed. Another alternative is use of tools like "PING" or "TRACEROUTE" to verify connectivity between nodes, but this does not detect application server internal issues. There are also tools available that require installation of probes, that in turn sends the information to a central or distributed management platform, but this requires commercial agreements across many administrative domains. The traffic generated by the probes, might get lost, and also have a volume that consumes unnecessary bandwidth on the communication links. Some of the probe mechanisms do require dedicated communication ports, requiring changes to firewalls and communication links, and in some administrative domains this is not a feasible path, due to restrict policies on configurations of firewalls and communication links.

The Web Service world on the Internet makes extensive use of servlet techniques, to be able to split applications in multiple tiers. Typically they are deployed with a HTTP front-end, an application logic layer, and a database layer, hence the term 3 layer architecture. The easies way to accomplish this is with WAR files, which are deployed into J2EE servers. Making the token passing as a simple service in this way will make use of the front-end and the application logic, and hence cover a minimalistic, yet functional set of that particular server. This works equally well with virtualized servers. One reason for choosing HTTP as token bearer is that most firewalls are open for HTTP traffic, and no network configuration is required.

3.2 The Token

The data structure that is passed around is a list of stings and a time reference, a reference to the next node, which is circulated. The data structure is shown in Fig. 2. Where the first server is Server D and the last server on the route is Server A.

— **Token identity number**
— **Token dispatch time**
— **Pointer to next URL**
— **http://serverD.example.com/Echo_Token**
— **http://serverB.example.com/Echo_Token**
— **http://serverC.example.com/Echo_Token**
— **http://serverE.example.com/Echo_Token**
— **http://serverA.example.com/Receive_Token**

Fig. 2. The internal structure of a token

This data structure is the passed around, and the pointer is incremented with one, for each node that receives the list. A typical length of this list is 5-10 URL's, that make up the service in question, that one wishes to monitor.

3.3 Token Flow

The token is created in the generator, and then passed through a set of nodes called the echo instances, according to the list, and finally ends up in a receiver node, see Fig. 3. The current version of the algorithm is implemented in Java, and writes data to log files locally on each node. The data contains a timestamp, available free memory in the Java virtual machine, HTTP status of the request, and the time it used to do the trip end to end. The data collected on the sender side is shown in Fig. 6.

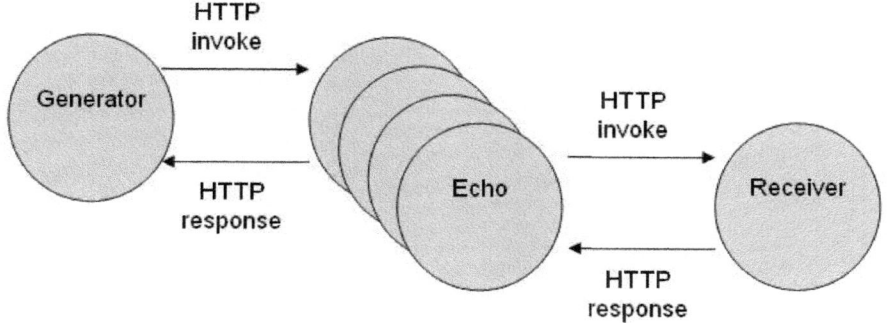

Fig. 3. Passing the token end-to-end

3.3.1 Generator

The token is dispatched at fixed 30 second intervals, in order to minimize the bandwidth consumption of the token itself. Since the assumption is that 10 seconds would be the maximum tolerable delay for a user, we measure the trip time, as part of the token passing. The use of the HTTP protocol built on top of TCP protocol, means that the invocation is robust and capable of internal retries in case of network faults. However if the IP link does not return to a operational state, the link will eventually time out, and signal to the invoking service that communication could not be established, and resulting in error messages on the generator side.

3.3.2 Echo

As shown in Fig. 1, servers are marked B, C, D and E, have Web Services capable of the executing the Echo function. These intermediate nodes will then upon reception of a token inspect the data structure shown in Fig. 2, and increment the pointer, and then forward the token to the next server on the list, by invoking its URL in the token. This means that the TCP connection is held open over all nodes, until the last node on the token list is accessed.

3.3.3 Receiver

The receiving endpoint could be any HTTP capable node, but in order to measure the trip time for the token, the receiver resides on the same physical node as the sender, and hence is using the same time reference. It is therefore easy to compute both the round-trip time, as well as the end-to-en time.

3.4 Experiment Evaluation

From the individual log files that are collected, the interested data can be extracted and plotted. In Fig. 4 shows a plot for the month of January. The tokens have a sequence number included, and if we plot the difference between two consequent tokens, we get a plot on how many have been lost. The plot shows multiple occasions where large numbers of token are lost, and this translate several minutes of down time in a given service. Since the token arrival time is known, a watchdog mechanisms dispatches SMS's to the author when a few consequent tokens are lost.

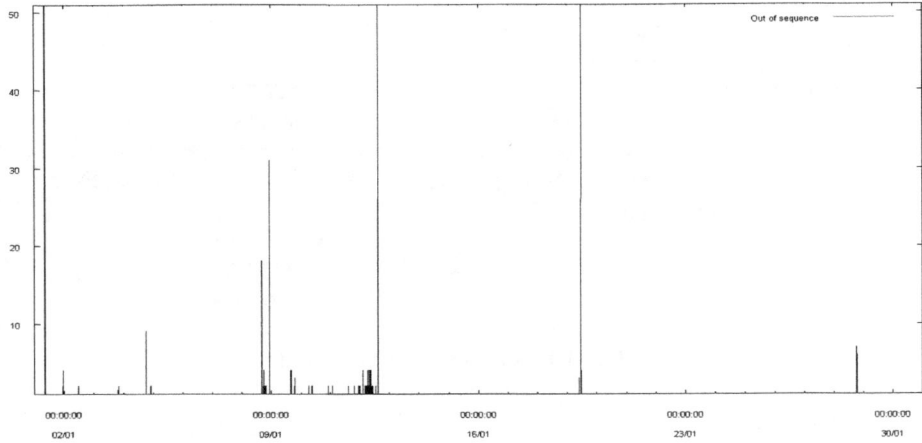

Fig. 4. Tokens on receiver side

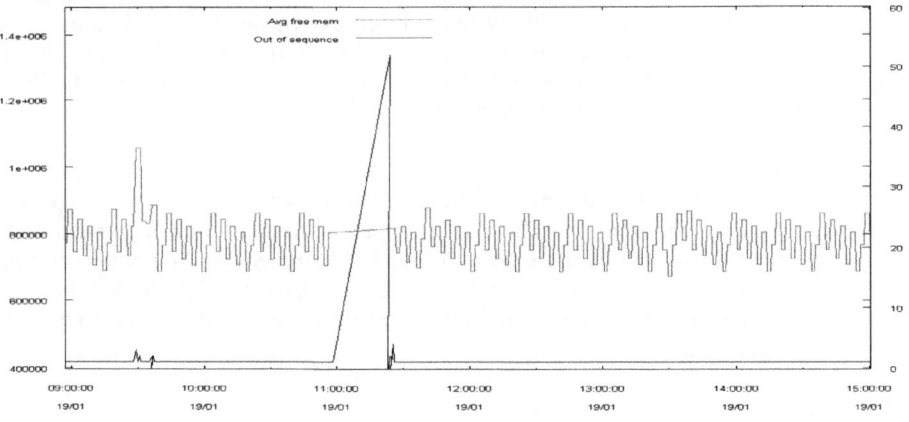

Fig. 5. Zoom in on token reception

Since log files are created at all nodes, one can do post processing of the data and analyze in grater detail on what goes wrong. Zooming in on the incident on the 19th of January, as shown in Fig. 5, the plot shows in blue that at 11:00, 50 tokens are lost.

A few tokens are also lost at 9:30. The cyan plot is the available free memory. The saw tooth shape of the cyan plot illustrates the garbage collection in the JVM. However in a real situation the data from the end node might not be accessible if it resides in a different administrative domain. Therefore it is of interest to look at the data generated at the sender side.

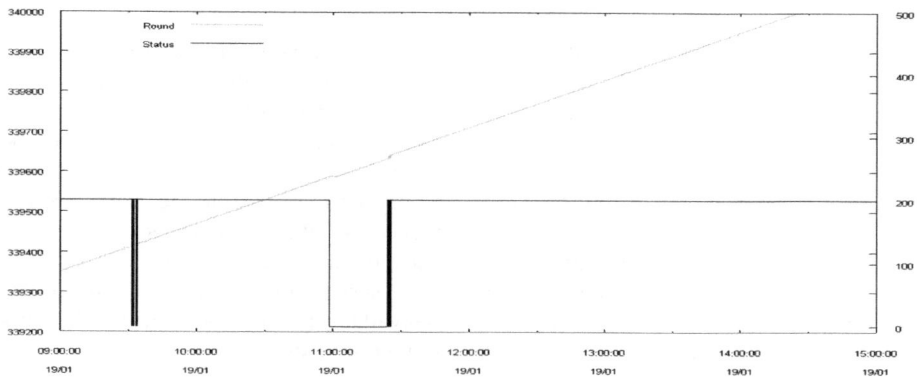

Fig. 6. Status at sender side

The black graph is the HTTP status, and the gray graph is the token sequence number. When the black graph stays at the value "200", we have normal operation, otherwise there is some kind of failure, like when the graph shows "-1" indicating that there is no route or connection to the next application server. The large dip in the plot at 11:00 indicates a particular incident, which was due to human failure. A system administrator replaced a network cable to a switch at 9:30, and then later at 11:00 decided to reconfigure a router, but accessed the wrong unit, causing the network to fail, and then it took 27 minutes to restore the operation.

4   The Trial System

The system is build from a set of servers running various flavors of Unix operating systems, Tomcat application servers [5] and JVM [6]. The system has run continuously since November 2008. The servers span over three different administrative domains, and in one of the domains the server is installed a virtualized node. None of the organizations have automated monitoring of their servers. Due to the sequential behavior of the algorithm, all port numbers are cycled in a short time period, and miss configurations in routers and firewalls are quickly detected, and administrative action can be taken. For the month of January 2010, 85116 tokens were dispatched from the generator. Of these 1459 tokens experienced a delay for more than 5 second, and 253 tokens failed, while only 3 tokens got other error messages. The incident on the 19th of January accounts for 52 lost tokens. Of all the tokens in the period 215 tokens were out of sequence, meaning that they were delayed more than 30 seconds. This is due to the robustness of TCP protocol, the tokens are eventually retransmitted, and not lost until the timeout of roughly 200 seconds is exceeded. None of the servers had any downtime in the entire period, so the unavailability is only due to network issues.

5 Conclusion and Further Work

This paper has briefly outlined a method to detect unavailability, and described one trial system. Only a small excerpt of the data is shown in this paper, but there are several interesting observations in the data that are under investigation. The variations in available free memory around the times when tokens are lost, and the distribution of the round trip time, need further investigation. Also when one has detected that the service indeed is down, different service recovery strategies should be trial as well.

However this approach shows that it is indeed possible to detect downtime across different administrative domains, and as a result maintenance action can be taken. Further work is to investigate how alternative servers or nodes could improve the availability, and how the knowledge of the unavailability could be distributed in order to trigger reconfiguration at the different administrative domains.

Acknowledgments

I would like to thank Professors Rolv Bræk and Bjarne Helvik at Department of Telecommunication at NTNU, for their advice and guidance in my research work.

References

1. George, P.: Improving Distributed Application Reliability with End-to-End Datapath Tracing, PhD at Electrical Engineering and Computer Sciences, University of California, Berkeley. Technical Report No. UCB/EECS-2008-68, May 22 (2008)
   http://www.eecs.berkeley.edu/Pubs/TechRpts/2008/EECS-2008-68.html
2. Bouricius, W.G., Carter, W.C., Schneider, P.R.: Reliability Modeling Techniques for Self-Repairing Computer Systems. In: Proceedings 24th National Conference ACM (1969)
3. Avižienis, A., Laprie, J., Randell, B., Landwehr, C.: Basic Concepts and Taxonomy of Dependable and Secure Computing. IEEE Trans. Dependable and Secure Computing 1(1), 11–33 (2004)
4. Java virtual machine, http://java.sun.com
5. Tomcat, http://tomcat.apache.org/
6. TCPDUMP tool, http://en.wikipedia.org/wiki/Tcpdump

Distributed Resource Reservation for Beacon Based MAC Protocols

Frank Leipold[1] and Jörg Eberspächer[2]

[1] Sensors, Electronics and Systems Integration
EADS Innovation Works Germany
frank.leipold@eads.net
[2] Institute of Communication Networks
Technische Universität München
joerg.eberspaecher@tum.de

Abstract. Wireless connected devices become increasingly popular in a large variety of applications. Consumer electronics most certainly is the field with the most wireless innovations in the past years; but also other areas, such as medical equipment, vehicular on-board networks or maintenance services, experience an increasing demand for wireless communication. Additionally the networks should *just work* and require as little maintenance as possible. Hence future WLAN and WPAN must be self-configuring, self-healing and distributed to provide flexible usage.

Assuming a homogeneous distributed MAC protocol with a beacon based reservation mechanism, a radio resource reservation algorithm is developed to fulfil the delay and data rate requirements from the devices. It uses a game theoretic approach to achieve infrastructure-less design and still provides fair resource allocation. Changes in the radio channel, failing devices or links and mobile nodes are detected and a reorganisation of resources is calculated.

1 Introduction

The constantly increasing performance demands for wireless high speed connections is the motor to further enhance existing technologies and to invent new and innovative communication interfaces. For wireless local area networks (WLAN) several candidates have been developed in the past years. The recently approved IEEE 802.11n standard achieves up to 600 Mbits/s, by using MIMO (multiple-input-multiple-output) technology.

High data rates can also be easily achieved by using ultra wideband (UWB) as the bandwidth of more than 500 MHz per channel provides plenty resources. The WiMedia standard [1] is capable of 480 Mbit/s with one single channel. The recently published upgrade even allows 1024 Mbit/s. This is done with one single antenna (hence no MIMO), which implies less complex transceivers.

One significant difference between WiFi and high-speed UWB is the transmit range. Due to the strict power limitations of at most -41.3 dBm/MHz for UWB, defined from the FCC [2], and mostly followed by other regulatory organisations,

F.A. Aagesen and S.J. Knapskog (Eds.): EUNICE 2010, LNCS 6164, pp. 217–225, 2010.

high data rate UWB has a typical range of about 10 meters. It also underlies more intense spatial limitations given by walls and other obstacles. This is often interpreted as a handicap of UWB, but it actually also implies two significant advantages. First it reduces the congestion in the wireless channel. For instance the IEEE 802.11b/g technology is currently often used in home applications. But the standard has only three non-overlapping channels (respectively four in Japan). Thus in apartment houses, where most flats have their own WiFi access point (AP), the channels are very crowded as the technology can achieve 20 to 100 meters transmit range even in indoor scenarios. With UWB the radio signals are almost confined to the individual apartments and the contentions for the wireless channels get relaxed. There will be less wireless systems sharing the common resources. The second advantage is the reduced risks of being eavesdropped. The attacker must be in very close proximity to successfully receive the UWB signals. Eavesdropping becomes much more difficult and the risk of falling victim to an attack declines.

Even though WiMedia is a fully distributed algorithm, without any coordinating nodes such as a WiFi-access-points, the network still has *service providing access points*, for instance a node that is connected to a LAN and operates as a bridge between the LAN and the WiMedia network to enable access to the Internet. These WiMedia (or UWB) APs do not have any special role in terms of the PHY or MAC protocol.

Using UWB for WLAN applications requires a larger number of APs. For the office scenario about each room requires at least one AP, depending on the size of the room. This makes the network management more complex. For now usually the WiFi networks are managed by maintenance personnel, but with a large number of APs to manage, this becomes a very complicated task. Several automated resource management algorithms for WLAN systems have been proposed yet. For WiMedia little work has been done on this subject. With its completely distributed algorithms the existing approaches can not be used efficiently.

In an earlier work [3] an integer optimization algorithm was presented to calculate the optimal AP placement and resource allocation plan for WiMedia networks. It is very complex and time consuming algorithm that will run during the design phase of the network. But during the operation of the network changes to the wireless channel will occur and a fast adaptation to the new conditions must be made. Depending on the size of the network a centralised optimisation may take too long. A distributed approach is preferred.

Our applications target wireless on-board networks for public transportation vehicles, such as aircraft. Cabin management systems, containing reading lights, signs, speakers, small displays, shall be connected wireless to reduce production and maintenance costs and to make the cabin layout more flexible. UWB currently is the most promising technology that provides a very robust radio channel for the aircraft environment [4], high data-rates and no licensing problems. Furthermore the network shall have self-healing capabilities and adapt quickly to failures and changes. Therefore the resource management algorithm should be distributed and enable the network to operate with the best possible configuration even when some parts are failing completely.

The network management of the public transportation scenario is not very different from an office environment or home consumer electronics. Therefore the gained results in this paper can be easily applied to other situations, where devices must be connected to service access points.

The rest of this paper is organised as followed. Section 2 describes features of WiMedia relevant to this work. Section 3 describes briefly game theory and some related research. In section 4 the distributed resource management algorithm is presented. Finally section 5 concludes this paper.

2   Unique WiMedia Features

The WiMedia standard and also other possible standards that use a distributed beacon based MAC implementation, has some unique features, which are of value for the resource management.

WiMedia is completely infrastructure-less, meaning all nodes have the same physical and MAC capabilities. Unlike in other common protocols there is no coordinator or MAC level access point. The protocol uses periodic superframes containing a beacon period and a data period; see Figure 1(a). All nodes allocate a beacon slot in the beacon period, regardless if they have to transmit user data or not. This slot is fix over time and changes only in rare conditions. For DRP (distributed reservation protocol) channel access the beacons are used to announce transmissions and reserve parts of the data period, the so called MAS (media access slots). Conflicts are identified early and collisions can be mostly avoided. The only collisions may occur in the beacon slot allocation process.

The beacons contain details of all upcoming transmission in the data period. This means a single node knows when a neighbour communicates with another node. This feature makes WiMedia not susceptible to the *hidden station problem* and it does not need RTS/CTS messages.

With the beaconing a hard limitation of WiMedia applies. The beacon period has only 96 beacon slots with two slots reserved for special functions. This means no more than 94 nodes must be within range on the same channel. The group of nodes within transmit range of a subject node is called *beacon group* (BG). Each node includes the members of the BG and the respective beacon slot ids into its own beacon; hence it is broadcasting this information to the surrounding nodes. Furthermore the standard also defines the *extended beacon group* (EBG), which is the set of nodes representing the neighbour's neighbours. Figure 1(b) shows the BG and the EBG of a node. A newly activated node picks a beacon slot that is not occupied by a node of the BG nor from the EBG. This rule implies that for any given node the beacon slot ids of the BG members must be unique, otherwise they would cause collisions. For the EBG members the ids can be identical, because the subject node can not listen to them, but must not

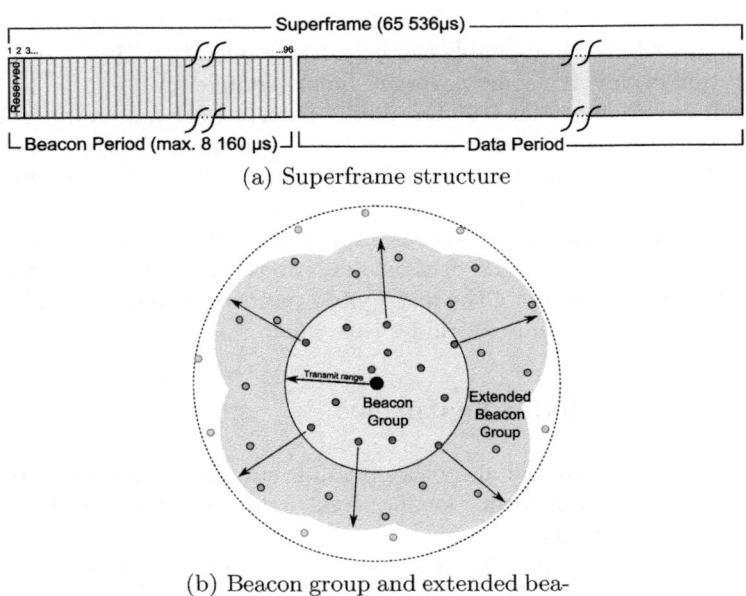

(a) Superframe structure

(b) Beacon group and extended beacon group

Fig. 1. WiMedia beacon groups: Figure (a) shows the beacon period and data period of a superframe. In (b) the relation of the beacon group and the extended beacon group are shown.

transmit at the same time as it would cause collision at the node that has the subject node and the EBG-node in its BG.

The consequence of overlapping areas in WiMedia is the extension of the EBG. When a border node is within range of two APs, it will add parts of the second AP nodes to its BG, which again appear in the EBG of the first AP. Hence overlaps increase the extended beacon group size and therewith limit the maximal node density.

The beacon group size is directly related to the node density and transmission range as slots may only be occupied by one node. The extended beacon group size additionally depends on the activation sequence of the network, as for a single node the neighbour's neighbours beacon slots may be identical.

3   Game Theory for Resource Management

For the distributed resource management algorithm presented in this work, one key feature is to have a fair resource allocation. Nodes should not act selfishness and occupy wastefully available resources. The objective of the algorithm is to maximize the overall efficiency of the network. Hence nodes must not only take their own situation into account, but also those of the surrounding nodes.

The game theory is capable of defining fair and distributed algorithms. It originates from economics in the 40's, but has also been applied to biology, engineering, political science, computer science, and philosophy. The problem is defined in a strategic game with a set of rational players. Each player tries to maximise its payoff function by choosing the best strategy and also by taking the actions of the other players into account. Usually the players strategies will result in an equilibrium, for instance the Nash equilibrium, where no player can change the own strategy without decreasing its payoff. Many different types of games have been developed: cooperative or non-cooperative, symmetric or asymmetric, zero-sum or non-zero-sum, simultaneous or sequential, just to name a few.

Game theory has been proposed for resource management in wireless networks for all kinds of technologies. For IEEE 802.11 the authors of [5] present a game to share the available radio channels. The payoff function based on transmission delay, channel access length and throughput. In [6] an algorithm for OFDM based communication is described that minimizes the transmission power, while still achieving the QoS requirements of the system. For IEEE 802.16 networks [7] presents a game definition to control the amount of bandwidth given to new connections, with respect to delay, throughput and other QoS parameters in the network.

Even though the amount of available resource management algorithms based on game theory is huge, no one handles a comparable system to that of a Wi-Media network. The distributed beaconing mechanism and DRP channel access scheme is unique. The priority for a WiMedia network lies in the efficient MAS allocation. To utilise the WiMedia features a new resource management algorithm is required.

4 Algorithm

The resource management algorithm assumes APs with fix locations. Furthermore the network consists of a wired backbone that connects the APs. The APs provide the service for wireless end devices to get access to the wired back bone network.

The MAC mechanism for the WiMedia nodes shall be DRP. For the aircraft scenario DRP is essential, as it provides guaranteed resources to the end devices. The QoS requirements for end devices are defined in required data rate and maximum time delay. Assuming only one MAS is reserved per end device and taken into account the 64 ms of a superframe, the delay of an message over the UWB link can be nearly 64 ms. For applications with smaller delay requirements, the resource management algorithm must reserve two or more MAS per end device and per superframe with defined maximum gaps between the MAS.

Despite the mentioned DRP reservation scheme the network may also use the alternative channel access PCA (prioritized contention access) for non critical applications. MAS that are not completely used by the owing device can be released and used for PCA, a CSMA like access scheme. This way the unused MAS sections can still be used for applications where collisions can be tolerated.

The following two lists show input and output parameters of the algorithm:

Input parameters
- Link quality
- Devices per AP
- BG/EBG size
- Bandwidth utilisation
- Delay requirements

Output parameters
- Channel allocation
- MAS reservation
- Beacon slot reservation

The link quality is given in RSSI (received signal strength indicator) or LQE (link quality estimator). Both are defined in the WiMedia standard and should be accessible from the application layer. To provide the required information of the surrounding nodes, the QoS requirements, neighbourhood relation and currently superframe structure is exchanged on a two-hop distance. This is mostly already present in the WiMedia protocol and requires only little extra signalling.

Efficiency in a WiMedia network using DRP can be defined as the configuration with the minimal used MAS, which again implies the minimal usage of resources and maximum remaining bandwidth. The straight forward approach then would be to choose the AP only by signal quality. A node would therefore always try to connect to the AP with the best signal. But for an unlucky network topology, as is Figure 2, this could mean that many nodes try to connect to the same AP while a second AP close by is ignored, because it is further away. If these are too many nodes, the network performance decreases at this AP. The goal of the algorithm is to enable some nodes to choose a different AP, if the actually best AP has a lot of load.

From the protocol perspective two limiting factors exist. The first one is the number of beacon slots. As described earlier, only 94 slots are available for the BG and the EBG. Not more than 94 nodes can be simultaneously in transmit range. To allow new nodes to connect to the network, a small amount of slots should be kept unoccupied.

The second limiting factor is the number of MAS. 256 MAS exist in one superframe, which must be shared among the nodes in transmit range. The

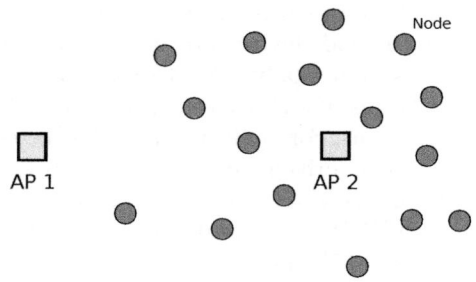

Fig. 2. Network topology where most nodes would try to connect to the same AP, if the selection is only based on signal quality or distance

MAS reservation depends on the data rate and delay requirements. Again a few MAS should be constantly available to allow new nodes to join the network.

In contrast to the MAS slots stands the data rate between an AP and an end device. An increased data rate requires less MAS reservations and enables more nodes to join the network.

The reservations between an end device and AP are made from the end device. It periodically calculates the utility function U_n for all AP in range. It is defined as:

$$U_n = B_n^* + M_n^* + D_n^* \tag{1}$$

where B_n^* contains the beacon slot utilisation, M_n^* the MAS slot utilisation and D_n^* the possible data rate. All three values are normalised to values between 0 and 1, where 0 means the best achievable value and 1 the worst value. Each end device periodically calculates the utility value for each AP and initiates an AP change, if a better solution was found than the currently one.

B_n^* is defined as the number of reserved beacon slots b_a at the AP a, divided by the maximal number of beacon slots b_{max} to the power of v:

$$B_n^* = \left(\frac{b_a}{b_{max}} \right)^v \tag{2}$$

Analogous to this, M_n^* is defined as:

$$M_n^* = \left(\frac{m_a}{m_{max}} \right)^v \tag{3}$$

where m_a are the number of reserved MAS at the AP and m_{max} the number of maximal available MAS slots. m_a is obtained by calculating the own number of required MAS, depending on the delay and data rate requirements, and adding it with the already used MAS by other nodes. The MAS reservation algorithm.

B_n^* or M_n^* describe the utilisation of the beacon slots or MAS. They should only get a value close to 1 if the number of utilised slots becomes close to the maximum. By modifying v one can control the threshold.

Finally D_n is a linear relation of the possible data rate between the end device and the AP $d_{n,a}$ and the maximal possible data rate d_{max}:

$$D_n^* = \left(1 - \frac{d_{n,a}}{d_{max}} \right) * T \tag{4}$$

D_n^* must degrade over time if B_n^* or M_n^* are above the threshold. Therefore T is used; it usually has a value of 1, unless the threshold is reached. Then T will increase for each round (when the utilisation values are recalculated), until the node has changed the AP or B_n^* and M_n^* drop below the threshold, due to other nodes disconnecting from the AP. With this feature one can achieve to let the nodes closest to the other AP (in Figure 2) first change the AP. D_n^* for the overloaded AP increases for all nodes in its range. The nodes close to the other AP will have a relative low D_n^* for the less used AP, because they are closer. They will be the first to change.

All three addend in Equation 1 are influenced by the current reservations of surrounding nodes. The algorithm considers the actions of the other nodes and derives its own actions. After some rounds an equilibrium will be found where each node can not improve its situation.

For the APs themselves only the channel selection and beacon slot reservation must be done. For them the procedure is to choose a channel that is not used by another AP in its BG or EBG. If no free channel can be found the AP should choose the channel with the least nodes assigned to it.

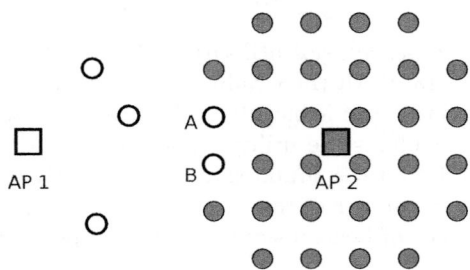

Fig. 3. Final results of the algorithm, after 9 rounds. The transmit range is larger than the show field; hence all nodes are in range of each other. The filled nodes are on one channel and the unfilled nodes on another one. Nodes A and B, close to the left AP, initiated an AP change after 8 round.

In Figure 3 the results of the algorithm are shown. To demonstrate the results in a smaller scale, the maximal beacon slots number and MAS were set to smaller values as defined in the standard. After the first round of calculating the utilisation values, each node chose the closest AP. This resulted in a situation where the threshold for the beacon slots was exceeded. Hence for the following rounds the parameter T for calculating the utilisation of AP 2 increased, until the utilisation value of AP 1 for nodes A and B became smaller as for AP 2. After node A and B having changed the AP, the threshold is no longer exceeded and the network is in an equilibrium.

5   Conclusions

In this work a resource management algorithm based on game theory for high data rate UWB networks was presented. The network consists of service access points, which provide access for wireless end devices to a wired network, such as LAN or Internet. The algorithm achieves a fair distribution of resources and adapts to changes in the radio channel or topology. Network efficiency is based on the number of used MAS slots, which also encompass bandwidth and timing requirements. The MAS reservation algorithm is essential for scenarios with high MAS utilisation, as the efficiency of the overall algorithm is tightly coupled to the efficiency of the MAS reservation for large loads.

The distributed approach does not require a dedicated node to compute the network parameters. Hence whatever parts of the network are failing, the network still tries to connect as many nodes to the APs as possible. The algorithm can be used in various applications: on-board networks, office and home WLAN or ad-hoc communication.

References

1. ECMA International, ECMA-368: High Rate Ultra Wideband PHY and MAC Standard, 3rd edn. (December 2008)
2. FCC, Report and Order. FCC 02-48 (April 2002)
3. Leipold, F., Bovelli, S.: Requirements for Radio Resource Management in Multi-cell WiMedia Networks. In: ICT-MobileSummit (2009)
4. Chuang, J., Xin, N., Huang, H., Chiu, S., Michelson, D.: UWB Radiowave Propagation within the Passenger Cabin of a Boeing 737-200 Aircraft. In: IEEE 65th Vehicular Technology Conference, VTC 2007-Spring (April 2007)
5. Berlemann, L., Hiertz, G., Walke, B., Mangold, S.: Radio resource sharing games: enabling QoS support in unlicensed bands. IEEE Network 19 (July-August 2005)
6. Han, Z., Ji, Z., Liu, K.: Power minimization for multi-cell OFDM networks using distributed non-cooperative game approach. In: IEEE Global Telecommunications Conference, GLOBECOM 2004, November-3 December 2004, vol. 6 (2004)
7. Niyato, D., Hossain, E.: Radio resource management games in wireless networks: an approach to bandwidth allocation and admission control for polling service in IEEE 802.16. IEEE Wireless Communications 14 (February 2007)

On Runtime Adaptation of Application-Layer Multicast Protocol Parameters[*]

Christian Hübsch, Christoph P. Mayer, and Oliver P. Waldhorst

Institute of Telematics, Karlsruhe Institute of Technology (KIT), Germany
{huebsch,mayer,waldhorst}@kit.edu

Reasonable choice of protocol parameters is crucial for the successful deployment of overlay networks fulfilling given service quality requirements in next generation networks. Unfortunately, changing network conditions, as well as changing application and user requirements may invalidate an initial parameter choice during the lifetime of an overlay. To this end, runtime adaptation of protocol parameters seems to be a promising solution—however, it is not clear if protocol parameters can be adjusted dynamically at runtime in a distributed setting. In this paper, we show—using the NICE application layer multicast protocol as an example—that runtime adaptation of protocol parameters is indeed feasible. We propose an algorithm for adapting the NICE clustersize parameter k dynamically at runtime and discuss the impact on service quality. Our simulations show that runtime adaptation of NICE protocol parameters is promising for service improvement and that data latencies can be optimized by up to 25% without increasing the overhead significantly for most of the nodes.

1 Introduction

During the evolution towards next generation networks, the Internet architecture faces a number of limitations due to upcoming requirements like mobility, multi-homing, or multicast communication. E. g., a clean implementation of multicast communication within the Internet architecture has been proposed for years but never experienced widespread end-to-end deployment. One reason for the rare deployment of IP-based multicast is its inflexibility and requirement for provider trust. Therefore, deployment of overlay networks attracted great interest throughout the scientific community as a feasible approach for adding new services like multicast communication on top of the Internet architecture in an unintrusive and flexible way. A large number of protocols have since been proposed that enable the overlay-based deployment of new services like multicast, or distributed hash tables [4].

As services implemented by application-layer overlay networks typically provide worse service quality than native implementations in lower layers, careful

---

[*] This work was partially funded as part of the *Spontaneous Virtual Networks* (*SpoVNet*) project by the Landesstiftung Baden-Württemberg within the BW-FIT program and as part of the Young Investigator Group *Controlling Heterogeneous and Dynamic Mobile Grid and Peer-to-Peer Systems* (*CoMoGriP*) by the *Concept for the Future* of Karlsruhe Institute of Technology (*KIT*) within the framework of the German Excellence Initiative.

F.A. Aagesen and S.J. Knapskog (Eds.): EUNICE 2010, LNCS 6164, pp. 226–235, 2010.
© IFIP International Federation for Information Processing 2010

parameterization and fine-tuning of the overlay protocol becomes essential. Unfortunately, runtime changes of network conditions, as well as application and user requirements can make a good parameterization turn worse during the lifetime of an overlay. This requires re-adjustment of overlay parameters during runtime to prevent degradation of service quality—at best without any service downtime.

In this paper we show by the example of the NICE application-layer multicast (ALM) protocol that runtime adaptation of parameters is indeed feasible. For this purpose, we conduct in-depth simulation experiments using a NICE implementation in the P2P simulator OverSim [2]. As service quality measures, we use message latency and resulting protocol overhead. We identify the size of the node clusters maintained by NICE, the thresholds for cluster refinement and the interval of the heartbeats exchanged by members of a cluster as relevant parameters for tuning the service quality. Subsequently, we show that these parameters can be adapted at runtime by providing an algorithm for dynamically setting the cluster size as an example. We discuss the (positive and negative) impact of an adaptation on service quality measures. Our simulation experiments indicate that runtime adaptation of the NICE protocol parameters is promising for service quality as it may improve latency up to 25%.

The remainder of this paper is structured as follows: Related work is presented in Section 2. To make the paper self contained we introduce the NICE ALM protocol in Section 3. We then present an in-depth analysis of the NICE behavior with respect to parameters that determine service quality in Section 4. In Section 5 we discuss how parameters can be adapted at runtime and illustrate the impact of runtime adaptation on service quality measures. Finally, concluding remarks are given in Section 6.

2   Related Work

Work in overlay adaptation and optimization can be roughly divided into two groups: *structure optimization* and *underlay optimization*. Structure optimization focuses on performance of the overlay structure in itself, whereas underlay optimization tries to adapt the overlay structure to the underlay, aiming to overcome performance disadvantages of overlay-based networks. Our work can be categorized as structure optimization. Work that has been performed under aspects of overlay robustness and churn [7] is explicitly excluded in our work.

Our work is most closely related to [6]: The authors present the DHT protocol Accordion that adjusts itself to different network sizes by adaptation of the routing table size. While their focus is on performance and robustness under churn in a DHT protocol, we work on adaptation in ALM and focus on service quality. Furthermore, we concentrate on feasibility using the NICE protocol as an example. Earlier work by the same authors [7] presented an analytical parameter-space evaluation that focuses on systems under churn. The authors identify the routing table size for DHT protocols as the most important parameter. This is similar to the cluster size parameter of the NICE protocol that we use in this work for dynamic adaptation.

Fan and Ammar describe reconfiguration policies for adaptation of overlay topologies [3]. The authors consider design problems for static and dynamic overlay networks, the dynamic being based on occupancy cost and reconfiguration cost. Their work provides insight into general reconfiguration and the question of when to perform reconfiguration, whereas our work focuses on the concrete effects of parameter adaptation.

Our work is based on a defined overlay structure with adaptation within the parameter-space. Jelasity and Babaoglu [5] in contrast use a topology-space to develop a protocol for topology structure adaptation, called T-MAN. The T-MAN protocol allows for runtime variation of overlay topologies.

The authors of the NICE protocol performed analysis on static parametrization of the protocol, with focus on the k parameter [1]. In their work they look at stretch and stress and how they are affected by selecting a constant parameter k at design time. The ZIGZAG protocol [9] for application-layer multicast is similar to NICE in its layered clustering structure, but outperforms NICE with respect to node degree, and failure recovery. ZIGZAG, too, defines a constant parameter k, similar to NICE. Therefore, we see applicability of our work to ZIGZAG.

Interesting work has been published by Mao et al. on the MOSAIC system for dynamic overlay composition at runtime [8]. The MOSAIC system can dynamically compose a set of overlay protocols to include/exclude specific properties—like mobility, or performance features—provided by the respective overlay. In contrast, our work tries to optimize behavior inside a single overlay protocol.

3   NICE Protocol

The NICE protocol [1] is an early approach for ALM that implements an overlay aiming at scalability by establishing a cluster hierarchy among participating member nodes. In the following, we give a short description of the protocol.

3.1   Basic Protocol

NICE divides all participating nodes into a set of clusters. Protocol traffic is mainly exchanged between nodes residing in the same cluster, leading to good scalability. In each cluster, a cluster-leader is determined that is responsible for maintenance and refinement in that cluster. Furthermore, all cluster-leaders themselves form a new set of logical clusters in a higher layer, exchanging protocol data. Respective cluster-leaders are determined from one layer for the next higher layer. This process is iteratively repeated until a single cluster-leader in the topmost cluster is left, resulting in a layered hierarchy of clusters. Each cluster holds between k and $(\alpha k - 1)$ nodes, α and k being protocol parameters. In case of size bound violations, a cluster is split, or merged with a nearby cluster. Clusters are formed on the basis of a "distance" evaluation between nodes, where distance is basically given by network latency. NICE aims at combining "near" nodes in the same cluster. Cluster-leader election is accomplished by determining the node nearest to the graph-theoretic center of that cluster. Nodes in the same cluster periodically exchange heartbeat messages to indicate their liveliness and report measurements of mutual distance to other nodes in that cluster. Cluster-leaders decide on splitting and merging of clusters as they are aware of the current cluster size and all distances between nodes inside their cluster.

The objective of NICE is to scalably maintain the hierarchy as new nodes join and existing nodes depart. Conformance tests and rearrangements are performed by NICE periodically. Based on the hierarchical clustering structure, paths for data dissemination in NICE are defined implicitly. A node intending to send out multicast data sends its data to all nodes in all clusters it currently resides in. A node receiving data from inside its cluster forwards the packet to clusters it is part of except the one it received it from. This leads to each participant implicitly employing a dissemination tree to all other nodes in the structure. To analyze the effects of parameter adjustment, we implemented NICE in the open-source overlay simulation framework OverSim [2] based on the technical descriptions given in [1].

4   Static Parameter Selection

In this section we analyze the protocol parameters of NICE with focus on their impact on service quality.

4.1   Protocol Parameters and Service Quality Measures

The protocol behavior of NICE is adjustable by a variety of parameters. For completeness, we mention them here shortly in order to focus on the most relevant in the following. In our implementation of NICE we find the parameters α, k, HBI, min_CL_Dist and min_SC_Dist, triggering cluster sizes, interval length between heartbeat messages, and decision bounds for clusterleader estimations, respectively. Furthermore, the protocol employs several timers to detect failures in communication or structure. The *Maintenance Interval* determines the interval in which a node checks for protocol invariants. *Peer Timout* is defined to be the period of time after which a node assumes another node has failed or gone. It is a configurable multiplicity of HBI. *Structure Timeout* is the period of time after which a node assumes to be partitioned from the structure and attempts to reconnect. The *Query Timeout* detects lost queries in NICE for initiation of retransmissions (while NICE is typically soft-state, some messages have to be assured to be received in order to work properly).

As service quality measures, we consider data latency and protocol overhead. Data latency is given by the average time that elapses between sending a data packet via multicast and receiving it. Protocol overhead is measured by the average bandwidth used by a node for sending control messages.

Table 1. Protocol and simulation parameters

NICE-specific		Simulation-specific	
Parameter	Value	Parameter	Value
α, k	3	Number of nodes	512
HBI	5s	Offset after last join	60 s
Maintenance Interval	3.3 s	Measurement phase	600 s
Peer Timeout	2 HBI	Joins	∼every 3 s
Query Timeout	2 s	Data Interval	6 s
Structure Timeout	3 HBI		
min_CL_Dist	30%		
min_SC_Dist	30%		

Several pre-evaluations have shown that the major impact is limited to 3 relevant parameters in NICE:

- the clustersize parameter k
- the refinement of node cluster memberships, in the following referred to as *inter-cluster refinement* (especially by adjusting min_SC_Dist)
- the rate of protocol heartbeat messages (HBI)

Due to space limitations we only focus on k in the remainder of this paper.

4.2   Experimental Setup

Our experiments are conducted using the peer-to-peer simulation framework OverSim [2]. As network model we chose OverSim's *SimpleUnderlay* and use the protocol-specific and simulation-specific parameters given in Table 1.

In our simulations we analyze a total of 512 nodes in NICE. From simulation start, a new node joins the network every 3 s. After the last node joined we employ a backoff time of 60 s to stabilize the hierarchy. Then, every node starts sending a data packet using the multicast structure every 6 s. After 10 min of data exchange, we again employ a backoff of 60 s before finishing the simulation run. All simulation settings have been conducted with 30 different seeds of the random number generator and mean values have been calculated for all performance measures.

4.3   Clustersize Parameter k

The clustersize parameter k determines the thresholds of cluster sizes that trigger splitting and merging a cluster with neighboring clusters. As all nodes in a cluster directly exchange protocol messages, increasing k will intuitively increase per-node overhead. In contrast, larger clustersizes also lead to fewer layers in the hierarchy. Therefore, data packets have to traverse less overlay hops, leading to lower overall data latencies. Thus, adjustment of k trades off protocol overhead against data packet latencies.

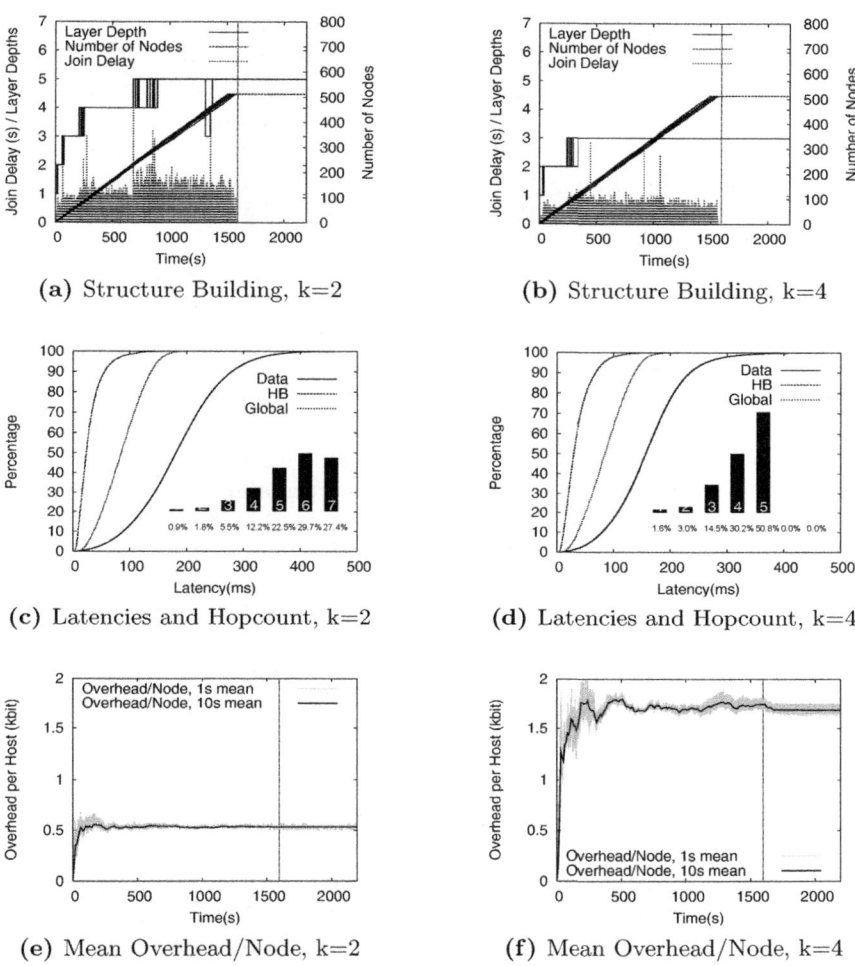

Fig. 1. Sensitivity to Clustersize Parameter k

Figure 1 gives an overview of the impact of different choices of k. Here, the rows reflect the resulting NICE structure, latencies and overlay hopcounts, and overhead per node, respectively. For the individual columns we set k to be 2 and 4, respectively. Concerning the hierarchy structure (Figures 1a and 1b) we find that small changes of k already have mentionable impact on protocol properties. The figures show the number of nodes in the structure and the resulting number of layers as a function of simulation time. For each joining node it also shows the *join delay*, i.e., the time that has passed between first contacting the RP and finally joining a cluster in layer L_0. Each figure visualizes the values for 10 out of 30 different seeds each.

In general, incrementing k leads to a decrementation in the hierarchy depth. While the final structure with $k = 2$ converges to five layers, it converges to three with $k = 4$. Each additional layer also increases the mean join delays that joining nodes take to become part of the structure. Figures 1c and 1d show the resulting latencies and hopcount distributions for each k, acquired after the structure finished its building process. The figures show three aspects of latencies: (1) global network latencies between all nodes as they have been placed randomly in the simulation field, (2) latencies inside the clusters built (i. e. the heartbeat round-trip times), and (3) data latencies that the data packets experience when being routed through the overlay. We also show the distribution of hopcounts for the data packets, meaning how many overlay nodes they pass before reaching all nodes in NICE. It is clearly visible that the NICE clustering process combines nodes with low network latencies in the same cluster. Decreasing hopcounts lead to lower latencies. Finally, Figures 1e and 1f compare the resulting mean overhead per node for each k. As the number of neighbors in a cluster increases with k, the overhead due to heartbeat and other signaling also grows. Note that adjusting k also influences the impact on the underlay by trading off between resulting stress and stretch [1]. Due to space limitations, we will not discuss these aspects in this paper.

We conclude from these experiments that adjusting the clustersize parameter k enables trading off data latency and control overhead. Thus, it is promising to adapt k in a scheme for runtime parameter adaption. However, it is not clear whether k can be changed for an existing NICE overlay at runtime. We will discuss this issue in more detail in the following Section 5.

5   Runtime Parameter Adaptation

Instead of choosing parameters for overlays at design time, we propose to enable the protocol to adjust them during runtime. In the following, we provide an algorithm for choosing the clustersize parameter k dynamically and illustrate its impact on the service quality measures.

5.1   Adaptive Selection of the Clustersize Parameter k

In Section 4.3 we stated how k implicitly trades off overhead against data latencies in NICE. Assuming the protocol has knowledge about the desired latency constraints on the one hand and the tolerable resulting overhead on the other hand, it may adaptively re-adjust k during runtime to provide desired data latencies without exceeding its overhead bounds. As data packets in NICE traverse the whole structure through its hierarchy layers, latencies will increase with the depth of the hierarchy.

Given a NICE structure of depth d (d being the number of hierarchy layers), to decrease the overall depth by one, k has to be chosen such that all nodes in layer $d - 1$ become part of one bigger cluster. In that case, there will be only one cluster leader left on layer d, i. e. this cluster is eliminated. To determine a

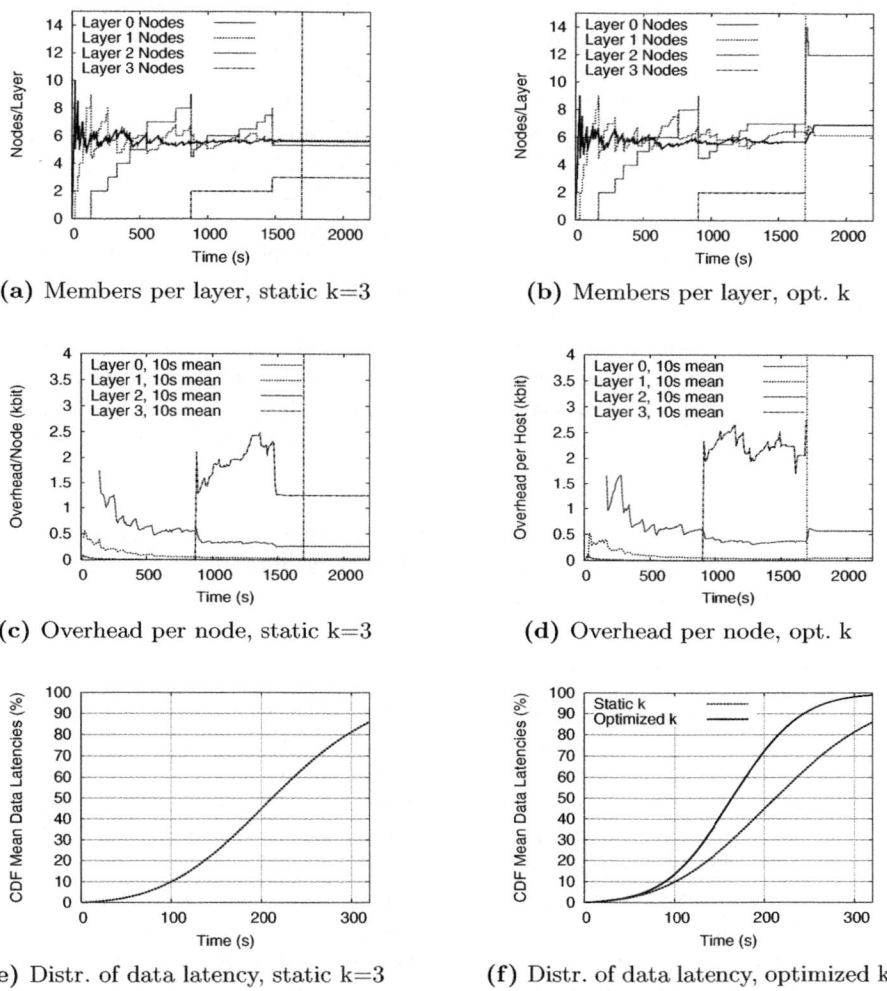

Fig. 2. Clustersize k Runtime Adaptation Schemes (512 nodes)

suitable k under a worst-case assumption, the cluster-leader of the highest cluster in layer d must assume that every cluster in layer $d-1$ holds its maximum node number of $\alpha k - 1$ nodes. As the highest cluster-leader knows the number x of nodes in layer d (i.e. its direct neighbors in the single highest cluster), it may determine the worst case number of nodes in layer $d-1$ to be $x(\alpha k - 1)$. Based on this information the cluster-leader calculates a new value k_{new} as follows:

$$k_{new} = (x * (\alpha * k - 1) + 1)/\alpha$$

This ensures that all nodes in layer $d-1$ will fit in a single cluster, resulting in a decrease of one layer in the hierarchy structure. After calculating k_{new}, the cluster-leader instructs all nodes in layer d to merge their $d-1$-clusters with him, so that he stays the last node in layer d which is equivalent to eliminating the highest layer cluster. Furthermore, it propagates k_{new} to its new cluster members by including the new value in its periodic heartbeat messages. A node receiving a changed value k will update its own cluster it is leader of, but with a randomized bounded backoff to prevent all nodes from refining their structural part at the same time. Note that the worst case estimation may raise k to a value that potentially induces much more overhead to the participating nodes than really necessary. Thus, the protocol may be optimized by the following addition: If the highest cluster-leader knows the exact number of nodes in layer $d-1$ it can choose k to be just big enough to hold all such nodes in one cluster. To gain this knowledge, all nodes in cluster layer d tell their specific current number of $d-1$ cluster nodes to the leader of the highest cluster by including this information in their periodic heartbeat messages. At the time of changing k, the highest leader may sum up these numbers to find a value num that satisfies the following:

$$(\alpha * k) \leq num \leq x * (\alpha * k - 1)$$

In general this value will be smaller than the one computed using the worst case assumption, leading to a choice for k that is significantly smaller. We will illustrate this fact in the following Section.

5.2   Performance Results

Figure 2 gives insights in the dynamic adjustment of k during runtime. Each column covers a different case, being (1) no runtime adaptation of k at all, (2) optimized adaptation of k to decrease the hierarchy depth. In case (2) the adjustment of k is triggered actively 300 seconds after the structure has stabilized, i. e., the new value of k is propagated and the merge of all clusters in $d-1$ is triggered. The first column compares the development of the average number of members in clusters of a specific layer in the two cases (1) – (2), respectively. Figure 2a shows that with a static value of k, the clusters in each layer show a comparable size, with more layers being created with growing number of participants. When adapting k at runtime (Figure 2b), the highest layer is eliminated, while the next lower layer grows notably.

Looking at the overhead per node, Figures 2c and 2d show that the overhead per node naturally grows with the highest layer the node resides in. In case of adapting k, all nodes in the highest layer after the adaptation have higher overhead due to the higher number of cluster participants they have to exchange protocol messages with. While the overhead grows, it still remains manageable.

Comparing data latencies, we see that latency is significantly reduced as shown in Figure 2e and 2f. To illustrate the gain of the adaptation more clearly, Figure 2f compares the latencies before and after the runtime adaptation of k. The figure shows a decrease in mean data latency by 25%. We conclude that the clustersize parameter k can be adapted during runtime in order to trade off latency against protocol overhead. How to determine an optimal new value for k is subject to future research.

6   Conclusion

Careful selection of overlay parameters is crucial for the successful deployment of overlay-based services in next generation networks. As conditions may change due to dynamics in the network or overlay structure, we propose the self-adaptation of parameters at runtime to adapt to service- and user-requirements. As a first step towards this autonomous behavior we identified parameters with high impact and showed the feasibility of adapting the clustersize parameter k during runtime using the exemplary NICE ALM protocol. By extensive simulation we presented the impact of protocol parameters and behavior during parameter changes. We have shown that runtime adaptation of k is feasible and can reduce data latencies by up to 25%.

References

1. Banerjee, S., Bhattacharjee, B., Kommareddy, C.: Scalable Application Layer Multicast. In: Proc. Conf. on Applications, Technologies, Architectures, and Protocols for Computer Communications (SIGCOMM 2002), October 2002, vol. 32, pp. 205–217 (2002)
2. Baumgart, I., Heep, B., Krause, S.: OverSim: A Flexible Overlay Network Simulation Framework. In: Proc. 10th IEEE Global Internet Symp. (GI 2007) in conjunction with IEEE INFOCOM, May 2007, pp. 79–84 (2007)
3. Fan, J., Ammar, M.H.: Dynamic Topology Configuration in Service Overlay Networks: A Study of Reconfiguration Policies. In: Proc. 25th IEEE Int. Conf. on Computer Communications (INFOCOM 2006), April 2006, pp. 1–12 (2006)
4. Hosseini, M., Ahmed, D.T., Shirmohammadi, S., Georganas, N.D.: A Survey of Application-Layer Multicast Protocols. IEEE Communications Surveys & Tutorials 9(3), 58–74 (2007)
5. Jelasity, M., Babaoglu, O.: T-Man: Gossip-based Overlay Topology Management. In: Brueckner, S.A., Di Marzo Serugendo, G., Hales, D., Zambonelli, F. (eds.) ESOA 2005. LNCS (LNAI), vol. 3910, pp. 1–15. Springer, Heidelberg (2006)
6. Li, J., Stribling, J., Morris, R., Kaashoek, F.M.: Bandwidth-efficient Management of DHT Routing Tables. In: Proc. 2nd conference on Symp. on Networked Systems Design and Implementation (NDSI 2005), May 2005, vol. 2, pp. 99–114 (2005)
7. Li, J., Stribling, J., Morris, R., Kaashoek, F.M., Gil, T.M.: A Performance vs. Cost Framework for Evaluating DHT Design Tradeoffs under Churn. In: Proc. 24th IEEE Int. Conf. on Computer Communications (INFOCOM 2004), August 2005, vol. 1, pp. 225–236 (2005)
8. Mao, Y., Loo, B.T., Ives, Z., Smith, J.M.: Mosaic: Unified declarative platform for dynamic overlay composition. In: Proc. Int. Conf. on Emerging Networking Experiments and Technologies (CoNEXT 2008), December 2008, pp. 883–895 (2008)
9. Tran, D.A., Hua, K., Do, T.: ZIGZAG: An Efficient Peer-to-Peer Scheme for Media Streaming. In: Proc. 22th IEEE Int. Conf. on Computer Communications (INFOCOM 2003), March 2003, vol. 2, pp. 1283–1292 (2003)

A Framework with Proactive Nodes for Scheduling and Optimizing Distributed Embedded Systems

Adrián Noguero[1] and Isidro Calvo[2]

[1] European Software Institute,
Parque Tecnológico de Zamudio, #204, 48170, Zamudio, Spain
adrian.noguero@esi.es
[2] DISA (University of the Basque Country),
E.U.I. de Vitoria-Gasteiz, C/Nieves Cano, 12, 01006, Vitoria, Spain
isidro.calvo@ehu.es

Abstract. A new generation of distributed embedded systems (DES) is coming up in which several heterogeneous networked devices execute distributed applications. Such heterogeneity may apply to size, physical boundaries as well as functional and non-functional requirements. Typically, these systems are immersed in changing environments that produce dynamic requirements to which they must adapt. In this scenario, many complex issues that must be solved arise, such as remote task preemptions, keeping task precedence dependencies, etc. This paper presents a framework aimed at DES in which a central node, the Global Scheduler (GS), orchestrates the execution of all tasks in a DES. The distributed nodes take a proactive role by notifying the GS when they are capable of executing new tasks. The proposed approach requires from the underlying technology support for task migrations and local preemption at the distributed nodes level.

Keywords: Distributed embedded systems, Framework, Reconfigurable Architectures, Middleware.

1 Introduction

Nowadays, there is a clear trend in the software industry to create distributed systems from already designed components. This trend, which is especially relevant in the embedded systems industry, has promoted the use of middleware technologies such as Java-RMI, CORBA/e, CORBA-RT, ICE, DDS or even SOA architectures. Indeed, distributed computing has proven its added value especially in some application domains, such as multimedia telecommunications, manufacturing, avionics or automotive [1], [2].

Typically, middleware technologies abstract the details of underlying devices facilitating the creation of distributed applications by providing uniform, standard, high-level interfaces to developers and integrators. Another objective of middleware technologies is supplying services that perform general purpose functions in order to avoid duplicating efforts and facilitating collaboration among applications. This paper sticks to the second objective. Namely, it presents a middleware framework for DES

F.A. Aagesen and S.J. Knapskog (Eds.): EUNICE 2010, LNCS 6164, pp. 236–245, 2010.
© IFIP International Federation for Information Processing 2010

aimed to manage the deployment and execution of a set of tasks in a set of distributed nodes, so the time requirements of the overall system are met and the use of the resources of the distributed nodes is optimized (e.g. CPU, volatile / non-volatile memory or battery). The proposed framework allows deploying tasks to the nodes in runtime for achieving a better overall optimization.

The framework uses a set of entities, namely the Global Scheduler (GS) and the Remote Servers (RS), which provide the infrastructure to execute the tasks at the distributed nodes. In particular, the GS orchestrates the execution of all tasks at the DES according to scheduling and optimization policies implemented as pluggable components. The RS act as local managers responsible for executing application tasks at the distributed nodes.

Even though typically DES must be configured at start-time, the framework allows dynamic reconfigurations of the system at run-time, providing a certain degree of flexibility and adaptability to changing requirements. These reconfigurations include modifications of both functional requirements at run-time, such as updating the executable code of the applications, as well as non-functional requirements, such as changing QoS parameters or introducing new nodes or removing existing ones without impacting the functionality of the system. These characteristics, which allow optimizing the computational load of a distributed system by adding or removing tasks from the system without changing the underlying hardware, may be applicable in certain application domains such as distributed multimedia applications or home automation, in which changing the hardware may become a complex issue.

A centralized scheduling approach has been selected for the sake of flexibility since concentrating all the information of the system in one single node facilitates the coordination of the distributed nodes. However, this approach has some drawbacks since a single GS may become a critical point of failure. In the future the authors intend to introduce replicated GS in the framework for improving fault tolerance.

The layout of the paper is as follows: section 2 presents a description of some relevant works on this topic; section 3 describes the proposed software architecture; and lastly, in section 4 some preliminary conclusions are drawn and the future work on the topic is described.

2   Related Work

The implementation of DES has been a very important research topic in the last decades. Examples of early investigations on the field can be found in [3] and [4], where some of the first solutions applicable to DES were described.

More recent works on the field have explored the implementation of complex scheduling frameworks for distributed systems on top of popular middleware architectures, such as CORBA or Java RMI. Their main advantage lies in the flexibility and the implementation simplicity provided by the middleware layers, which enables developers to abstract from low level details of the distributed system, such as communication protocols, operating systems, etc. These works vary in vision and scope. For example, references [5] and [6] focus on the timing requirements of a DES, providing a framework able to orchestrate task activations in time and, therefore, allowing DES designers to easily decouple the execution of periodic tasks that make use of the same resources. Other works, such as [7] and [8] implement control loops to change the timing characteristics of the distributed tasks in order to achieve schedulability and improve the overall performance of a DES. Lastly, some works focus on specific characteristics of DES, such as the management of aperiodic tasks, admission control strategies or task migrations strategies [9] [10].

The framework proposed in this paper addresses some of the same objectives as previous works, but it introduces some innovative aspects. Firstly, the proposed framework aims at merging scheduling and resource optimization in the same DES structure. Secondly, it follows a different approach based on proactive distributed nodes, instead of reactive nodes fully dependent on the decisions of a global scheduler. Also, the proposed framework allows dynamic reconfigurations in run-time (e.g. changes on the number of nodes in the DES, software updates in the code of the tasks or changes in the tasks' parameters). Finally, the proposed framework includes mechanisms to manage some of the complexities of DES, namely, precedence dependencies between tasks and remote preemptions.

3 Framework Description

The proposed framework is composed of two component types: one Global Scheduler (GS), responsible for deploying and activating the tasks according to a predefined application graph (see Fig. 1) as well as an optimization criterion and several Remote Servers (RS), which encapsulate the distributed processors and execute the tasks. In this work applications are defined as the execution of a set of tasks following an application graph. As shown in the figure the tasks that compose an application may be executed in different nodes, being migrated from a node to another following to the decisions of the GS. Tasks are encapsulated with a special structure, known as Task Wrapper (TW), which contains not only the executable code, but also a set of parameters that characterize them.

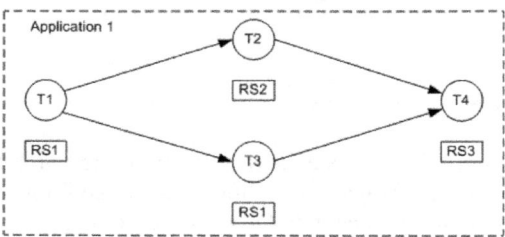

Fig. 1. Example graph of an application

The framework is to be implemented as a middleware layer that enables the users to easily configure a distributed system; however, due to its characteristics, the nodes hosting the components of this framework must provide the following services to the middleware layer:

- **Task migration.** The target software platform must support task migration between different nodes of the system. This can be achieved by explicitly sending the executable code of the application to the nodes of the system (e.g. by using Java serialization). Sometimes, this restriction can be relaxed, reducing the flexibility of the framework, by previously deploying the tasks at the distributed nodes so they are activated by the central node.

- **Local schedulers.** In order to implement remote preemptions the framework requires the use of preemptive local schedulers. This role may be assumed either by a local preemptive OSs, or by dedicated local schedulers which must be able to preempt local tasks when required by the framework.

Any DES whose nodes meet the latter requirements is candidate for the implementation of the proposed framework. Task migrations may represent a considerable overhead when the required execution times of the tasks are similar to their migration times. Hence, the framework is applicable to applications where the execution times are much longer than migration times. Possible target applications may be found in the multimedia applications domain. As a matter of example, a Java based intelligent surveillance system formed by different types of nodes such as IP cameras, network video recorders, control centers and video analysis and streaming nodes, could be candidate for applying this approach since Java technology allows task migration via object serialization and the distributed nodes are capable of using local schedulers. Also, this kind of system requires an extensive use of CPU that may justify task migrations. Moreover, the optimal use of physical resources, such as CPU, memory or battery, justifies the implementation of optimization policies to improve the overall performance of the applications.

3.1 Functional Overview

Briefly, the proposed architecture works as follows. The GS activates the tasks that compose the applications and orders them according to the application graphs and the scheduling and optimization policy. Since the GS is in charge of the activation of the periodic tasks, it needs a global timing reference for the whole DES, which defines a minimum granularity of the invocations at the DES. This timing reference unit will be called the *Elementary Cycle* (EC) and it will described below.

RS abstract the distributed processors in charge of executing tasks. Whenever an RS completes the execution of all assigned tasks, it notifies the GS its availability with a *WorkCompleted* message as depicted in Fig 2. The GS reacts by selecting the next task in the queue and deploying it to the free RS.

Deploying a task to an RS typically involves migrating a task from the GS to this RS; however, since task migrations have a great impact on the overall performance of the framework each RS is equipped with a sort of cache memory. So, depending on the status of the cache of the target RS, the GS may carry out three different deployment types. If (1) the deployed task is not in the cache of the target RS and it has enough memory for holding the new TW a regular deployment is performed and the TW object is physically transmitted to the RS. Else, (2) if the deployed TW is already in the cache of the target RS, the TW object is not sent; instead, the GS only sends the input parameters to run the task. Lastly, (3) if the deployed task is not in the cache of the target RS and it has not enough memory to keep it, an overwrite deployment takes place. In such case, the TW is sent to the RS, and it overwrites an older TW from the cache of the RS to store the new task.

To implement this functionality the GS keeps a set of ordered task queues, associated to every RS of a DES, that are updated when changes are produced. After each update, the task located in the first position of the queue is considered by the GS as the highest priority task to be deployed to the RS associated with that queue.

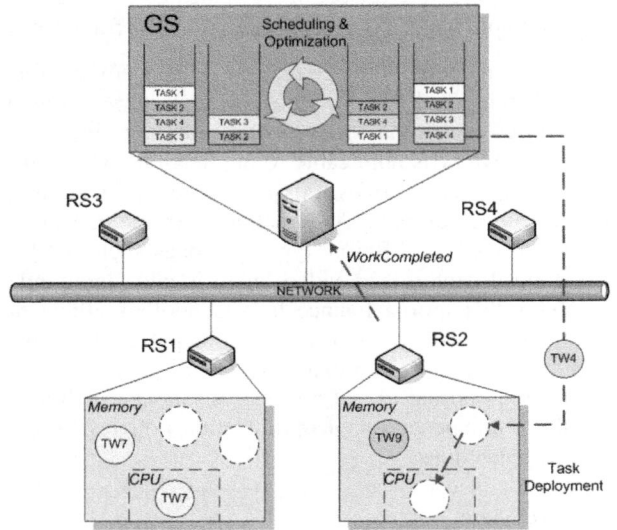

Fig. 2. Overview of the proposed architecture

Often, distributed applications are formed by a combination of tasks that are executed in a concrete sequence. This implies that every task may have one or several predecessors and successors. Tasks may also require inputs from other tasks to perform their work. This information is modeled in the GS by an application graph which is built from the TW. The GS uses it in run-time to compose the applications and to select the more appropriate nodes for optimizing the use of the resources according to the selected criteria. The communications between a task and its successors are centralized by the GS which receives the outputs of the completed tasks from the RS and hand them over as inputs to the next tasks in the graph. Predecessor and successor tasks along with input and output elements are included in the TW structure, as it will be further explained along with the system characterization.

The proposed framework assumes that tasks can be executed in any RS of the DES; however there will be cases in which some tasks will only be executable in certain nodes, e.g., when a specific library or hardware is required. The framework proposes the use of node bindings for modeling these requirements that will specify the list of nodes where a task may be executed.

In order to achieve a soft real-time behavior, it is necessary to implement a remote task preemption mechanism. The proposed framework relies on the local schedulers of the distributed nodes to implement a simple preemption mechanism. Preemptions are triggered by the GS when a critical situation is detected, that is, when the laxity of a task, defined as its time to deadline minus its remaining execution time, becomes too short. If so happens the GS triggers a special deployment routine called *Preemption Deployment*, which deploys the critical task to a non-free RS. This kind of deployment stops the running task and executes the newly deployed one.

One of the key benefits of the proposed framework is its high level of dynamism. The GS provides a reconfiguration interface to dynamically add, remove or change the tasks in the system. The GS is equipped with an admission control system to prevent changes that could lead the system to unschedulable situations. This approach allows changes in the applications ensuring certain QoS parameters.

Certain characteristics of the presented framework have direct implications in its behavior. For example, the authors have chosen a centralized approach because it provides more flexibility and higher scalability even though it may reduce fault tolerance. In future works the authors will introduce replicated GS in order to improve fault tolerance. Task migrations are also a key challenge since they may affect negatively the performance of the system. In future works the authors will quantify the impact of task migrations if different scenarios. Lastly, the presented framework does not consider shared resources dependencies among tasks as this issue can be worked around splitting the tasks and using precedence dependencies.

3.2 System Characterization

The middleware framework described in the previous section requires keeping in memory a model of the tasks in the DES. Indeed, both the GS and the RS use a special structure to abstract the concept of a task, the Task Wrapper (TW). A TW not only models the task timing characteristics and precedence dependencies but also contains the logic of the application. Fig. 3 depicts this structure using UML notation and Java-like types for simplicity.

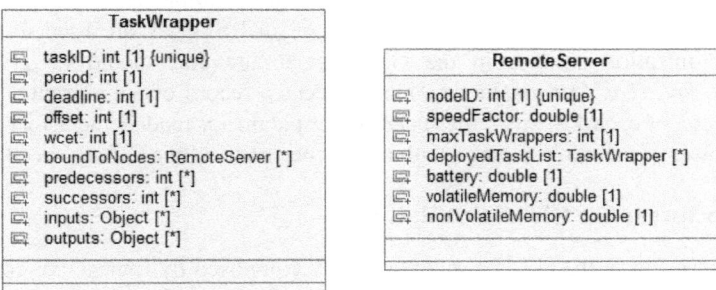

Fig. 3. Characterization of the Remote Servers and Tasks

The proposed task model requires a unified timing reference to be used in the entire DES. For the sake of consistency, all time measurements have been referred to an abstract time unit, the Elementary Cycle (EC), which is defined in a centralized way in the Global Scheduler as the minimum time between two subsequent task activations and specifies the time granularity of the system. The EC must be configured during the start-up phase of the GS and it cannot be modified by any means in run-time. As a consequence, all the timing parameters of the TW model are referred to this parameter.

The task model is composed by the following parameters per TW:

- **Task identifier.** Used to identify univocally a task in the distributed system.
- **Timing parameters.** This set of parameters characterizes the timing properties of a TW. They are all defined as integer multiples of the EC. These parameters include: *period, deadline, offset* and *worst-case execution time (wcet)*. Note that the *wcet* parameter is referred to a node with a unitary speed factor as detailed in the RS description section below.
- **Precedence graph management parameters.** This group of parameters includes information related to the application graph such as *predecessors, successors, input data (inputs)* and *outcomes (outputs)* of one task. The GS uses these parameters to build the precedence graph in order to activate the tasks in the graph in order and manage the inputs and outputs involved.
- **Bound to nodes.** This parameter is used to attach one TW to a specific node or list of nodes. It may be used when a node owns a specific hardware resource or software component (e.g. a library) required for the execution of a task.

The framework also requires the GS to keep information about the status of the RS in the DES, since this information is essential for applying optimization policies. Therefore, the GS maintains a model in memory that represents the DES using a *Remote-Server* data structure per RS (also included in Fig. 3). Regarding the RS model, the following parameters are considered:

- **Node identifier.** Integer value that identifies univocally an RS in the distributed system.
- **Physical parameters.** Namely *speedFactor, maximumTaskWrappers, deployedTaskList, battery* and *volatile / non-volatile memory.* This group of parameters models the physical status of an RS. They are used to implement optimization policies in the GS. Special attention should be given to the *deployedTaskList* parameter, since it keeps a record of the current status of the cache of each RS. Also, the *speedFactor* parameter models the actual processing power of a node, compared with a reference node with a unitary speed factor.

3.3   Structure of the Global Scheduler

As shown in Fig. 4 the GS is a modular entity composed by four active components; namely (1) *Activator*, (2) *RSInterface*, (3) *ReconfigurationInterface* and (4) *PreemptionManager*. Along with these components the GS also uses several data structures that represent the current status of the DES.

For each connected RS, the *TaskQueuesManager* (see element #5 of Fig. 4) keeps an ordered queue with all the tasks that must be activated at that node. These tasks are ordered according to a scheduling policy and then, the ordering is refined according to an optimization policy. A reordering is committed every time a new task is placed in a queue. As shown in the figure, the framework is open to be used with different scheduling policies such as RMS, EDF or MUF, which are connected to the queues as pluggable components. Similarly different optimization policies can be connected to the queues.

Additionally, the GS maintains a *SystemModel* (element #6), which is a data structure updated every EC by the *Activator* (element #1). This model is used to keep track of the remaining times for the next activation of each periodic task, the current laxities of the active tasks and the current status of the connected RS.

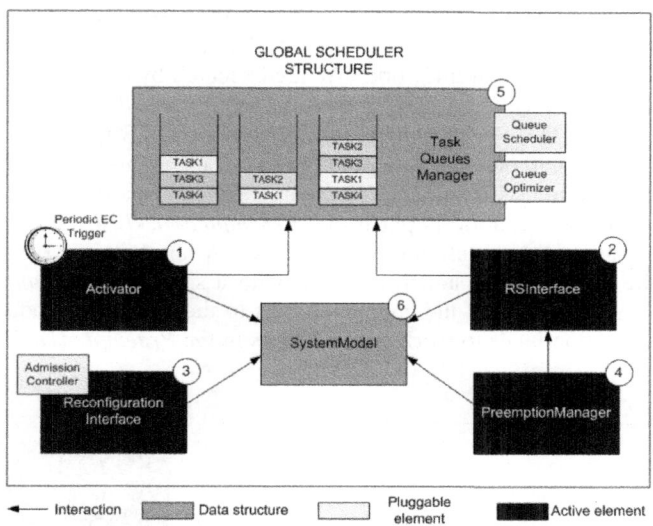

Fig. 4. Structure of the Global Scheduler

Apart from updating the *SystemModel*, the *Activator* is in charge of adding the tasks activated every EC to the queues. Therefore, it also manages the precedence dependencies between tasks.

All communications between the GS and the RS are handled via the *RSInterface* (element #2). Additionally, whenever the *RSInterface* receives a *WorkCompleted* message from any RS, it is the *RSInterface* itself who deploys the next task in the queue to the requesting RS and updates the *SystemModel* accordingly.

Users and applications may interact with the GS to require dynamic reconfigurations at runtime through the *ReconfigurationInterface* (element #3). This interface allows changes in the task parameters (period, deadline, etc.) as well as adding or removing tasks to the system. All changes are recorded in the *SystemModel*; however, before any changes are committed this component executes an admission control test that checks whether the new configuration is feasible. This functionality is provided by a pluggable component that may be exchanged to use different admission policies.

Finally, the GS implements a *PreemptionManager* (element #4) module whose objective is to prevent tasks from missing their deadline. It uses the tables in the *SystemModel* to detect potential deadline misses. Should any problems be detected, the *PreemptionManager* would instruct the *RSInterface* to activate a preemption deployment routine to one or more RS.

3.4 Structure of the Remote Servers

An RS is an entity that manages only one processor of the DES (see Fig. 5). The main role of an RS is to execute the tasks deployed by the GS and, when all assigned tasks are completed, declare its availability to the GS via a *WorkCompleted* message.

Communications with the GS are handled via the *GSInterface* component (element #1). When a new TW is received it is stored in the *DeployedTaskList* (element #3). This element plays a similar role to the caches in processors, allowing the GS to reduce deployment times when a TW is already loaded in an RS, and improving the overall performance of the framework. TW are executed by the *TaskExecutor* component (element #2), which is capable of starting, stopping and resuming the execution of a TW and sends the *WorkCompleted* messages to the GS with the results of each task when they are completed. It also implements a mechanism that allows preempting a TW in execution with another. When an RS receives a preemption deployment message the TW in execution is put in the *PreemptedTaskList* (element #4) and the *TaskExecutor* starts the execution of the new task. When the latter task terminates its execution, the task executor notifies the GS with a special *TaskCompleted* message which includes the results of the completed task to the GS to be handed over to successor tasks, and continues to work until all tasks in the *PreemptedTaskList* have been executed.

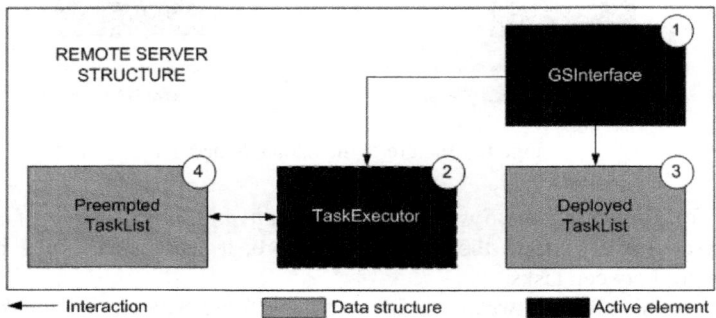

Fig. 5. Structure of the Remote Server (RS)

4 Conclusions and Future Work

This paper presents a middleware framework for DES that supplies a set of services aimed to manage the deployment and execution of a set of tasks in a set of distributed nodes, so the time requirements of the overall system are met and the use of the resources is optimized according to different criteria (e.g. use of CPU, memory or battery). In particular, it provides a timing reference for the whole DES, support for executing in order the tasks of a DES, a certain degree of dynamism in run-time to adapt to changing requirements (e.g. task parameters, code updates) and a reconfiguration interface to request these changes. This framework is aimed at DES that allow task migration and that may use local schedulers at the nodes. It uses a central modular component (known as GS) that orchestrates the system, another component type

(RS) to abstract the individual processors involved in the DES and a special structure that abstracts the concept of task (TW). The GS is built in a modular way so the designers of the applications may easily choose from different scheduling and optimization policies the one that suits best to their applications.

In the future the authors will introduce replicated GS in the framework to improve fault tolerance and will evaluate the performance of the framework as well as its behavior using the real-time Java implementation. Special care will be taken regarding task migration costs and memory consumption issues.

Acknowledgments. This work has been supported by the ARTEMIS JU through the iLAND project 2008/10026, the Basque Government through the TEReTRANS project IE08-221 and by the University of the Basque Country (UPV/EHU) through grant EHU09/29.

References

1. OROCOS, The OROCOS project – Smarter control in robotics and automation,
   http://www.orocos.org/
2. Real-Time CORBA with TAO, ACE and TAO success stories,
   http://www.cs.wustl.edu/~schmidt/ACE-users.html
3. Levi, S., Tripathi, S.K., Carson, S.D., Agrawala, A.K.: The MARUTI Hard Real-Time Operating System. SIGOPS Operating Systems Review 23(3), 90–105 (1989)
4. Stankovic, J.A., Ramamritham, K., Humphrey, M., Wallace, G.: The Spring System: Integrated Support for Complex Real-Time Systems. International Journal of Time-Critical Computing Systems 16, 223–251 (1999)
5. Calvo, I., Almeida, L., Noguero, A.: A Novel Synchronous Scheduling Service for CORBA-RT Applications. In: Proceedings of the 10th IEEE International Symposium on Object and Component-Oriented Real-Time Distributed Computing, ISORC 2007, pp. 181–188 (2007)
6. Basanta-Val, P., Estévez-Ayres, I., García-Valls, M., Almeida, L.: A synchronous scheduling service for distributed real-time Java. IEEE Transactions on Parallel and Distributed Systems (2009) (Accepted for future publication)
7. Wang, X., Lu, C., Gill, C.: FCS/nORB: A feedback control real-time scheduling service for embedded ORB middleware. Microprocessors and Microsystems (June 2008)
8. Kalogeraki, V., Melliar-Smith, P.M., Moser, L.E., Drougas, Y.: Resource management using multiple feedback loops in soft real-time distributed object systems. The Journal of Systems and Software 81, 1144–1162 (2008)
9. Zhang, Y., Lu, C., Gill, C., Lardieri, P., Thaker, G.: Middleware Support for Aperiodic Tasks in Distributed Real-Time Systems. In: Proceedings of the 13th IEEE Real Time and Embedded Technology and Applications Symposium, RTAS, pp. 113–122 (2007)
10. Zhang, Y., Gill, C., Lu, C.: Reconfigurable Real-Time Middleware for Distributed Cyber-Physical Systems with Aperiodic Events. In: Proceedings of the 28th International Conference on Distributed Computing Systems ICDCS 2008, pp. 581–588 (2008)

Resource Adaptive Distributed Information Sharing

Hans Vatne Hansen[1], Vera Goebel[1],
Thomas Plagemann[1], and Matti Siekkinen[2]

[1] University of Oslo, Department of Informatics
{hansvh,goebel,plageman}@ifi.uio.no
[2] Aalto University School of Science and Technology,
Department of Computer Science and Engineering
matti.siekkinen@tkk.fi

Abstract. We have designed, implemented and evaluated a resource adaptive distributed information sharing system where automatic adjustments are made internally in our information sharing system in order to cope with varying resource consumption. CPU load is monitored and a light-weight trigger mechanism is used to avoid overload situations on a per-machine basis. Additional improvements are obtained by calculating what we call a utility score to better determine how the data structures in the system should be arranged. Our results show that resource adaptation is an efficient way of improving query throughput, and that it is most effective when the number of stored data items in the system is large or many queries are performed concurrently. By applying resource adaptation, we are able to significantly improve the performance of our information sharing system.

Keywords: autonomic networks, self-optimization, resource adaptation.

1   Introduction

Data centric networking has become an important networking paradigm. Focus on content instead of its location has started to drive the design of communication systems, and over the last decade much research efforts have been invested into developing solutions for networking data instead of hosts. These activities have addressed several kinds of approaches ranging from building overlay type of solutions [13,10,7] to designing complete data-centric network architectures from scratch [6,4].

It is desirable to have general purpose information sharing solutions that are capable of storing and querying various kinds of data. This is true for both the current situation as well as for the future. Many applications today rely on specific overlay solutions, and the different solutions are usually incompatible, which introduces significant amount of unnecessary overhead and complexity. Future Internet approaches such as ANA (Autonomic Network Architecture) [1] would benefit from such a general purpose information sharing system as an

F.A. Aagesen and S.J. Knapskog (Eds.): EUNICE 2010, LNCS 6164, pp. 246–255, 2010.

integral and reusable core component of the architecture. We have designed and implemented a fully distributed system called MCIS (Multi-Compartment Information Sharing) which is based on a distributed hash table (DHT) type of structured peer-to-peer system. The core functionality of MCIS is storing and querying different types of data. We define a data type as a set of attributes and a data item as a tuple of attribute values. Within the entire MCIS system, each different data type is formed by separate, logical units for data management called attribute hubs. In these attribute hubs, data items and queries are routed independently from each other. Furthermore, the system supports multi-attribute range queries.

One of the challenges in deploying such a fully distributed system that relies on cooperation of every node, is that these nodes can have significantly different amounts of resources available, such as CPU, memory, storage space and energy. The amount of available resources depends on the one hand on the device itself, e.g. mobile devices have scarce resources compared to desktop computers, and on the other hand on the current workload caused by the applications running on that node. A single overloaded node can negatively impact the performance of the overall system. Thus, there are potentially several bottlenecks in the system in form of overloaded nodes. Therefore, in order for such a system to perform well, individual nodes should handle overload situations by dynamically adjusting to the varying resource consumption.

In this paper, we describe our MCIS system enhanced with an automated resource adaptation mechanism. MCIS is designed in such a way that the different attribute hubs introduce replication of data in order to gain performance in query processing. We take advantage of this design in the resource adaptation mechanism so that there is a trade-off in the level of replication and the query processing efficiency against resource consumption and the number of active attribute hubs. We introduce a utility metric to identify the hubs with the most positive effect on the query processing efficiency and give them priority when reducing the number of active hubs.

We evaluate the performance and behavior of the resource adaptive MCIS by storing and retrieving real Internet traffic traces from the Cooperative Association for Internet Data Analysis (CAIDA) [15]. The evaluation results do not only demonstrate the feasibility of resource adaptation in MCIS, but also the superior performance of a resource adaptive MCIS compared to a non-adaptive MCIS. This work is done as part of the ANA project and the complete source code of our system is available at the project website, http://ana-project.org/.

The remainder of this paper is structured as follows. Section 2 introduces MCIS. A detailed description of the resource adaptation scheme is given in Section 3. In Section 4, we evaluate the system behavior. In Section 5, we draw conclusions and explain future work.

2 Multi-Compartment Information Sharing

MCIS is a fully distributed store and query system. It is based on Mercury [2] which is a DHT-based system that supports multi-attribute range queries, such as `size` > 100 MB, `size` < 300 MB when looking up data items with size between 100 MB and 300 MB. As a DHT-based system, MCIS is fully decentralized.

Each data type in MCIS is administrated by an instance of a Mercury system. The data types can have several attributes like `string name`, `int size`, and `string type`; and the specification of the collection of these attributes is called a schema. An example of a data item for this schema is ''example, 150, MKV''. The data types are managed by attribute hubs which are logical ring structures, one per attribute in the schema. These attribute hubs, like `Hname`, `Hsize` and `Htype` in Figure 1, organize and route the data items related to the specific attribute independently of each other. This is done in the following way: Data items are replicated and stored in all hubs while a query is only forwarded to the hub where it is expected to be executed most efficiently. A wild card attribute in a query needs to be evaluated at each node in the hub corresponding to that attribute, while a small range for another attribute narrows the search down to only few nodes in that other hub. In this way, replication is introduced to improve query processing efficiency. Figure 1 shows how a data item is stored in all attribute hubs, i.e. `Hname`, `Hsize` and `Htype`, and how a potential query could most efficiently retrieve the data from `Hsize`.

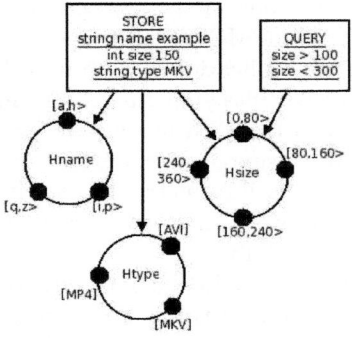

Fig. 1. MCIS hub structure

It should be noted that hubs do not necessarily need to be actively used for all the attributes in the schema, but they help to decrease hop count in a system with a large number of nodes. In other words, hubs can be joined or left at any time in order to adapt to resource consumption, but with the expense of potentially heavier query processing and longer query response times when few hubs are active. This potential expense is the trade-off our resource adaptation mechanism relies on. As shown in Figure 1, there are four nodes in the system. All of these nodes participate in `Hsize`, but only three of them participate in `Hname` and `Htype`.

3   Resource Adaptation

Distributed systems like MCIS can have a high degree of churn, and one reason for this is failing nodes [11]. There is a risk that one failing node can harm the entire system, and even with data replication, the overall performance decreases with node failures. The main problem in distributed systems is that the number of concurrent users can become too high for the system to handle. Our solution to this problem is self-optimization, which in MCIS means to automatically detect certain changes in the individual nodes and adapt to these changes in order to improve the service.

Available memory and processing power are critical system resources that vary depending on all running processes on the machine. Our strategy for self-optimization is to identify when resource consumption is at a level where MCIS is unable to function properly or to service its users, make internal changes to adapt to this, and consequently improve performance. It is possible to adapt to variance in almost all resources such as memory, storage space and energy. We call this resource adaptation, and it consists of two distinct functions in addition to the actual MCIS application: 1) The resource consumption is measured in order to determine when the consumption is at a critical level. We focus on CPU load, but any other node resource could be measured. 2) The measurement data is analyzed, and nodes join or leave attribute hubs when pre-defined thresholds are reached.

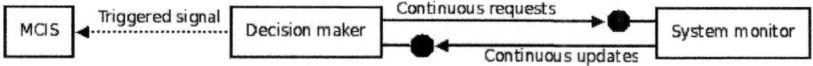

Fig. 2. Feedback control system

Figure 2 shows our feedback control system for resource adaptation of MCIS. The two resource adaptation modules dictate how many active attribute hubs each data type has, based on CPU load and pre-defined thresholds. The **System monitor** inspects system resources at fixed intervals and makes the obtained information available to the **Decision maker**. We use processor load as the resource adaptation trigger because calculating query routes in the attribute hubs is a CPU intensive task. Without sufficient available CPU capacity the queries will not be successfully executed. The processor load is calculated based on CPU queue length, i.e. the number of processes in the waiting queue when idle and blocking processes are omitted. This load number is calculated by the Linux kernel [9] as the exponentially weighted moving average within a one minute window. It is found to perform better than utilization indices when doing dynamic load balancing in distributed systems [3]. We normalize this measure by factor 100 such that an idle computer has a load of 0 and a fully loaded CPU has a load above 100. We have also investigated memory utilization, but found that it did not influence performance enough to be used as trigger.

The adjustment made in each MCIS node is whether it should join or leave a certain attribute hub. This is not a coordinated event, but made on a per machine basis. Individual nodes can leave a hub while the remaining nodes can continue to use it if they have enough available resources. Leaving an attribute hub leads to less routing calculation with a potential drawback of extra hops and higher response times. By minimizing calculations of where to retrieve data we expect that the MCIS node is able to answer more queries. This kind of optimization is especially valuable for resource constrained devices like PDAs or systems where resource demanding applications use a large percentage of the available processing capacity.

We choose which hub to join and leave based on what we call a *utility score* which ranks the different hubs based on how profitable they are when performing queries. Data items are always replicated and sent to each hub, but queries are only forwarded to the most selective hub, which is then responsible for providing the results. Hence, leaving the attribute hub where fewest queries are forwarded minimizes routing overhead. The utility score for each hub is calculated in the following way: If a hub stores a data item, its utility score is decreased by one, but if it forwards a query, the score is reset to zero. The result of this formula is that hubs that answer many queries will have a utility score close to zero, while less utilized hubs have negative scores. When MCIS is told to leave a hub, it chooses the hub with the lowest score. The reasoning behind this strategy is to first leave the hubs that add to resource usage by routing data items, but never route queries.

Table 1. Example time-line of utility scores

	T0	T1	T2	T3	T4
Hname	0	-3	-3	-5	0
Hsize	0	-3	0	-2	-2
Htype	0	-3	-3	-5	-5

Table 1 shows an example of how utility scores are calculated in a given situation. T0 through T4 are points in time, ranging from oldest to newest. At T1, 3 data items are stored and all the hubs decrease their utility score. At T2, a query is forwarded to `Hsize` making it reset its score to 0 while the other two remain at -3. At T3, 2 new data elements are stored. At T4, a query is forwarded to `Hname` giving the hubs three different utility scores. The most valuable hub at this point in time is `Hname`. `Hsize` is less important, while `Htype` is least important and the first hub an overloaded node leaves.

In summary, our resource adaptation extension is a simple and light-weight solution where the introduced overhead is too small to measure correctly. It monitors CPU load and calculates utility scores, but does not require global coordination or introduce extra networking traffic. All changes are made on a per-machine basis, and the potential result is fewer node failures and more queries performed.

4   Evaluation

The objective of the evaluation is to investigate the differences between our MCIS system with and without self-optimization to quantify the resulting performance improvement when adapting to changing resource consumption. The desirable outcome of the evaluation is higher efficiency and an increased number of queries performed when resource adaptation is applied. The number of simultaneous queries that MCIS can perform is known as throughput rate and is expressed as queries per time unit. We choose to evaluate our system based on throughput as it reflects the demands and requirements that applications using MCIS have for performance.

In our studies, the number of active hubs in MCIS is one of the key parameters. It varies according to resource consumption and influences how routing is done. External load is also an important parameter, and we use synthetic load to have total control over the quantity and when the load is applied. It is generated by two programs with the purpose of using a predetermined, fixed amount of processing capacity. We aim at keeping our evaluation realistic and choose data items and queries that represent real world use-cases. The data items we use in our studies are real Internet traffic traces gathered by CAIDA [15], and a typical investigation of these traces is anomaly detection. One example of our queries is locating traffic on a range of ports, known to be used by malicious programs. These queries also invoke route calculations and might not be issued if the system has insufficient resources. Other queries have also been tested.

The parameters we vary explicitly are the number of *stored data elements (DE)* and the number of *queries per minute (QPM)*. We choose these parameters because they correspond well to the internal load of MCIS. By changing one or both of these parameters we can investigate the implications and understand in which situations, if any, resource adaptation can improve MCIS.

We conduct our evaluation experiments in two phases with one local and one distributed test. In the first test, we use one MCIS node and experiment with a wide range of parameters. This test is an attempt to narrow our parameter values and prepare for the second, distributed test.

We repeat the experiments three times and use the same configuration on all machines to ensure that queries are processed under the same conditions, independently of which MCIS node they are forwarded to. The machines have 2.60

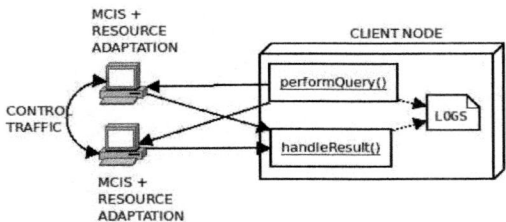

Fig. 3. Distributed evaluation setup

GHz processors and 1 GB of memory. In the distributed test, all the machines are connected directly to each other through the same switch. There are two MCIS nodes and one client node running, as shown in Figure 3.

In the **local test**, we see that MCIS has an average CPU load of 5 on the given hardware. This means that if the external load is below 95, MCIS has sufficient CPU. We conclude that a load of 95 can be the trigger for leaving an attribute hub as this is the critical level where MCIS can function properly. With different hardware specifications the load numbers will differ and the triggers must be adjusted accordingly. We determine all our parameter values based on what MCIS can handle on our hardware.

Table 2. Parameter values

	Minimum	Maximum
Data elements	1000	2000
Queries per minute	100	300
External CPU load	100	100

Table 2 shows our results from the local test and what parameter value ranges we use in the **distributed test**. For each of the varying query rates and number of stored data elements, we perform experiments with and without resource adaptation enabled. The improvement of the self-optimization is indicated when comparing throughput between the experiments with and without resource adaptation.

Figure 4 shows that the experiments are almost equal when resource adaptation is enabled, regardless of the number of stored data elements. MCIS is able to perform close to 100% of all the queries when both 100 and 200 queries are

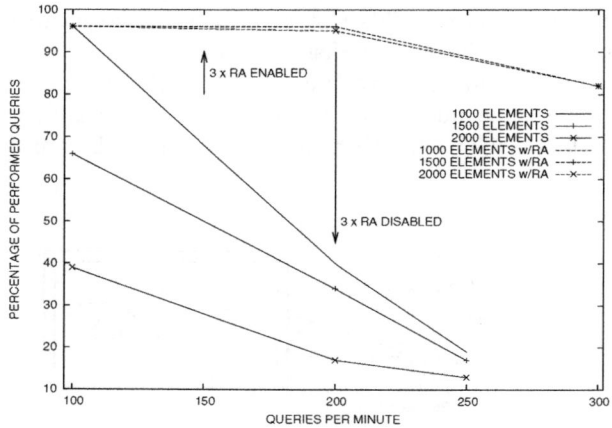

Fig. 4. Throughput comparison

performed per minute. This is not the case when adaptation is disabled. In the experiments where resource adaptation is disabled, the number of queries performed drops drastically when the rate is increased from 100 queries per minute to 200 queries per minute. This is especially true for the experiments with 1000 and 1500 stored data elements. In addition, we are unable to achieve consistent results when trying to query MCIS without adaptation 300 times per minute. The machines are fully saturated and we need to decrease the maximum query rate to 250 when resource adaptation is disabled.

Our analysis shows that resource adaptation has a positive effect on throughput and that it significantly improves the performance of MCIS. Most improvement is achieved when the number of stored data elements is large or when query rate is high. As expected, when a sufficient amount of resources are available the effect of resource adaptation is neglectable.

5   Related Work

Several scalable information systems that gather information about networked systems have been proposed in the literature such as in [14,16,8]. The first two approaches focus more on information aggregation aspects, while the last approach exhibits some self-configuration properties regarding resource utilization. InfoEye [8] is a self-configuring distributed information management system where management nodes communicate with monitoring sensors located at each overlay node in order to get information about the overlay nodes which execute application tasks. This information is provided via a query interface to applications. Since collecting information about every attribute of each overlay node is unfeasible in a large system, InfoEye self-configures in an optimal fashion the way it gathers the information from monitoring sensors, e.g. which attributes, from which nodes, and via push or pull mechanism. InfoEye with the centralized management nodes (one such node exists in [8]) is a quite different concept from MCIS which is a fully distributed system in which nodes make independent decisions.

Replication in distributed systems has been studied earlier. For example, a popularity/size-adaptive object replication degree customization scheme is described in [17]. In [5], the focus is on maximizing object availability in peer-to-peer communities under space constraints. Together with an additional large body of approaches, this work focuses on developing optimal models for replication in a system given certain constraints and objectives. Our scheme differs in that there is no system-wide coordination. Instead, individual nodes dynamically and independently make adjustments depending on their system state.

The closest match to our work that we could find is the concept of Elastic Routing Table (ERT) [12]. This mechanism is proposed to prevent overload situations at particular DHT nodes due to inherent load balancing problems in DHTs, the heterogeneity of network nodes, and the non-uniform and time varying popularity of content. ERT uses variable size routing tables for DHT nodes which are adjusted based on load at the particular node. The difference to our work is that we consider impacting the replication degree of data instead of the routing tables. In fact, ERT could be used as a complementary scheme in MCIS to adjust each node's routing tables for individual hubs.

6   Conclusions

In this paper we present MCIS, our distributed information sharing system which is able to self-adapt to changing levels of resource consumption when necessary. We achieve self-adaptation by altering internal data structures with the possible side-effect of extra hops and higher response times. This property allows the system to be deployed among heterogeneous nodes while alleviating the problem of individual nodes to become bottlenecks, which can decrease the performance of the entire system. Our evaluation results demonstrate that with the use of resource adaptation the system achieves higher throughput than without resource adaptation. These results are also relevant for other distributed systems that can trade-off response time for throughput.

This paper presents our first results in our pursue towards a fully self-organizing information sharing system. Hence, there are a number of challenges that we would like to address in our future work. The current version of MCIS attempts to handle overload situations in individual nodes. We want to benefit from the large body of analytical work on replication for high availability in order to drive the design of MCIS towards optimal resource adaptation. This work also includes replicating data through duplicate attribute hubs, which requires changes to the design of the underlying Mercury system. In addition, we plan to model the resource consumption to be able to evaluate the behavior of the MCIS system with a large number of nodes through simulations. Specifically, we are interested in understanding the quantitative impact of nodes leaving and joining hubs to the query response time in a large system. We are also interested in analyzing the trade-off between the number of joined hubs and query hop count.

Acknowledgments. This work has been funded by the ANA project (EU FP6-IST-27489). The authors would like to thank Stein Kristiansen for valuable feedback.

References

1. ANA Project. ANA Blueprint, sixth framework programme edition (February 2008)
2. Bharambe, A.R., Agrawal, M., Seshan, S.: Mercury: Supporting scalable multi-attribute range queries. In: ACM SIGCOMM (2004)
3. Ferrari, D., Zhou, S.: An empirical investigation of load indices for load balancing applications. Technical Report UCB/CSD-87-353, EECS Department, University of California, Berkeley (May 1987)
4. Jokela, P., Zahemszky, A., Rothenberg, C.E., Arianfar, S., Nikander, P.: Lipsin: line speed publish/subscribe inter-networking. In: SIGCOMM 2009: Proceedings of the ACM SIGCOMM 2009 conference on Data communication, pp. 195–206. ACM, New York (2009)

5. Kangasharju, J., Ross, K.W., Turner, D.A.: Optimizing file availability in peer-to-peer content distribution. In: IEEE INFOCOM 2007, 26th IEEE International Conference on Computer Communications, May 2007, pp. 1973–1981 (2007)
6. Koponen, T., Chawla, M., Chun, B.-G., Ermolinskiy, A., Kim, K.H., Shenker, S., Stoica, I.: A data-oriented (and beyond) network architecture. SIGCOMM Comput. Commun. Rev. 37(4), 181–192 (2007)
7. Li, X., Bian, F., Zhang, H., Diot, C., Govindan, R., Hong, W., Iannaccone, G.: Mind: A distributed multi-dimensional indexing system for network diagnosis. In: Proceedings of INFOCOM 2006, pp. 1–12 (2006)
8. Liang, J., Gu, X., Nahrstedt, K.: Self-configuring information management for large-scale service overlays. In: INFOCOM, pp. 472–480 (2007)
9. The Linux man-pages project. Linux Programmer's Manual. SYSINFO(2) (November 1997)
10. Ratnasamy, S., Francis, P., Handley, M., Karp, R., Schenker, S.: A scalable content-addressable network. In: Proceedings of SIGCOMM 2001, pp. 161–172 (2001)
11. Rhea, S., Geels, D., Roscoe, T., Kubiatowicz, J.: Handling churn in a dht. In: ATEC 2004: Proceedings of the annual conference on USENIX Annual Technical Conference, p.10. USENIX Association, Berkeley (2004)
12. Shen, H., Xu, C.-Z.: Elastic routing table with provable performance for congestion control in dht networks. In: ICDCS 2006: Proceedings of the 26th IEEE International Conference on Distributed Computing Systems, Washington, DC, USA, p. 15. IEEE Computer Society, Los Alamitos (2006)
13. Stoica, I., Morris, R., Karger, D., Frans Kaashoek, M., Balakrishnan, H.: Chord: A scalable peer-to-peer lookup service for internet applications. In: Proceedings of the ACM SIGCOMM 2001 Conference, San Diego, California, August 2001, pp. 149–160 (2001)
14. Van Renesse, R., Birman, K.P., Vogels, W.: Astrolabe: A robust and scalable technology for distributed system monitoring, management, and data mining. ACM Trans. Comput. Syst. 21(2), 164–206 (2003), doi.acm.org/10.1145/762483.762485
15. Walsworth, C., Aben, E., Claffy, K.C., Andersen, D.: The caida anonymized 2009 internet traces - equinix-chicago.dira.20090331-055905.utc (2009), http://www.caida.org/data/passive/passive_2009
16. Yalagandula, P., Dahlin, M.: A scalable distributed information management system. In: SIGCOMM 2004: Proceedings of the 2004 conference on Applications, technologies, architectures, and protocols for computer communications, pp. 379–390. ACM, New York (2004)
17. Zhong, M., Shen, K., Seiferas, J.: Replication degree customization for high availability. In: Eurosys 2008: Proceedings of the 3rd ACM SIGOPS/EuroSys European Conference on Computer Systems 2008, pp. 55–68. ACM, New York (2008)

Performance Impacts of Node Failures on a Chord-Based Hierarchical Peer-to-Peer Network

Quirin Hofstätter

Lehrstuhl für Kommunikationsnetze
Fakultät für Elektrotechnik und Informationstechnik
Technische Universität München, Germany
quirin.hofstaetter@tum.de

Abstract. Peer-to-Peer networks are designed to provide decentralized, fault-tolerant alternatives to traditional client-server architectures. In previous work, the the resilience of structured P2P networks has been evaluated analytically as well as by simulation. In this paper, we discuss the influence of peer failures on a core component of our *cost-optimized* hierarchical P2P protocol *Chordella*. We show that the used algorithms lead to a working system even if a high number of superpeers is subject to failures.

1   Introduction

Peer-to-Peer (P2P) networks are based on the construction of a an overlay network on the application layer, where the decentralized protocol allows the users to share certain resources. Current generations of P2P networks are based on Distributed Hash Tables (DHTs), used to maintain a consistent, global view of the available resources. Popular protocols are e.g. Chord [1] or Kademlia [2]. However, the stability of these overlay networks is directly affected by user behavior (joining or leaving the network) and the performance of the network connection. When many peers fail (e.g. mobile peers losing power supply or the wireless network link), the P2P network may be split into a number of disjoint networks or even break down completely. Hence a careful design of the used algorithms is crucial.

A number of previous works on the stability of P2P networks has focused the Chord protocol and its behavior in non-optimal conditions [3,4]. The authors examine the behavior of the overlay in settings with inconsistent routing table entries, the recovery mechanisms and their boundaries.

In our previous work [5,6], we extended standard Chord to a two-tier hierarchical P2P protocol for mobile use: *Chordella*. High performance nodes act as superpeers (SP) forming the DHT, nodes with less power (e.g. mobile phones) are attached to these superpeers as leaf nodes (LN). In order to drive the system in a cost-optimal state (in terms of network traffic), we introduce metrics and algorithms to determine, reach and maintain the optimal number of superpeers in the network in a fully distributed way.

F.A. Aagesen and S.J. Knapskog (Eds.): EUNICE 2010, LNCS 6164, pp. 256–258, 2010.

2   Impact on Chordella's Algorithms

In [6] we describe the used algorithm to achieve and maintain the cost-optimal ratio between superpeers and leaf nodes in the network. We show that it is able to drive the system in a state that the relative deviation between the theoretical optimum and the value measured in simulation does not exceed 5 % in a realistic churn scenario. In the worst case it stays within 10 %. These scenarios include random failures of peers in the network. In this work, we study the impact of non-random failures on the introduced algorithms. We use the same parameters for simulation as in [6], due to space limitations we refer to that publication for details. Of course, fhe failure scenario differs: We let a differently sized number of peers fail every twelve stabilization periods $T_{stab} = 5$ s. The removed peers build a block in the ID space, i.e. they are direct successors. Hence the Chord overlay experiences a worst-case failure. All results are averages of ten independent simulation runs, confidence intervals are too small to be shown.

Figure 1 shows the number of superpeers against the simulation time in a case where no peers fail (best case, solid line) and with equally distributed failures (as in [6], dashed line). You can see the three phases of the simulation: the join phase (0 to 7200 seconds) where the nodes join the network at a constant rate, the churn phase where nodes join and leave the network (normal operation mode) and the leave phase where all nodes leave the network at a constant rate. In this scenario the system is able to maintain a number of superpeers very close to the optimum. This random failure case is the reference for the following scenarios.

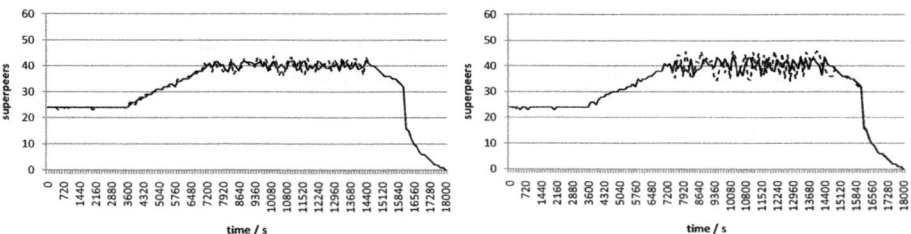

Fig. 1. Number of SPs: No failures (solid) / Equally distributed failures (dashed)   **Fig. 2.** Number of SPs: Random failures (solid) / 10% block (dashed)

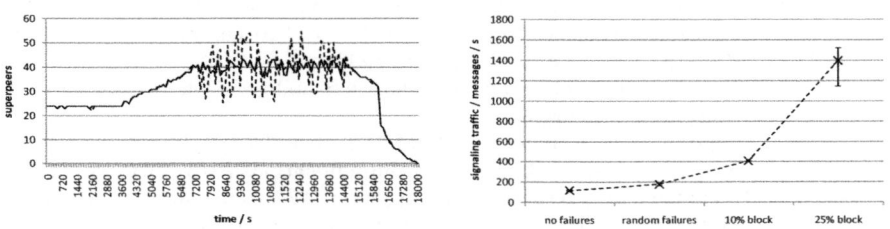

Fig. 3. Number of SPs: Random failures (solid) / 25% block (dashed)   **Fig. 4.** Signaling traffic in different failure situations: none/random/block

Figure 2 shows the case where in the operational phase (churn phase) a successional block of 10 % of all superpeers is selected every $12 \cdot T_{stab}$ and removed from the network compared to the random failure case. This leads to an immediate drop in the total number of superpeers. Although some overcompensations (due to the distributed nature of the used algorithm) are observable, the algorithm is able to compensate this error and regain a stable state within the twelve stabilization periods.

When the size of the block is increased to 25 % of the superpeers as shown in Figure 3, a similar result is obtained. Both, the drop and the overcompensation spikes are larger but again the system manages to return to a stable state.

Since the goal of the algorithms is a cost-optimal operation, it makes sense to also evaluate the costs (i.e. the signaling traffic) in the different failure scenarios. Figure 4 shows the comparison of signaling messages per second in the network. Obviously the introduction of failures leads more signaling traffic. Uniformly distributed failures only introduce a moderate increase compared to the best case without any disturbance. The introduced block-failures lead to 125 % more signaling messages compared to the random failures if 10 % of the nodes fail or even 670 % in the 25 %-block case.

3   Conclusion

The introduction of worst-case failures in the superpeer tier of the Chordella system leads to deviations from the optimal number of superpeers in the system. However, our algorithm is able to compensate these interferences within twelve stabilization periods. The number of messages needed to maintain a working system increases rapidly with the ratio of failed peers. Of course, the failures have also an impact on parameters like the query success and query duration. These results are are omitted due to space constraints.

References

1. Stoica, I., Morris, R., Karger, D., Kaashoek, M., Balakrishnan, H.: Chord: A scalable peer-to-peer lookup service for internet applications. In: Proceedings of the 2001 ACM SIGCOMM, p. 160. ACM, New York (2001)
2. Maymounkov, P., Mazieres, D.: Kademlia: A peer-to-peer information system based on the xor metric. In: Druschel, P., Kaashoek, M.F., Rowstron, A. (eds.) IPTPS 2002. LNCS, vol. 2429, p. 2. Springer, Heidelberg (2002)
3. Leonard, D., Rai, V., Loguinov, D.: On lifetime-based node failure and stochastic resilience of decentralized peer-to-peer networks. In: Proceedings of the 2005 ACM SIGMETRICS, pp. 26–37. ACM, New York (2005)
4. Binzenhöfer, A., Staehle, D., Henjes, R.: On the Stability of Chord-based P2P Systems. University of Würzburg 347 (2004)
5. Zoels, S., Despotovic, Z., Kellerer, W.: On hierarchical DHT systems–An analytical approach for optimal designs. Computer Communications 31(3), 576–590 (2008)
6. Zoels, S., Hofstätter, Q., Despotovic, Z., Kellerer, W.: Achieving and Maintaining Cost-Optimal Operation of a Hierarchical DHT System. In: Proceedings of the 2009 IEEE International Conference on Communications (ICC) - Next Generation Networking Symposium (2009)

A Low-Power Scheme for Localization in Wireless Sensor Networks

Jorge Juan Robles, Sebastian Tromer, Monica Quiroga, and Ralf Lehnert

Dresden University of Technology, Dresden, 01069 Germany
{robles,lehnert}@ifn.et.tu-dresden.de

Abstract. One of the most challenging issues in the design of system for Wireless Sensor Networks (WSN) is to keep the energy consumption of the sensor nodes as low as possible. Many localization systems require that the nodes keep the transceiver active during a long time consuming energy. In this work we propose a scheme to reduce the energy consumption of the mobile nodes that need to know their positions. Our strategy consists of decreasing the idle listening and an optimized allocation of the localization tasks on the nodes. Thus, the nodes that are externally powered calculate the position for the resource-constrained nodes. The scheme is based on a low-power IEEE 802.15.4 non-beacon enabled network.

Keywords: localization scheme, positioning algorithm, idle listening, wireless sensor networks.

1 Introduction

In a WSN the sensor node's position is used, e.g. in geographical routing, clustering techniques and context-based applications.

For the position estimation many localization algorithms utilize the information of the distance between nodes and the position of certain nodes, called anchors. The distance between two nodes can be calculated by measuring the received signal strength (RSS) at the receiver and considering an appropriate attenuation model. Unfortunately the RSS measurements have a big dispersion due to environment characteristics. Therefore many RSS measurements are required to achieve a good accuracy at expense of additional energy consumption. In our scenario the anchors are fixed and externally powered whereas the mobile nodes are battery-operated. The mobile nodes need to know their positions.

The goal of the proposed scheme is to provide, in an efficient way, the necessary information to the localization algorithm for the position calculation. Thus, RSS-based localization algorithms, such as Multilateration and Centroid [2] could be independently implemented on our scheme.

2 Idle Listening in the Localization Algorithms

Many localization systems, like in [3], are similar to the following simplified protocol:

F.A. Aagesen and S.J. Knapskog (Eds.): EUNICE 2010, LNCS 6164, pp. 259–262, 2010.
© IFIP International Federation for Information Processing 2010

A) The Mobile node (MN) broadcasts a localization request.
B) The anchors, which receive the request, send to the MN a certain number of packets containing relevant information for the position calculation.
C) The MN measures the RSS of each packet. Using these measurements and the received information, the MN executes a localization algorithm. It can also send the information to an anchor or a central unit.

Considering an IEEE 802.15.4 beaconless network [1], in the described basic protocol, the MN has to be active for a long time to receive enough packets from the anchors. This is due to the time that is required by the anchor for a successful transmission (CSMA/CA is used by the MAC layer). During the time between successful transmissions of the anchors the MN is listening (because it does not know when the next packet arrives) consuming a high amount of energy. This is called idle listening.

3 Low-Power Scheme for Localization

The low-power scheme tries to reduce the energy consumption of the MNs by reducing the idle listening period and incrementing the sleep periods. The network maintains a global synchronization. The anchors are active all the time and have the information related to the synchronization. The proposed scheme divides the localization task into four different phases (Fig. 1):

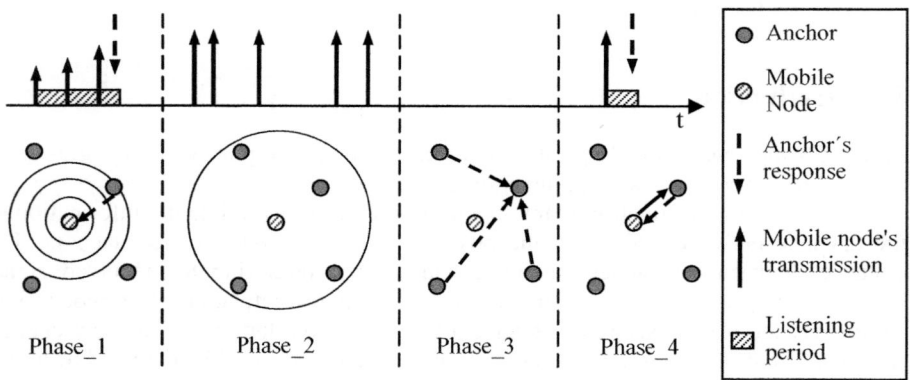

Fig. 1. Operation of the low-power scheme for a MN over the time

1) Phase_1: The MN broadcasts a localization request at minimal power transmission. If the anchors receive the request, they answer indicating the duration of the phases. If the MN does not receive any answer it tries to send another request at a higher transmission power. This process is to ensure that the nearest anchors to the MN can answer first. When the mobile node receives the first answer, it goes into sleep mode and waits for the following phase.

2) Phase_2: The MN broadcasts a certain number of packets at maximal power transmission. This process is random over the time decreasing the collision probability with other MNs. During the time between transmissions the MN goes into sleep mode in order to save energy. The anchors, which receive the packets, take the corresponding RSS measurements. The packets contain the information of the "selected anchor", which is the anchor that answered in the phase_1.

3) Phase_3: During all the phase_3 the mobile node is in sleep mode. The selected anchor is in charge of the position estimation of the MN. Thus, the anchors that have received packets in the phase_2, send their positions and RSS measurements to the selected anchor. The selected anchor executes a localization algorithm in order to estimate the MN position.

4) Phase_4: If the MN needs to know its position, it sends an information request to the selected anchor. The answer of the selected anchor contains the calculated position. After this task the MN goes into sleep mode.

If the anchors receive a localization request outside of phase_1 then they send a packet to the MN indicating when the phase_1starts.

The energy consumption of an IEEE 802.15.4 transceiver was studied for the basic protocol described in the section two and the low-power scheme (see Fig. 2). In our first analysis the following assumptions were taken:

1) The scenario consists of four anchors and one mobile node.
2) CSMA/CA needs 5 ms for a successful transmission [5].
3) The duration of each transmission is 1,5ms.
4) The information about the energy consumption in the different modes was obtained from the specifications of the transceiver AT86RF230 [4].
5) In the basic protocol each anchors transmits 5 packets to the MN.
6) In the basic protocol the MN listens during 150 ms and sleeps 850ms.
7) In the phase_1 the MN transmits 2 times waiting 10ms in each trial.
8) In the phase_2 the MN sends 5 packets.
9) In the phase_4 the MN receives a response after 10ms.
10) The duration of the localization period is 1000ms.

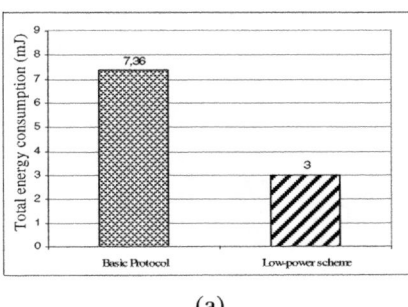
(a)

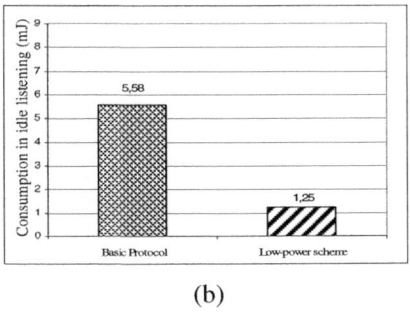
(b)

Fig. 2. a) Total energy consumption of the MN´s transceiver in one localization period. **b)** Energy consumption of the MN´s transceiver due to idle listening in one localization period.

4 Conclusions

We present the main operation of a novel low-power scheme for localization algorithms. It can reduce the energy consumption of the mobile nodes by decreasing the idle listening and increasing the sleep period. The future work will be focused on an extended performance evaluation of the proposed scheme.

References

1. IEEE standard 802.15.4-2006 (2006),
   http://standards.ieee.org/getieee802/802.15.html
2. Reichenbach, F., Blumenthal, J., Timmermann, D.: Comparing the Efficiency of Localization Algorithms with the Power-Error-Product (PEP). In: 28th International conference on Distributed Computing Systems Workshops, Beijing, China (2008)
3. Lorincz, K., Welsh, M.: MoteTrack: a robust, decentralized approach to RF-based location tracking. Personal and Ubiquitous Computing 11(6) (2007)
4. Datasheet IEEE 802.15.4 transceiver AT86RF240, http://www.atmel.com
5. Rohm, D., Goyal, M., Hosseini, H., Divjak, A., Bashir, Y.: Configuring Beaconless IEEE 802.15.4 Networks Under Different Traffic Loads. In: International Conference on Advanced Information Networking and Applications, Bradford, UK (2009)

Flow Aggregation Using Dynamic Packet State

Addisu Eshete and Yuming Jiang

Center for Quantifiable Quality of Service in Communication Systems
Norwegian University of Science and Technology
addisu.eshete@q2s.ntnu.no,
ymjiang@ieee.org

1 Introduction

While the Internet is tremendously scalable, it is based on simplistic design that provides only egalitarian best effort service for all traffic types. Two architectural extensions have been proposed, namely Intserv and Diffserv, to enhance the service architecture. While Integrated Services (Intserv) can provide hard Quality of Service (QoS) requirements (delay and bandwidth) to individual flows end to end, its per-flow mechanisms and signaling overhead make it too complex to be readily deployed in the Internet core. On the other hand, Differentiated Services (Diffserv) restores the scalability of the Internet by dropping per flow mechanisms and merging of flows into aggregates. The price paid is the lack of flexible and powerful services at flow granularity levels.

In this work, we plan to address the question: **can we *simultaneously* provide efficient and scalable QoS and also fulfil the requirements of individual constituent flows?**

2 The Techniques– Flow Aggregation and DPS

To provide scalability, we can employ flow aggregation which provides a host of benefits: (1) the number of flows in the core of networks are reduced, and so are the complexity associated with per flow management and operations at core routers, (2) scheduler efficiency of routers can be improved [2], and (3) when the reserved rate of a flow is coupled with delay as in guaranteed rate schedulers [3], flow aggregation can result in tighter bounds of the queueing delay [2,1]. However, there are known issues with flow aggregation. We outline two known problems: (1) FIFO aggregation of EF traffic in large arbitrary networks may explode the delay bound after a certain utilization threshold [1,4] and it is not possible to provide delay bounds for high utilization levels in these networks, and (2) flow aggregation usually needs to be nonwork-conserving to be fair to individual constituent flows. Continuous proliferation of very high capacity links means that the first issue may not be the major problem. The main challenge is the impact of (work-conserving) aggregation on the resulting QoS provision of the constituent flows of the aggregate.

F.A. Aagesen and S.J. Knapskog (Eds.): EUNICE 2010, LNCS 6164, pp. 263–265, 2010.

To provide efficient flow aggregation that also satisfies flow requirements, we plan to use *dynamic packet state* (DPS) [5]. Using DPS, per flow states or requirements are inserted into the packet headers at the edge as labels, freeing core routers from the task of maintaining flow states. The states to be encoded are dynamic since the current flow state is recomputed at each node and encoded as the packet traverses through the network. These labels are then used to coordinate the actions of distributed algorithms in the network domain. The labels are ultimately removed at the egress nodes for interoperability to existing architectures.

3   Our Approach

It is important to note we use flow aggregation for scalability, and DPS to make it efficient (work-conserving) and able to fulfill per flow requirements.

A schematic of our conceptual system model is shown in Fig. 1. Flows with similar characteristics are aggregated at the ingress node. We assume that routers have aggregation capability at their output ports. Core routers C_1 and C_2 are aware of the aggregate flow $f_{1,2}$ but they have no knowledge of the constituent flows f_1 and f_2. The challenge is to preserve the quality of service requirements of constituent flows in the network core without the core routers knowing the composition of the aggregate flow.

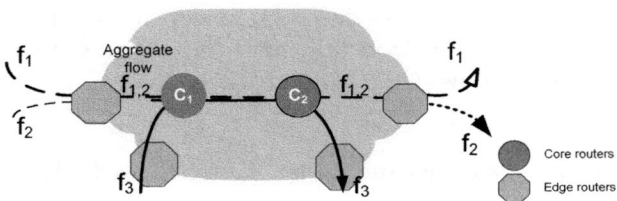

Fig. 1. Flow aggregation. f_1 and f_2 are constituents of flow $f_{1,2}$ while f_3 is cross traffic.

We propose a *virtual GR* which is a variant of GR (Guaranteed Rate) scheduling [3] modified for aggregation. At entrance to the domain, ingress node ($j = 1$), we start by encoding the $vGRC$ for packet i of the flow as follows:

$$e_1^i = max\left\{a_1^i, vGRC_1^{i-1}\right\} \quad \text{where } vGRC_1^i = e_1^i + l^i/r \tag{1}$$

At inner node j, when packet p^i arrives it is assigned:

$$e_j^i = vGRC_{j-1}^i$$
$$vGRC_j^i = e_j^i + l^i/r \tag{2}$$

where r is flow rate at simple scheduler and sum of flow rates at an aggregator. The QoS provided to the individual flows using this scheme should be comparable to the per flow schemes without aggregation. We believe the result of this work will be very useful to provide QoS guarantees to flows in a scalable manner in the future Internet.

References

1. Charny, A., Le Boudec, J.-Y.: Delay bounds in a network with aggregate scheduling. In: Proceedings of Quality of Future Internet Services, pp. 1–13 (2000)
2. Cobb, J.A.: Preserving quality of service guarantees in spite of flow aggregation. IEEE/ACM Trans. on Networking 10(1), 43–53 (2002)
3. Goyal, P., Lam, S.S., Vin, H.M.: Determining end-to-end delay bounds in heterogeneous networks. Multimedia System 5(3), 157–163 (1997)
4. Jiang, Y.: Delay bounds for a network of guaranteed rate servers with fifo aggregation. Computer Networks 40(6), 683–694 (2002)
5. Stoica, I.: Stateless Core: A Scalable Approach for Quality of Service in the Internet. PhD thesis, CMU, Pittsburgh, USA (December 2000)

Evaluating MDC with Incentives in P2PTV Systems[*]

Alberto J. Gonzalez[1,2], Andre Rios[1,2], Guillermo Enero[2],
Antoni Oller[1,2], and Jesus Alcober[1,2]

[1] Dept. of Telematics Engineering, Universitat Politecnica de Catalunya (UPC)
Barcelona, Spain
[2] i2Cat Foundation, Barcelona, Spain
{alberto.jose.gonzalez,andre.rios,antoni.oller,jesus.alcober}
@{upc.edu,i2cat.net},
guillermo.enero@i2cat.net

Abstract. The popularity of P2P video streaming is raising the interest of broadcasters, operators and service providers. Concretely, mesh-pull based P2P systems are the most extended ones. Despite these systems address scalability efficiently, they still present limitations that difficult them to offer the same user experience in comparison with traditional TV. These ones are mainly the free-riding effect, long start-up delays and the impact of churn and bandwidth heterogeneity. In this paper we study the performance of Multiple Description Coding (MDC) combined with the use of incentives for redistribution in order to mitigate some of them by means of simulations. Simulation results show that the use of MDC and incentive-based scheduling strategies improve the overall performance of the system. Moreover, an extended version of the P2PTVSim simulator has been developed to support MDC and incentives.

Keywords: P2P Streaming, MDC, Incentives, Continuity Index (CI), Delay.

1 Introduction

P2PTV streaming systems have become a popular service on the Internet (both at commercial and research level), with several successful deployments [1], and the most spread form of what is known as Internet TV. The use of these systems is promising because they offer the possibility to introduce added value to traditional TV broadcasting by providing flexibility, in terms of content delivery and interactive services. However, in order to become a truly successful application they need to be able to provide the same or even better user experience as TV broadcasting offers. P2PTV systems are expected to provide a high degree of scalability with different streaming rates and number of peers. They must also provide continuity under adverse churn conditions (especially in presence of flash-crowds) as well as ensuring delivery of data within a given deadline in order to provide smooth playback. Some aspects affecting the performance of these requirements are: free-riding (non-cooperation)

[*]This work was supported by MCyT (Spanish Ministry of Science and Technology) under the Project TSI2007-66637-C02-01, which is partially funded by FEDER.

F.A. Aagesen and S.J. Knapskog (Eds.): EUNICE 2010, LNCS 6164, pp. 266–269, 2010.

effect, long start-up delay and the impact of churn and bandwidth heterogeneity in the stability of the system. We tackle them in this work.

In this paper we propose an MDC-based system, which uses incentives for redistribution, in order to address the impact of losses in the Continuity Index (CI), delay and the performance problems due to the effect of free-riding. MDC is a technique designed to enhance error resilience and increase transmission robustness and scalability by means of splitting a stream into N different sub-streams (N≥2). Different splitting techniques can be used [2].

In order to validate the proposed solution we have deployed it in a simulatior called P2PTVSim [3]. However, we have extended it in order to be able to simulate the usage of MDC as well as the use of a specific incentive strategy (inspired by [4]). The obtained results show how the use of MDC provides a more robust behaviour against losses. Consequently, the Continuity Index of the system is improved. In addition, thanks to the use of incentives, the effect of free-riding is alleviated.

2 Proposed Solution

The simulated system is a mesh-pull-based P2P streaming system that uses MDC for providing robustness to the communication. In order to support MDC, the receiver-side scheduler was adapted. When the system uses MDC, four balanced descriptions are generated and distributed in chunks. It also introduces the use of incentives based on the contribution of the partners in the supplier-side scheduler. Incentives are used in order to encourage cooperation, so that those peers contributing more to their partners are more likely to receive more descriptions and therefore more quality.

2.1 Receiver-Side Scheduler

The Buffer Map consists of a matrix with as many rows as descriptors. Partners exchange their Buffer Maps periodically and they perform rounds of chunk requests to get the missing chunks considering the availability information provided by their neighbours. The schedule of these requests is critical to achieve an optimal result, retrieve the maximum number of descriptions and ensure the best CI and quality. Chunks that have already been received are marked with a B and chunk that have been requested with R. In the matrix, each chunk position has a number indicating the order of the requests. The scheduler looks, at the beginning of each round, for the first chunk index that does not have a buffered (B) or requested (R) chunk from any description. When it finds the first chunk index satisfying this condition it selects a random description (from the available ones according to the availability information provided by its neighbours) and makes the request. Then, it continues with the next chunk index that has non-buffered or requested chunk until there are no more chunk indexes fulfilling this condition. The scheduling algorithm continues by doing the same routine over chunk indexes with just one buffered or requested chunk, then with just two, and so on. The goal of this scheduling algorithm is to get the maximum number of descriptions for each chunk index time but always trying to avoid having high variation between the received number of descriptions from one time to another.

2.2 Supplier-Side Incentive-Based Scheduler

The supplier-side algorithm takes into account the contribution of their partners and serves them accordingly (incentivates). The effect of this supplier strategy is that peers that contribute more are more likely to receive a larger number of descriptions. As a supplier, each peer has a queue of requests from its partners and at each round it has to decide which of these requests is going to be served first. Instead of selecting a random one or handle them in a FIFO manner, a weighted selection is performed. The weight in this case is assigned by computing the percentage of chunks that have been provided by a specific neighbour from the total.

3   Simulation Results

Four different simulation scenarios were considered. The first scenario is the reference system, which is a mesh-pull P2P streaming system with single layered video and no incentives (default operation in P2PTVSim). Then, in a second scenario, we added incentives. The third one is an MDC-based (using 4 layers) mesh-pull P2P streaming system and the last scenario is a variation of the third one, adding incentives. Due to space limitations, we focus on the first and the last scenarios.

The parameters for the simulations are the following ones: a total of 1000 peers, with a mean degree of 10 partners and a 5 seconds buffer. Four different types of peers: class A, peers with 5Mbps of upload bandwidth (10% of the total); class B, peers with 1Mbps of upload bandwidth (40% of the total); class C, peers with 500Kbps of upload bandwidth (40% of the total); and class D, peers that act as free-riders, with 0Kbps of upload bandwidth (10% of the total). No download bandwidth constraint is assumed. The simulations last for the distribution of 2500 chunks and the chunk size was 1KB. For each scenario several simulations have been performed varying the packet loss rate according to a random loss model (Bernoulli).

The metrics that are measured for the evaluation of the techniques are the end-to-end delay, the CI and the average number of received descriptions.

3.1   Reference System

When considering the effect of losses (Fig. 1) in the reference system we can see that the CI decreases as the losses increase. The delay, as it can be seen in Fig. 2, increases as the percentage of losses increases and it is in the range of 1,5s to 3,5s.

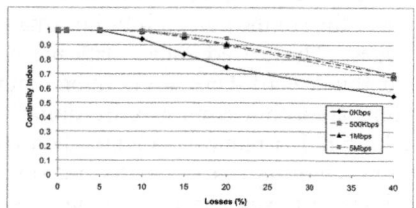
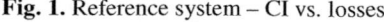
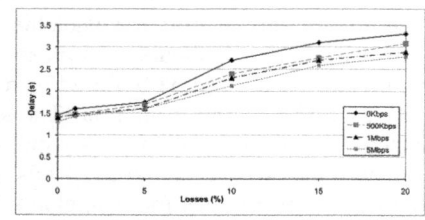

Fig. 1. Reference system – CI vs. losses   **Fig. 2.** Reference system – Delay vs. losses

3.2 MDC System with Incentives

Finally, we combined the use of MDC with incentives. Here, to the improvements introduced by MDC in terms of CI increase and low end-to-end delay we can add the overall performance boost introduced by the use of incentives. As Fig. 3 shows, the CI can be maintained at almost the maximum level even at high loss rates (40%). This combination allows a high level of playback continuity. The quality, in terms of average number of descriptions is also increased for the cooperating peers. Regarding to delay (Fig. 4) it is approximately the same delay that the MDC system provides without incentives.

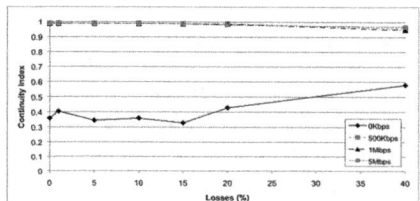   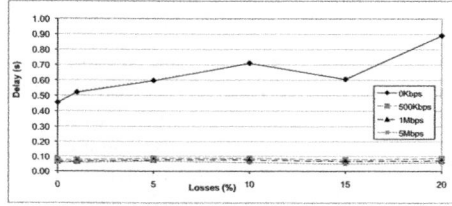

Fig. 3. MDC+incentives – CI vs. Losses   **Fig. 4.** MDC+incentives – Delay vs. Losses

4   Conclusions and Future Work

The aim of this work was to study the performance of MDC systems with incentive mechanisms in P2PTV systems. The evaluation of the solution was done by means of simulation. The gathered results show that the proposed solution clearly improves the considered metrics and the overall behaviour of the system compared to the performance of the reference system simulating a common P2PTV system. These results can be used as reference or guideline for further developments. As an additional outcome of this work, we have developed a simulation software (based on P2PTVSim) which provides a valid test-bed that can be used for future studies. Future work will include an analysis of the overhead introduced by MDC, PSNR measurement for quality estimation and the simulation considering churn effect. Start-up delay reduction will also be studied by means of MDC.

References

1. Xiaojun, H., Chao, L., Jian, L., Yong, L., Ross, K.: A Measurement Study of a Large-Scale P2P IPTV System. IEEE Transactions on Multimedia 9(8), 1672–1687 (2007)
2. Tillier, C., Crave, O., Pesquet-Popescu, B., Guillemot, C.: A Comparison of four Video Multiple Description Coding Schemes. In: EUSIPCO, Poznan (2007)
3. P2PTVSIM, http://www.napa-wine.eu/cgi-bin/twiki/view/Public/P2PTVSim
4. Zhengye, L., Yanming, S., Panwar, S.S., Ross, K.W., Wang, Y.: P2P Video Live Streaming with MDC: Providing Incentives for Redistribution. In: IEEE International Conference on Multimedia and Expo., pp. 48–51, 2–5 (2007)

Translation from UML to SPN Model: A Performance Modeling Framework

Razib Hayat Khan and Poul E. Heegaard

Norwegian University of Science & Technology
7491, Trondheim, Norway
{rkhan,poul.heegaard}@item.ntnu.no

Abstract. This work focuses on the delineating a performance modeling framework for a communication system that proposes a translation process from high level UML notation to Stochastic Petri Net model (SPN) and solves the model for relevant performance metrics. The framework utilizes UML collaborations, activity diagrams and deployment diagrams to be used for generating performance model for a communication system. The system dynamics will be captured by UML collaboration and activity diagram as reusable specification building blocks, while deployment diagram highlights the components of the system. The collaboration and activity show how reusable building blocks in the form of collaboration can compose together the service components through input and output pin by highlighting the behavior of the components. Later a mapping between collaboration and system component identified by deployment diagram will be demonstrated. Moreover the UML models are annotated to associate performance related quality of service (QoS) information for solving the performance model for relevant performance metrics to generate performance evaluation results.

Keywords: UML, SPN, Performance attributes.

1 Proposed Performance Modeling Framework

Our proposed performance modeling framework utilizes tool suite Arctis which is integrated as plug-ins into the eclipse IDE [1]. The proposed framework is shown in Fig.1 where steps 1 and 2 are the parts of Arctis tool suite. A developer first consults a library to check if an already existing collaboration block or a collaboration of several blocks solves a certain task. Missing blocks can also be created from scratch and stored in the library for later reuse. The building blocks are expressed as UML models. The structural aspect, for example the service component and their multiplicity, is expressed by means of UML 2.2 collaborations [2]. For the detailed behavior, we use UML 2.2 activities [2]. They express the local behavior of each of the service components as well as their necessary interactions in a compact and self-contained way using explicit control flows [1].

In the second step, building blocks are combined to more comprehensive service by composition. For this composition, we use UML 2.2 collaborations and activities as well. While collaborations provide a good overview of the structural aspect of the

F.A. Aagesen and S.J. Knapskog (Eds.): EUNICE 2010, LNCS 6164, pp. 270–271, 2010.

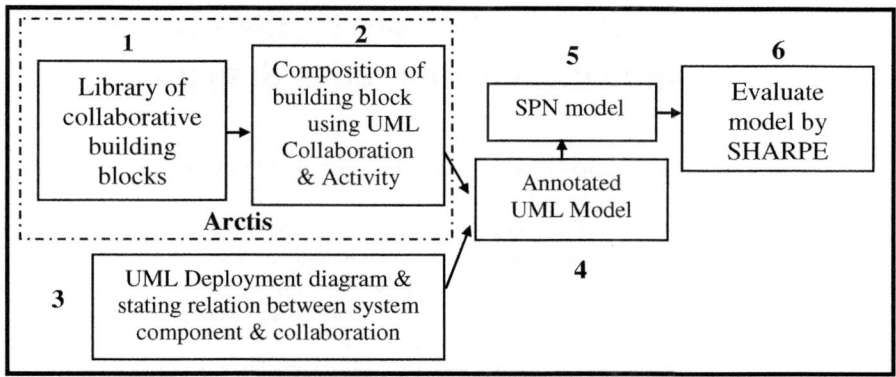

Fig. 1. Proposed performance modeling framework

composition, i.e., which sub-services are reused and how their collaboration roles are bound, activities express the detailed coupling of their respective behaviors [1]. Each sub-service is represented by a call behavior action referring to the respective activity of the building blocks. For each activity parameter node of the referred activity, a call behavior action declares a corresponding pin. By connecting the individual input and output pins of the call behavior actions, the events occurring in different collaborations can be coupled.

In the next step the UML deployment diagram of our proposed system as well as the relationship between system component and collaboration will be outlined to describe how service is delivered by the joint behavior of the system components [2].

Activity diagram and the deployment diagram are annotated in the following step according to the UML Profile for Schedulability, Performance and Time [2] to incorporate performance related quality of service information.

By considering all the above mentioned steps probable states and the parameter for triggering the change between the states of the performance model will be derived and later will be solved by the SHARPE tool [4].

2 Conclusion

However the size of the underlying reachability set is major limitation for large and complex system. Further work includes automating the whole translation process from our UML specification style to generate performance model and the way to solve the model through our proposed framework as well as tackling the state explosion problems of reachability marking for large system.

References

1. Kramer, F.A.: ARCTIS, Department of Telematics, NTNU,
   http://arctis.item.ntnu.no
2. OMG UML Superstructure, Version-2.2
3. OMG 2005, UML Profile for Schedulability, Performance, & Time Specification, V – 1.1 (2005)
4. Trivedi, K.S., Sahner, R.: SHARPE Performance evaluator. Duke University, Durham

An Open and Extensible Service Discovery for Ubiquitous Communication Systems

Nor Shahniza Kamal Bashah[1], Ivar Jørstad[2], and Do van Thanh[3]

[1] Norwegian University of Science and Technology, Department of Telematics,
O.S. Bragstads Plass 2E, NO-7491 Trondheim, Norway
nor@item.ntnu.no
[2] Ubisafe, Gamleveien 252, 2624 Lillehammer, Norway
ivar@ubisafe.no
[3] Telenor & NTNU, Snaroyveien 30, 1331 Fornebu, Norway
thanh-van.do@telenor.com

Abstract. In future ubiquitous communication systems, a service can be anything and introduced by anybody. Consequently, same or equivalent services may have different names and services with same name or type may be completely different. Existing service discovery systems are incapable of handling these situations. We propose a future service discovery, which is able to discover all these new service types. In addition, it is capable to find services that are not exact matches of the requested ones. More semantics are introduced through attributes like EquivalenceClass, ParentType and Keywords.

Keywords: Service advertisement, service discovery, service lookup, service matching, service registration.

1 Introduction

In this research, a future service discovery system is proposed to solve the problem of discovering services in a ubiquitous communication environment, where similar services can have different names in different languages and services with same name may not offer the same functions and capabilities. The requirements on future service discovery systems are derived and more details can be found in [1]. The future service discovery system consists of service registration, service advertisement and request and service matching.

For the **service registration** the service provider must supply the details about the service that are necessary to detect, identify and use the service as follows:

- **Service name** – to denote and recognise the service. We propose to extend the service name format of future service discovery to allow any language.
- **Service type** – to carry on a search and identification of the functions offered by the service. There are three options of service type which are:
  - o *Brand-new service type* – without relation with any existing service type. Hence the ParentType has to be set to nil and the EquivalentClass is then left empty.

F.A. Aagesen and S.J. Knapskog (Eds.): EUNICE 2010, LNCS 6164, pp. 272–273, 2010.
© IFIP International Federation for Information Processing 2010

    o *Equivalent service type* - the service type has a different name and may be another implementation but is equivalent to an existing service type. The service provider has to add the name of the existing service type and also all the known service type in different languages.

    o *Subtype of an existing service type* - The service type has all the functions and features of an existing service type but has also additional ones. The ParentType is hence indicated.

There are two alternatives for implementing **service advertisement:**

- The service registry or broker can take the initiative and broadcast the list of available services periodically.
- A client wanting to find services issues a request for a list of available services to the service registry.

However, both alternatives have their own advantages and disadvantages hence, selection of appropriate alternatives has to be considered and empirical methods have to be used. Anterior service advertisement may be a solution to be considered. Service advertisement can be made before the mobile user enters the location thus contribute to speeding up the service discovery and ensuring service continuity.

Instead of asking for the list of available services and carrying the service matching itself the client can **request** for a particular service or set of services. However, they may not have the opportunity to discover new and unknown services.

The **service matching** can be performed by the mobile client after the acquisition of the list of available services or by the network system upon receipt of a service request. We propose to extend the match operation on the Equivalence class, Service subtyping and Keywords of the requested service type.

2 Conclusions

This research explains the necessity of an innovative service discovery capable of handling similar services with different name and different services with same name. A prototype is currently under development and will be accomplished in the near future. Next, the experiments with various number of clients, number of services, different bandwidths and complicated service ontology will be carry out. The collected results will then be used to optimise the service discovery prototype. Another relevant and exciting future work will be the contribution to the extension of the 802.21 MIHF (Media Independent Handover Function) [2] for devices equipped with multiple wireless access technologies. Such an extension will ensure service continuity across multiple heterogeneous access networks.

References

1. Kamal Bashah, N.S., Jørstad, I., van Do, T.: Service Discovery in Future Open Mobile Environments. In: Proc. the Fourth International Conference on Digital Society (ICDS 2010), February 10-16, St. Maarten, Netherlands Antilles (2010), ISBN: 9780769539539
2. Media Independent Handover Services, Draft IEEE Standard for Local and Metropolitan Area Networks: Media Independent Handover Services, IEEE P802.21/D7.1 (August 2007)

Author Index